7th Southwest Process Technology Conference 2015

Galveston, Texas, USA
1-2 October 2015

ISBN: 978-1-5108-1769-2

Printed from e-media with permission by:

Curran Associates, Inc.
57 Morehouse Lane
Red Hook, NY 12571

CURRAN ASSOCIATES INC. proceedings.com

Some format issues inherent in the e-media version may also appear in this print version.

Printed by Curran Associates, Inc. (2016)

AIChE
120 Wall Street, FL 23
New York, NY 10005-4020

Phone: (800) 242-4363
Fax: (203) 775-5177

www.aiche.org

Additional copies of this publication are available from:

Curran Associates, Inc.
57 Morehouse Lane
Red Hook, NY 12571 USA
Phone: 845-758-0400
Fax: 845-758-2634
Email: curran@proceedings.com
Web: www.proceedings.com

TABLE OF CONTENTS

2015 CONFERENCE KEYNOTE

CATALYSIS & REACTION ENGINEERING

ENERGY EFFICIENCY IN PLANT EQUIPMENT

PROCESS CONTROL & OPTIMIZATION

CHEMICAL PROCESS TECHNOLOGY

DISTILLATION & SEPARATION

PROCESS SAFETY

STS DINNER PROGRAM AND SOCIAL HOUR

ENGINEERING ETHICS

THERMODYNAMICS & PROCESS SIMULATION

UPSTREAM PROCESS TECHNOLOGY

ICIS
Chemicals Energy Fertilizers
Prices | News | Analysis | Consulting

China, oil, and the US petrochemical project wave

Joseph Chang, Global Editor
ICIS Chemical Business

7th Southwest Process Technology Conference
Galveston, October 1, 2015

Agenda

- The big picture – commodities, macroeconomics, China
- Feedstocks and competitiveness
- US projects – ethylene
- US projects – propylene
- US projects – methanol
- Capital spending (CAPX) cycles
- Global context – Europe, Middle East, China

The big picture

ICIS

www.icis.com

Commodities meltdown

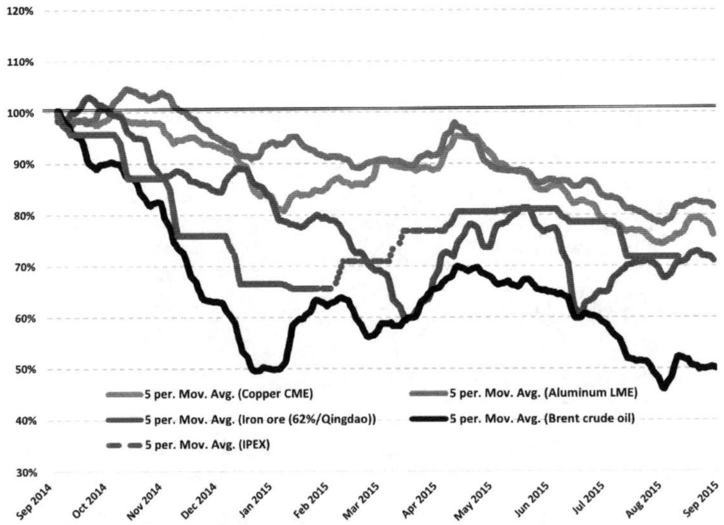

SOURCE: Bloomberg, ICIS

ICIS

www.icis.com

Global manufacturing PMI

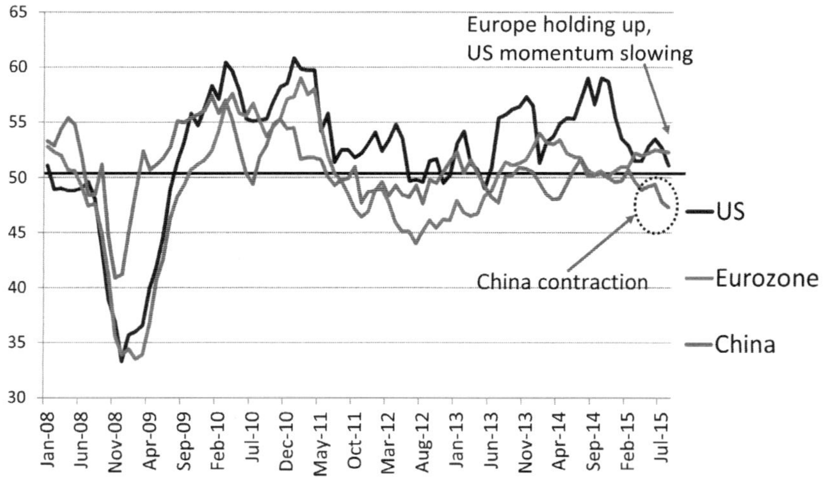

ICIS

Data to July 2015
SOURCE: US based on ISM, Eurozone on Markit, China on Caixin/Markit (formerly HSBC/Markit)

www.icis.com

China devaluation impact

August 11, 2015 – devaluation of yuan vs US dollar

- Points to extreme weakness in China. A symptom of the economic illness. More devaluations to come?
- Competitive devaluations in Asia elsewhere and spreading. Parallels to Asia financial crisis of 1997 (started with devaluations)
- **Deflationary** for commodities, which are priced in US dollars
- China exports more competitive – can export goods more cheaply (deflationary)
- IMPORTS into China less competitive (also deflationary). Don't count on China/emerging markets to revive profit growth
- Longer term, a weaker yuan could help lead to a stronger economy

ICIS

www.icis.com

Feedstocks and competitiveness

ICIS

www.icis.com

Oil-based naphtha converges on US NGLs

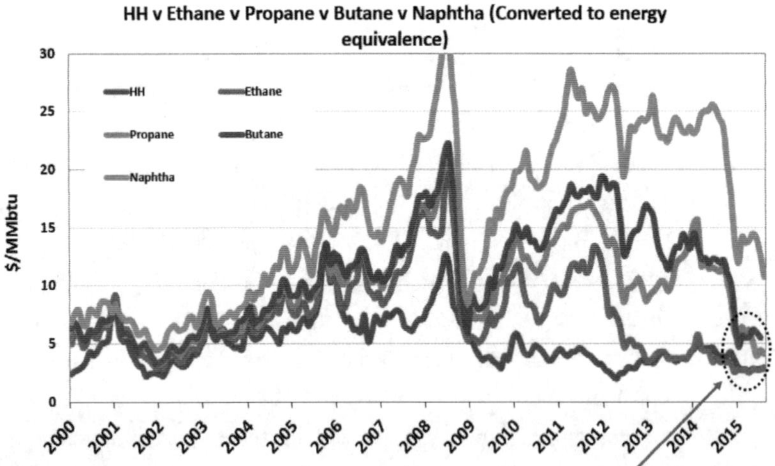

SOURCE: ICIS Consulting

ICIS

www.icis.com

US chemical prices track oil – NOT natural gas

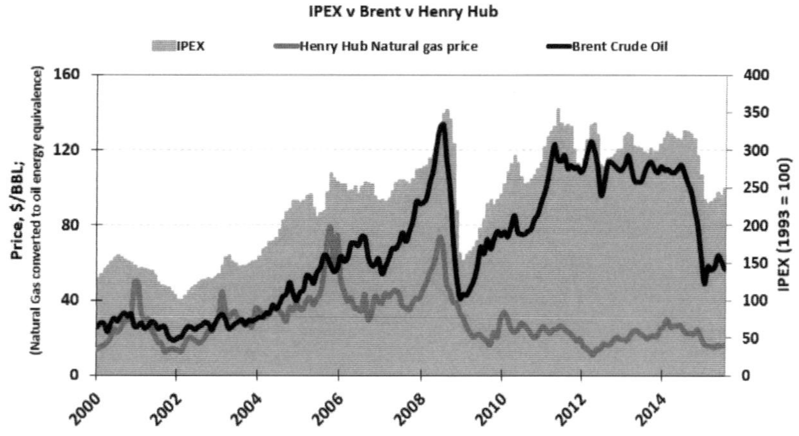

Source: ICIS, Data as of end-July 2015

ICIS

R2 US IPEX, natgas = 2.0%

R2 US IPEX, crude oil = 89.3%

www.icis.com

Cost curve flattens. More competitive!

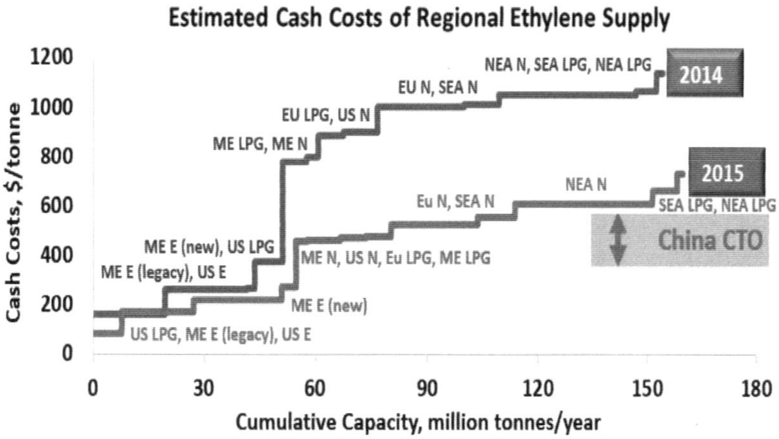

SOURCE: ICIS Consulting; 2015 data as of end-June 2015

ICIS

www.icis.com

US, Europe, Asia polyethylene margins

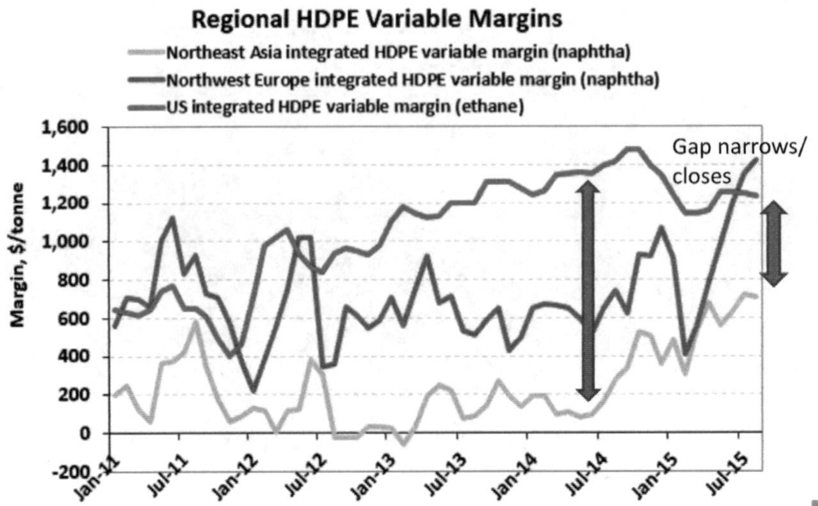

LyondellBasell – US vs Europe, H1 2015

O&P Americas	O&P Europe, Asia, Intl
EBITDA $2.05bn	EBITDA $849m
Sales $5.23bn	Sales $5.97bn
EBITDA margin = 39.1% 2014 = 30.0%	**EBITDA margin = 14.2%** 2014 = 9.3%

"The outlook for the third quarter remains positive for our portfolio. Natural gas and NGL remain well supplied and favorably priced."

– LyondellBasell CEO Bob Patel, July 2015

Polyolefins segment EBITDA margins (US or US asset heavy)

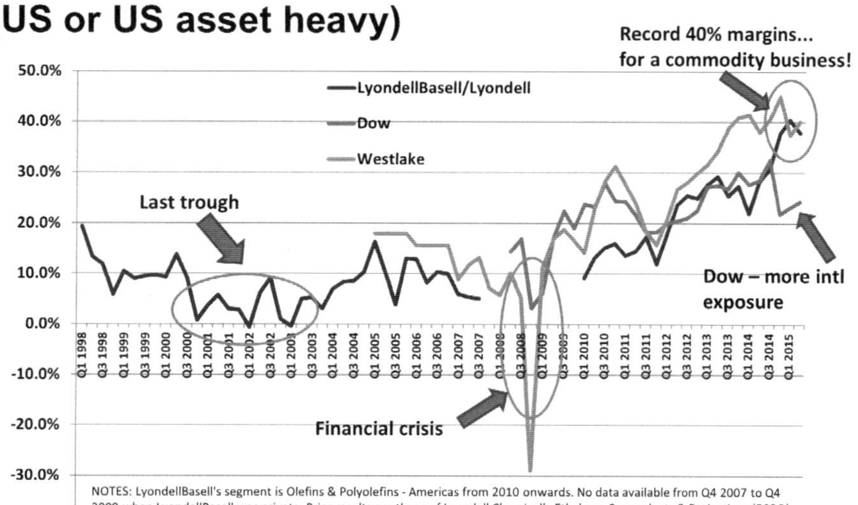

NOTES: LyondellBasell's segment is Olefins & Polyolefins - Americas from 2010 onwards. No data available from Q4 2007 to Q4 2009 when LyondellBasell was private. Prior results are those of Lyondell Chemical's Ethylene, Co-products & Derivatives (EC&D) segment. From Nov 2004 onwards, EC&D included acetyls from Millennium deal. Q2 2004 and prior, Equistar results for Lyondell
Dow's segment is Performance Plastics from Q4 2010 onwards. Prior to that, it is Dow's Basic Plastics segment. From Q1 2008 and prior, no EBITDA data for Basic Plastics is available.
Westlake's segment is Olefins, and EBITDA has been calculated from income from operations + D&A. The flat lines for 2005 and 2006 represent FULL YEAR margins, as quarterly data is unavailable.

ICIS

www.icis.com

US petrochemical projects - ethylene

ICIS

www.icis.com

New US ethane crackers under construction

Company	Capacity Ethylene, '000 tonnes/year	Downstream	Location	Start-up
Chevron Phillips Chemical	1,500	Bimodal HDPE (500kt), mLLDPE (500kt) - at Sweeny	Cedar Bayou, Texas	Mid-late 2017
Dow Chemical	1,500	ELITE PE (400kt), LDPE (350kt), elastomers (320kt), EPDM (200kt)	Freeport, Texas	H1 2017
ExxonMobil Chemical	1,500	mLLDPE plus LLDPE (650kt x 2) at Mont Belvieu	Baytown, Texas	H2 2017
Formosa Plastics	1,590	PE unspec (525kt), LDPE (625.5kt), MEG (1,000kt), plus DEG, TEG, PEG	Point Comfort, Texas	2017
Occidental Chemical/Mexichem	544	Feed into existing VCM plant of 1,050kt	Ingleside, Texas	Q1 2017
Sasol	1,500	LDPE (450kt), LLDPE (450kt), EO/EG (300kt), ethoxylates and detergent alcohols (300kt)	Lake Charles, Louisiana	Q1 2018

TOTAL CAPACITY: 8.13m tonnes/year

ICIS

www.icis.com

Planned expansions of existing US crackers

Company	Capacity Ethylene, '000 tonnes/year	Location	Start-up
LyondellBasell	113	Channelview, Texas	Aug-15
LyondellBasell	363	Corpus Christi, Texas	Q2 2016
Westlake Chemical	113	Lake Charles, Louisiana	H1 2016
LyondellBasell	250	Channelview, Texas	2017
Huntsman	19.3	Port Neches, Texas	NA

TOTAL CAPACITY: 858,000 tonnes/year
= about 1 new cracker

ICIS

www.icis.com

… plus planned new US crackers (to be built?)

Company	Capacity Ethylene, '000 tonnes/year	Downstream	Location	Start-up	Status
Shintech	500	VCM (300kt), PVC (300kt), caustic soda (200kt) - all Q2 2016	Plaquemine, Louisiana	2018	'Committed'
Axiall/Lotte	1,000	MEG (Lotte), feed into existing PVC for Axiall	St Charles, Louisiana	Q4 2018	FID H2 2015
Shell	1,500	HDPE/LLDPE (550kt x 2), HDPE (500kt)	Monaca, Pennsylvania	2018	Permits obtained, FEED stage
Total	1,000	None - meet derivative needs	Port Arthur, Texas	End-2019	Permitting, FEED awarded
Odebrecht/ Braskem	1,050*	PE (3 units)	Wood County, West Virginia	Post 2020	Feasibility, FID 2016
PTT Global/ Marubeni	1,000	HDPE (700kt), MEG (500kt), EO (100kt)	Belmont County, Ohio	2021	Feasibility, FID 2016
Formosa Plastics	1,200	LDPE, HDPE, EG	Louisiana	NA	Feasibility
Williams	1,500	NA	Geismar, Louisiana	NA	Feasibility

* ICIS Chemical Business estimate based on statements by Braskem

TOTAL CAPACITY: 8.75m tonnes/year, but don't count on all of it!

ICIS

www.icis.com

…and something old/new. 'Vampire crackers'

Company	Capacity Ethylene, '000 tonnes/year	Downstream	Location	Start-up	Status
Indorama	370	None – to meet derivative needs in EO, MEG, PET	Lake Charles, Louisiana	End 2017	Bought idled plant

- Sep 24 – Indorama Ventures buys Equistar plant closed in 2001
- Additional investment of $175m to revive cracker
- Start-up targeted by end of 2017
- Indorama will achieve goal of back integration in North America
- Dow restarted an idled 380kt cracker in Louisiana at the end of 2012

ICIS

www.icis.com

US cracker projects – recalibrate expectations

- 6 new crackers under construction = 8.13m tonnes/year
- Plus expansions at existing facilities, restart = 1.23m tonnes/year
- TOTAL of 9.3m tonnes/year of new capacity, or **33% of existing US capacity base by 2018. CONSERVATIVE SCENARIO**
- IF ALL the ones being planned also get built – 8 more new crackers totaling 8.75m tonnes/year, then **64% capacity increase**
- Companies recalibrating project economics for low crude oil. Expect delays and cancellations
- **2nd wave of expansions will be much weaker**

ICIS

www.icis.com

Timing of US ethylene expansions

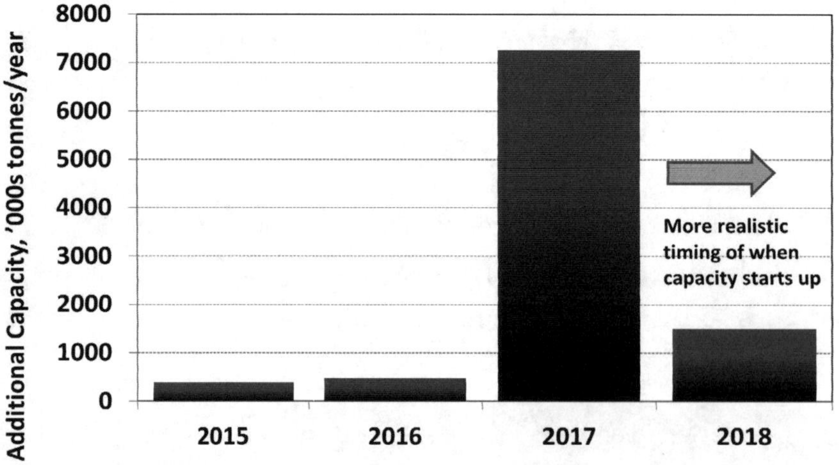

New crackers under construction, plus planned expansions of existing plants. Excludes new crackers in planning stage

ICIS

www.icis.com

Expected US C2 capacity vs past cycles

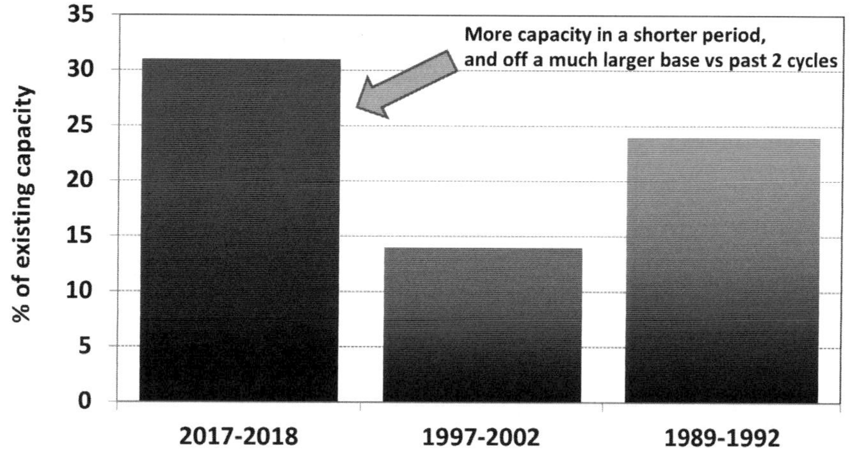

Note: 1997-2002 excludes NOVA's cracker in Joffre, Alberta, Canada

ICIS

www.icis.com

US petrochemical projects - propylene

ICIS

www.icis.com

On-purpose propylene plants – PDH, plus BASF's MTP

Company	Capacity '000 tonnes/year	Location	Start-up	Status
Dow Chemical	750	Freeport, Texas	H2 2015	Mechanically complete
Enterprise Products	750	Mont Belvieu, Texas	H2 2016	Under construction
Formosa Plastics	658	Point Comfort, Texas	2017	Under construction
Ascend Performance Materials	1,000	Chocolate Bayou, Texas	Mid-2019	FEED complete, FID Q1 2016
BASF (MTP)	475	Freeport, Texas	2019	FEED stage, FID mid-2016
Sunoco	New plant	Pennsylvania	NA	Evaluation
Dow Chemical	New plant	US	NA	Unknown
Enterprise Products	New plant	Texas, US	Shelved	Cancelled
REXtac	300	Odessa, Texas	Cancelled	Cancelled

ICIS

www.icis.com

Global propylene prices

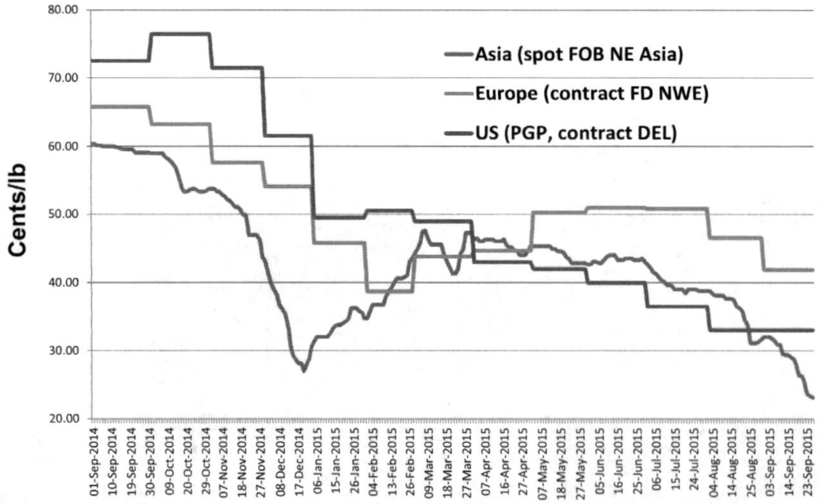

ICIS

www.icis.com

US MTP/PDH – not looking great at $50 oil

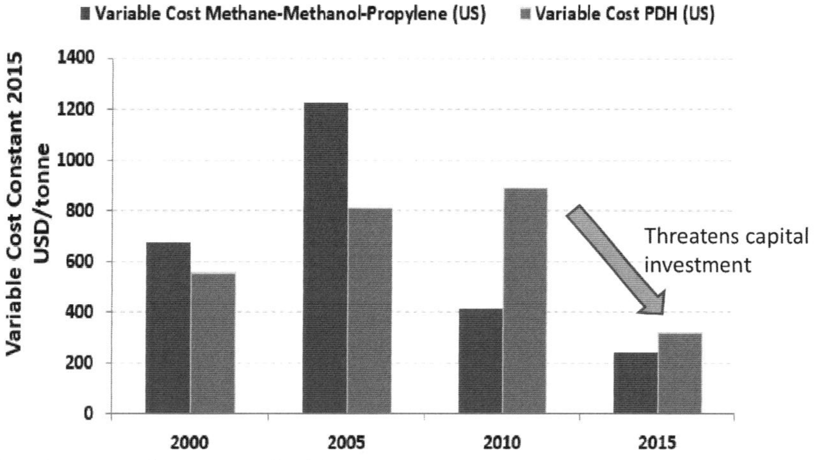

SOURCE: ICIS Consulting; 2015 until end-August

ICIS

www.icis.com

US petrochemical projects - methanol

ICIS

www.icis.com

Planned new US methanol projects

Company	Capacity '000 tonnes/year	Location	Start-up	Status
Methanex	1,000	Geismar, Louisiana	End 2015	Under construction
Celanese/Mitsui	1,400	Clear Lake, Texas	Oct 2015	Under construction
Yuhuang Chemical	1,700	St James Parish, Louisiana	2018	Construction starts Q4 2015
G2X Energy	1,400	Lake Charles, Louisiana	NA	Construction starts end 2015
South Louisiana Methanol	1,900	St James Parish, Louisiana	NA	Air permit acquired
Celanese/Mistui	1,300	Bishop, Texas	NA	Permits filed, FID 2016
OCI (Natgasoline)	1,750	Beaumont, Texas	NA	NA
Valero	1,600	St Charles, Louisiana	NA	NA
Castleton Commodities International	1,800	Plaquemines Parish, Louisiana	NA	NA
Northwest Innovation Works	1,600	Tacoma, Washington	NA	NA
Northwest Innovation Works	3,200 (2 units)	Oregon	NA	NA
Connell Group/Sino Life Insurance	7,200 (2 units)	Texas or Louisiana	NA	NA

ICIS

www.icis.com

Global methanol prices

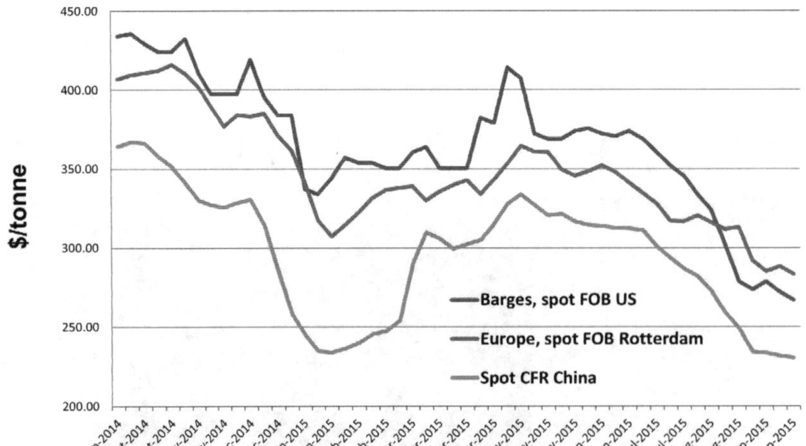

ICIS

www.icis.com

Capital spending cycles

ICIS

www.icis.com

US chemical industry CAPX
$bn

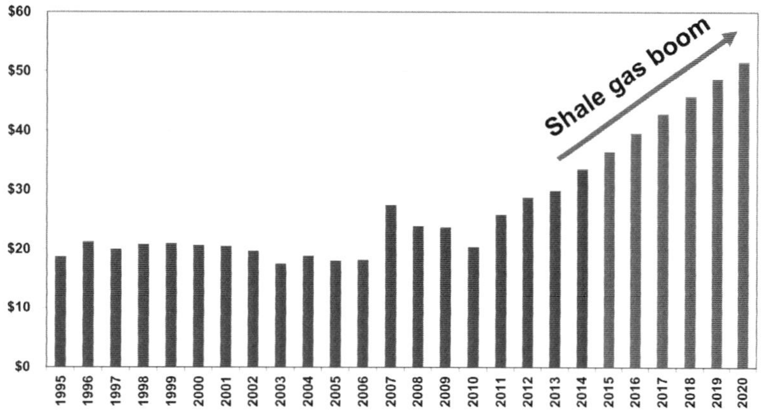

SOURCE: American Chemistry Council

ICIS

www.icis.com

Capital spending as % of depreciation

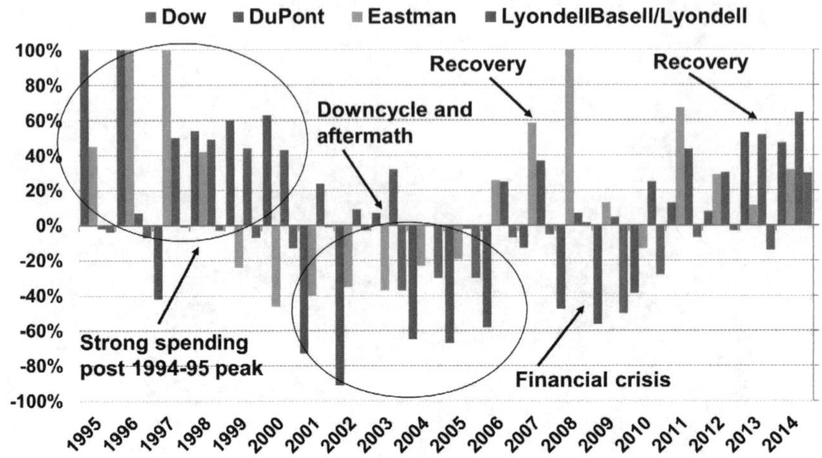

NOTE: Above 0 = % CAPX higher than depr, below 0 = % CAPX below depr, Bars shown are capped at +100% and -100%

ICIS

www.icis.com

Global context

Europe
Middle East
China

ICIS

www.icis.com

Europe – competitiveness improves on oil

Feedstock strategy in flux

- Closures in Europe have already happened – crackers, PE plants
- Naphtha-based economics have improved significantly on oil decline.
- Companies announced plans to import US ethane for feedstock – INEOS, Borealis, SABIC, Versalis
- But some will reevaluate import plans, in light lower oil prices = better naphtha competitiveness

ICIS

www.icis.com

Europe, Asia to import cheap US ethane

Company	Location	Amount/year	Target date
INEOS	Rafnes, Norway	800,000 tonnes	2015
INEOS	Grangemouth, UK	NA	H2 2016
Borealis	Stenungsund, Sweden	240,000 tonnes	Q3 2016
Reliance	India	1.5m tonnes	Q4 2016
SABIC	Teesside, UK	NA	2016
Versalis	Dunkirk, France	NA	2016

ICIS

www.icis.com

Middle East – the Iran factor

Potential lifting of sanctions by end 2015/2016

- Iran has massive unrealized potential, for oil and gas, and petrochemical production. 2nd largest natural gas reserves
- Needs technology, expertise and financing
- It will take time – maximizing oil production will be the 1st priority. Then later petrochemical operating rates. Then new projects.
- Elsewhere, Sadara (Saudi Aramco, Dow) and Petro Rabigh II (Saudi Aramco, Sumitomo) to start up in 2016. Borouge 3 in Abu Dhabi ramping up olefins and PE, PP production now. But not much else planned. 2 Qatar cracker projects cancelled.

ICIS

www.icis.com

Iran's planned petrochemical projects (to 2018)

Product	Capacity '000 tonnes/year
Methanol	19,140
Urea	10,750
Ammonia	7,445
Ethylene	5,318
HDPE	1,940
Propylene	1,582
MEG	1,500
LLDPE/HDPE	1,200
PP	750

SOURCE: National Petrochemical Company (Iran), March 2015

ICIS

www.icis.com

China CTO/MTO projects – delays/cancellations

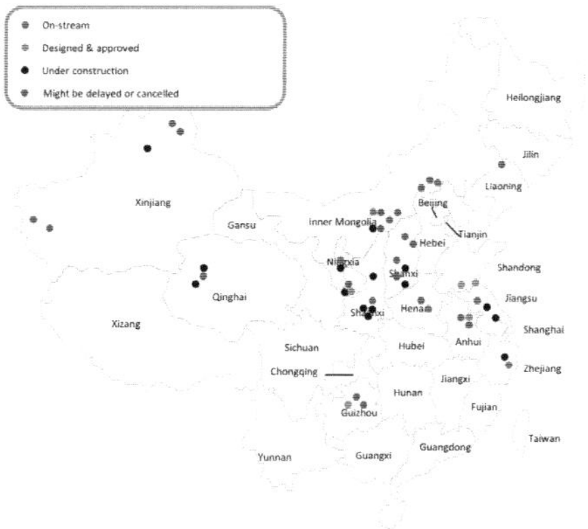

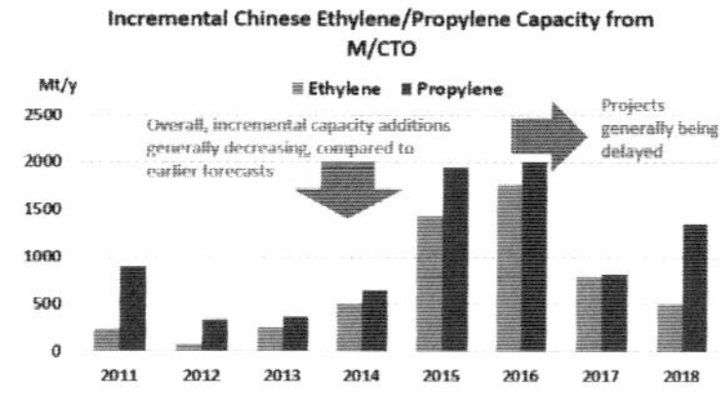

- Stricter environmental standards
- Big doubts over growth in the coal-to-olefins industry. Water is an issue for CTO
- Operationally competitive; Capitally & Regionally handicapped
- For coastal-based MTO, cost and feedstock availability challenges

SOURCE: ICIS Consulting and ICIS Supply and Demand Database

ICIS

www.icis.com

China ethylene – solid gains

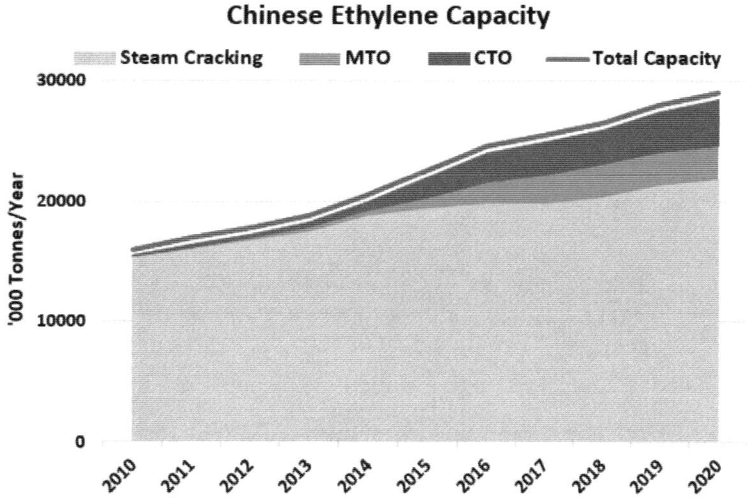

SOURCE: ICIS Consulting

ICIS

www.icis.com

China propylene – much more!

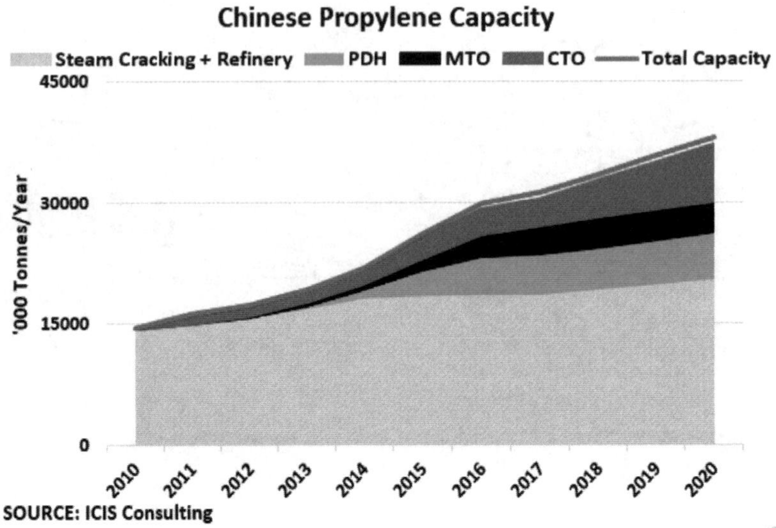

SOURCE: ICIS Consulting

ICIS

www.icis.com

Conclusion

China, oil change dynamics

- China weakness and devaluation to pressure commodities, including petrochemicals. Macro outlook more challenging.
- Crude oil collapse and volatility puts many planned projects at risk – in the US, Asia and even the Middle East.
- Yet US footprint in global petrochemicals will grow. Likely 6-8 new crackers and derivatives projects, plus on-purpose propylene, methanol.
- **Overcapacity is still a real threat.** Major capacity is coming on, especially in the US, in a challenging global macro environment. But low crude will blunt the next wave.

ICIS

www.icis.com

Thank you!

ICIS Chemical Business
Joseph Chang, Global Editor
joseph.chang@icis.com
+1 212 791 4224

Subscriptions:
Individual and bulk, print and digital
www.icis.com/chemicalbusiness

ICIS Consulting
Karl Bartholomew (Americas)
karl.bartholomew@icis.com
+1 713 525 2605

ICIS

www.icis.com

Turning a Tier 3 Profit

Esteban, J.,
Hartman, M.,

Originally presented on: 10/1/2015 9:30:00 - 10:00:00

Currently refiners are seeking options to address reductions in gasoline pool sulfur as required by the US EPA in 2017 as part of the Tier 3 regulations. However, several customers working with Criterion have already made strides to both meet the future Tier 3 regulations and are reaping the benefits of early success. With advantages in volume swell, hydrogen upgrade, and operating cost structure these refineries are turning a profit from the new stringent regulations. These profits have been realized as increases in gasoline volume from improved FCC feed quality and operations, distillate (specifically ULSD) volume via synergies in co-processing, as well as overall refining capacity improvements. While profitability as it relates to Tier 3 upgrades are contingent on refinery configuration, this paper explores the options and solutions employed with a case study from Criterion's extensive portfolio demonstrating these benefits.

Effect of Catalyst Shape on Yield and Performance of Packed Beds

Gregor D. Wehinger†, Thomas Eppinger‡ & Getnet Agegnehu §, Ravi Aglave§

†Chemical & Process Engineering, Technische Universität Berlin
‡ CD-adapco, Nuremberg, §CD-adapco, Houston

AIChE
Southwest Process Technology Conference

7th AIChE Southwest Process Technology Conference | October 1-2, 2015 | Galveston, TX

CD-adapco

1. **Dry reforming of methane (DRM)**
2. **Motivation for spatially resolved CFD simulations**
3. **Effect of catalyst shape on yield and performance of DRM**
4. **Outlook to catalytic foams**
5. **Conclusions**

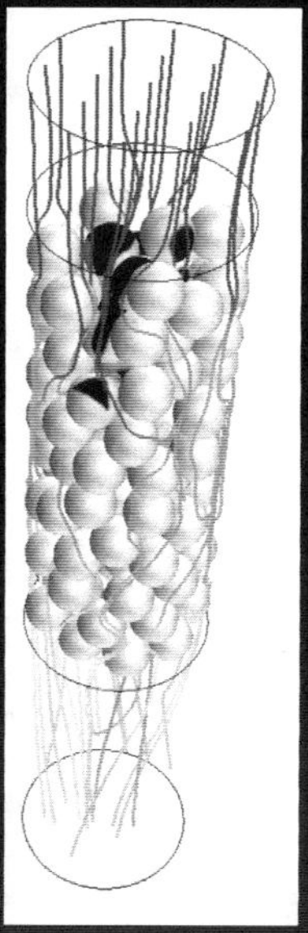

Dry reforming of methane

CD-adapco

- Dramatic increase of atmospheric concentrations of "greenhouse" gases during last decades
- Chemical engineering's field of interest:
  - Process with CH_4 and CO_2 → synthesis gas

➢ **Dry reforming of methane (DRM):**

- Catalytic endothermic process:
  - $CH_4 + CO_2 \leftrightharpoons 2H_2 + 2CO \quad \Delta H \approx 260\ kJ/mol$
- Mainly Nickel-based catalysts
- Coke formation as drawback
- Performed at 700-1000°C

Fixed-bed reactors for DRM

CD-adapco

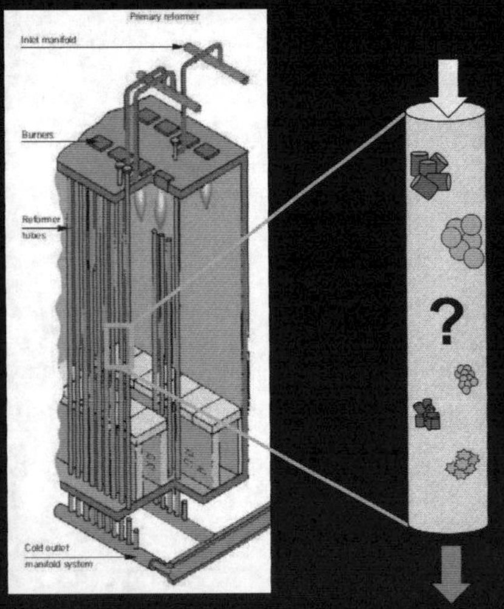

- **DRM: highly endothermic reaction**
- **Heat transfer into the reactor → small tube diameter**
- **High gas velocities and small pressure drop → large particles**

➢ **Small tube-to-particle-diameter-ratios (*N* < 15)**

- **Conventional description based on plug flow and pseudo-homogeneous kinetics fail for these low *N* beds**

➢ **Adequate modeling with full CFD and detailed chemical models**

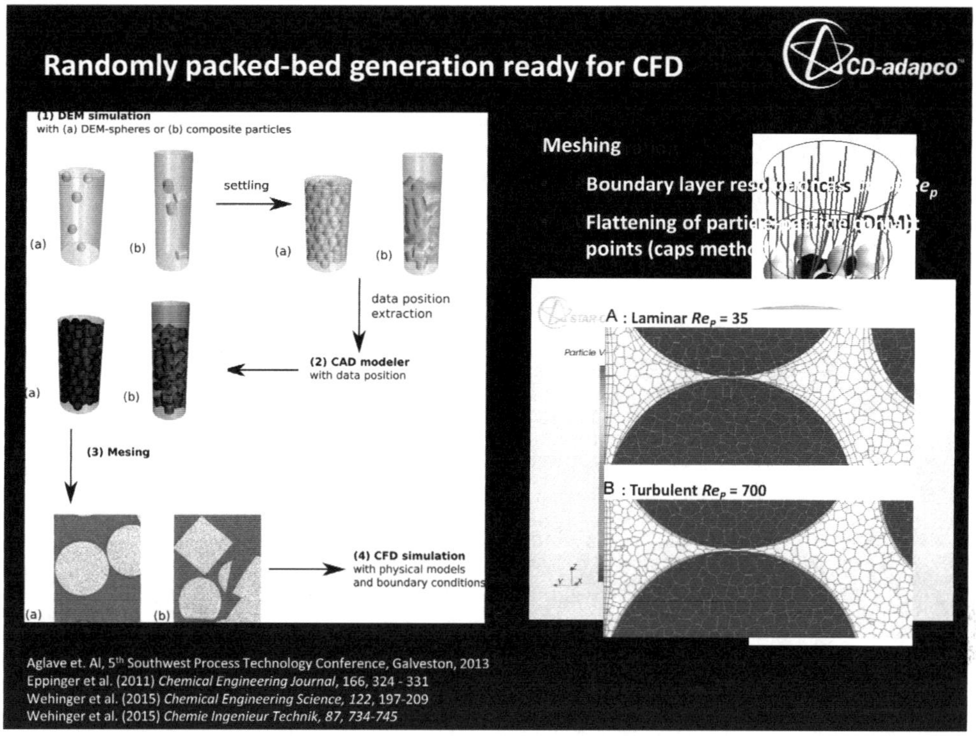

Comparing different particle shapes

CD-adapco

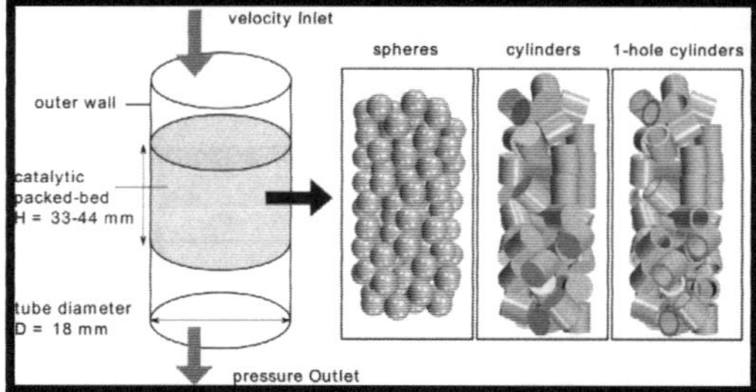

	spheres	cylinders	1-hole cylinders
Bed height H_{bed} (mm)	33	46	46
Tube diameter D (mm)	18	18	18
Particle diameter d_p (mm)	4	5	5
Particle height (mm)	-	5	5
Inner diameter (mm)	-	-	3.7
H/d_p ratio	4.5	3.6	3.6
Specific area a (m²/m³)	718	556	754
$F_{cat/geo}$	90	116	85
Mean porosity $\bar{\varepsilon}$	0.51	0.45	0.73
Superficial velocity v_{in} (m/s)	1.42	1.66	2.66
Reynolds number Re_p (-)	50	62	62
Inlet/Wall temperature (K)	1123	1123	1123
Feed $x_{CH4}/x_{CO2}/x_{N2}$	0.475/0.475/0.05		

- **Microkinetics from McGuire et al. (2011)**
- **Similar hydraulic residence time: 0.012 s (=1/*GHSV*)**
- **Similar catalytic active area: $F_{cat/geo}*A_{geo} = const$**
- **Three different particle shapes**

McGuire et al. (2011) Applied Catalysis A: General, 394, 257 - 265
Wehinger et al. (2015) *Chemie Ingenieur Technik, 87, 734-745*

Porosity and intersititital velocity

CD-adapco™

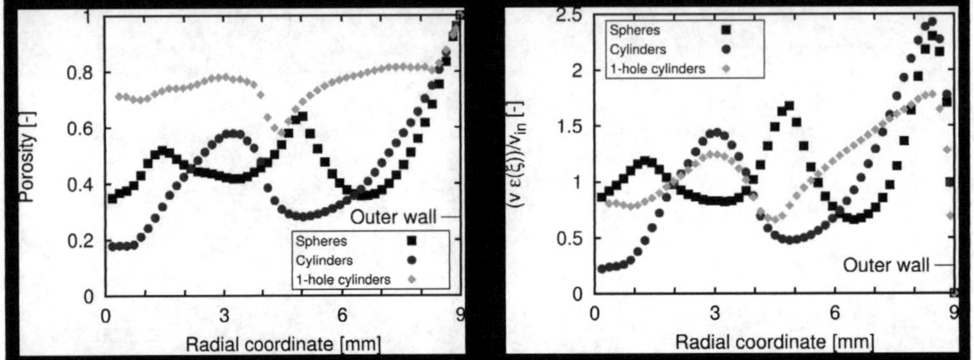

Flow field and hydrogen production

CD-adapco™

1-hole cylinders: particle diameter: 5 mm, $Re_p = 62$

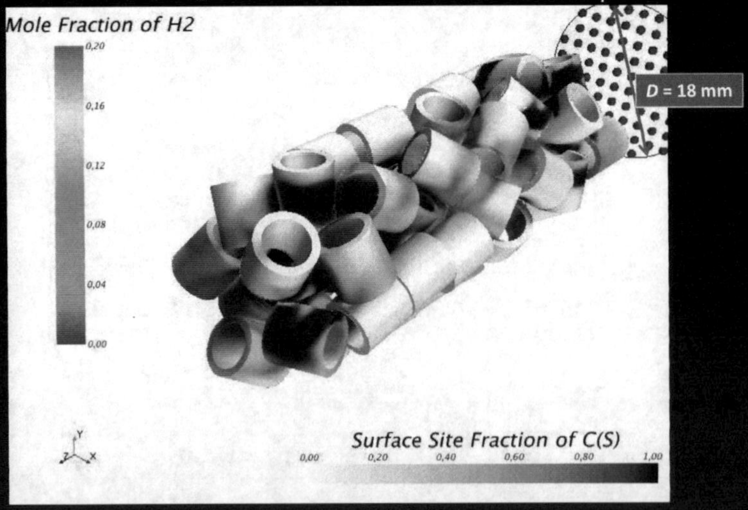

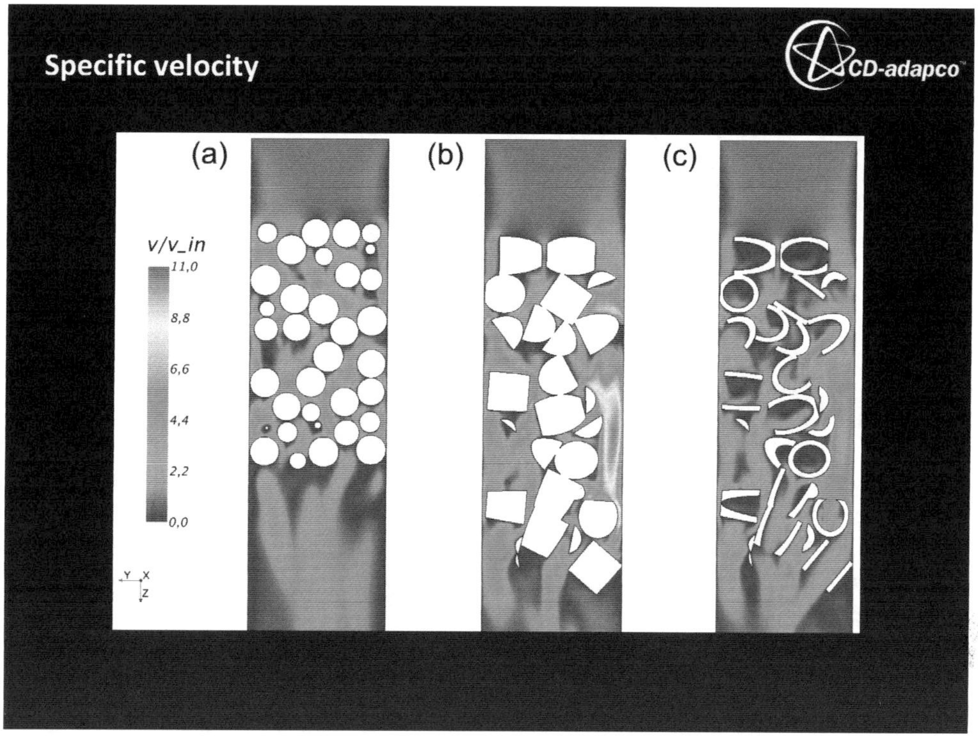
Specific velocity
CD-adapco
(a)
(b)
(c)
v/v_in
11,0
8,8
6,6
4,4
2,2
0,0

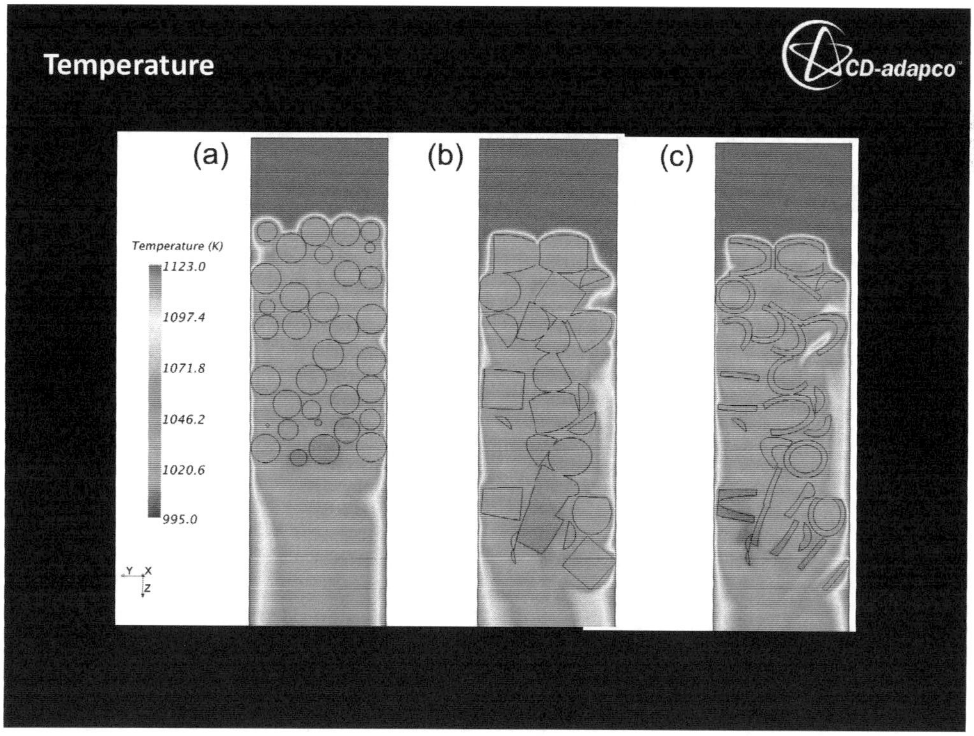
Temperature
CD-adapco
(a)
(b)
(c)
Temperature (K)
1123.0
1097.4
1071.8
1046.2
1020.6
995.0

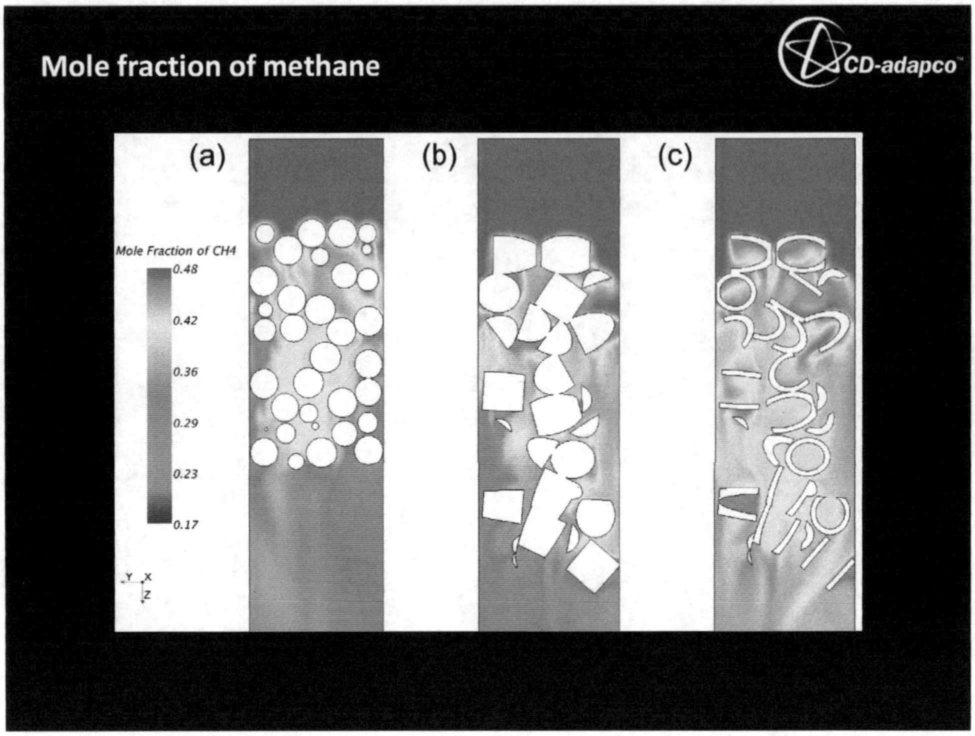
Mole fraction of methane
CD-adapco
(a)
(b)
(c)
Mole Fraction of CH4
0.48
0.42
0.36
0.29
0.23
0.17
Y
X
Z

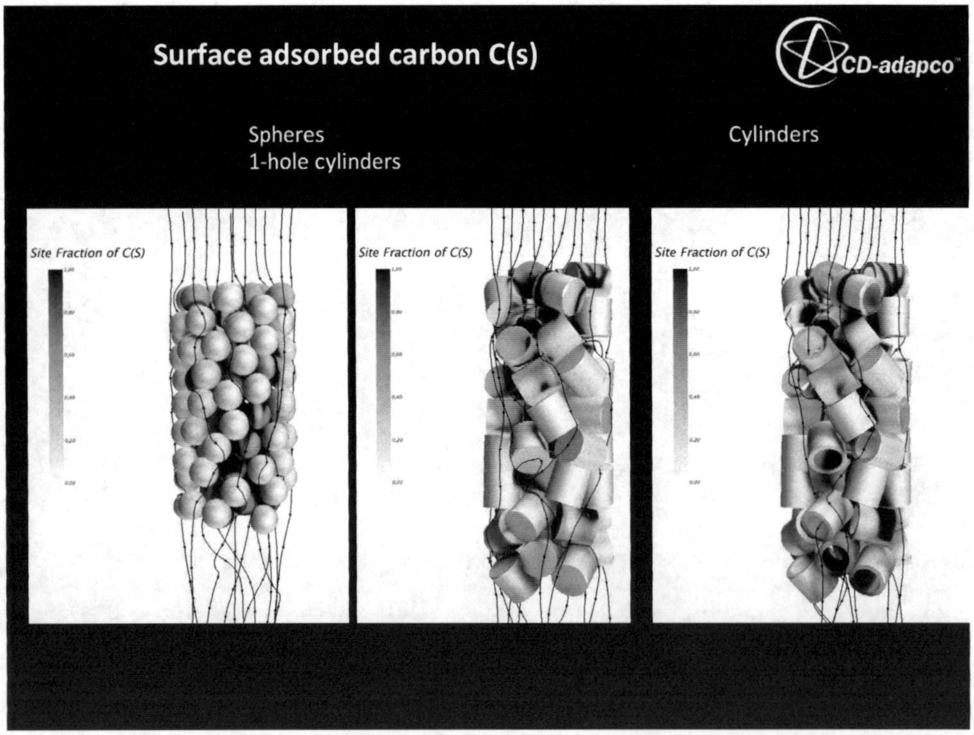
Surface adsorbed carbon C(s)
CD-adapco
Spheres
1-hole cylinders
Cylinders
Site Fraction of C(S)
Site Fraction of C(S)
Site Fraction of C(S)

Cross-section averaged values

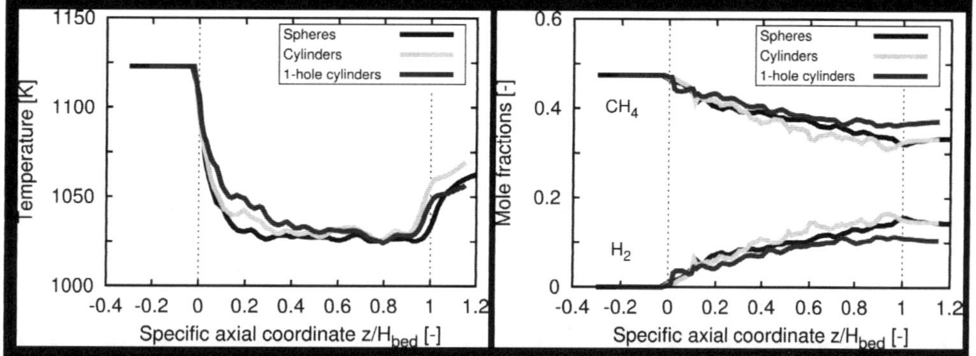

Cumulative curve of residence time

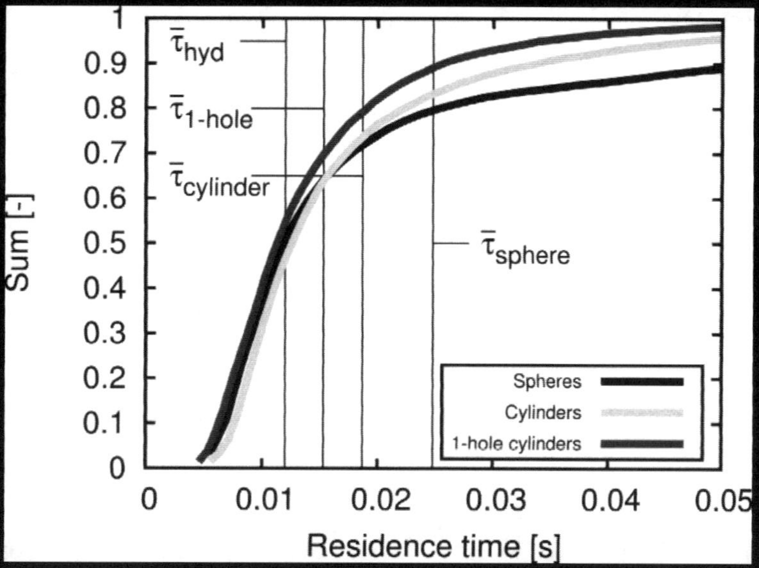

Quantitative comparison

Fixed-bed type	Mean residence time	X_{CH4}	X_{CO2}	Y_{CO}	Y_{H2}	Y_{H2O}	$\bar{\Theta}_{CO*}$	$\bar{\Theta}_{C*}$
	(s)	(%)	(%)	(%)	(%)	(%)	(-)	(-)
Spheres	0.0248	28.2	28.4	14.6	14.3	0.1	0.238	0.516
Cylinders	0.0187	30.2	30.5	15.6	15.3	0.1	0.287	0.447
One-hole cylinders	0.0153	21.8	22.0	11.3	11.0	0.1	0.221	0.539

PART II: Outlook to catalytic foams

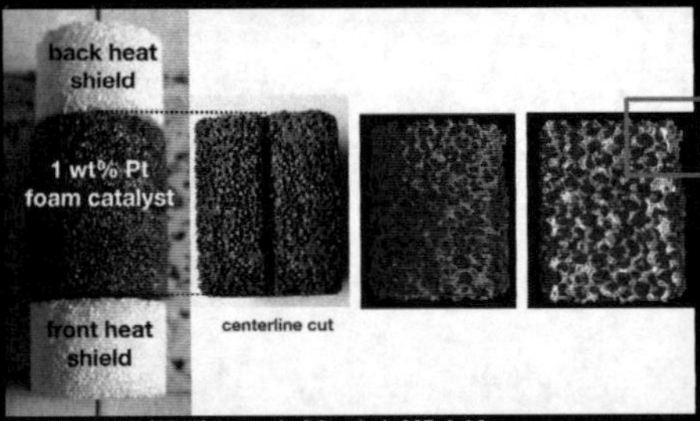

Korup et al. (2013) *Journal of Catalysis 297*, 1-16.

- **Fully computer-aided parameterized geometric generation (catFM)**
  - Basis: DEM packed-bed
  - Parameters: porosity, spec. surface area, strut profiles
- **Coupling with detailed reaction mechanism**
- **Heat transfer through struts**

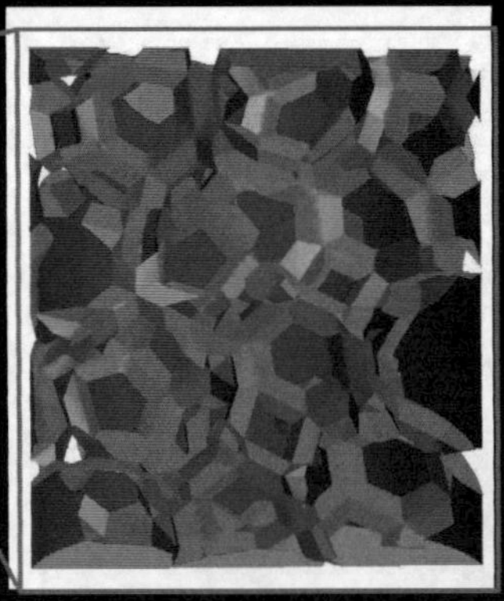

CAD data for CFD simulations

Wehinger et al. (2015) Chemical Engineering Journal, *accepted*

Experiments from Dalle Nogare et al. (2008)

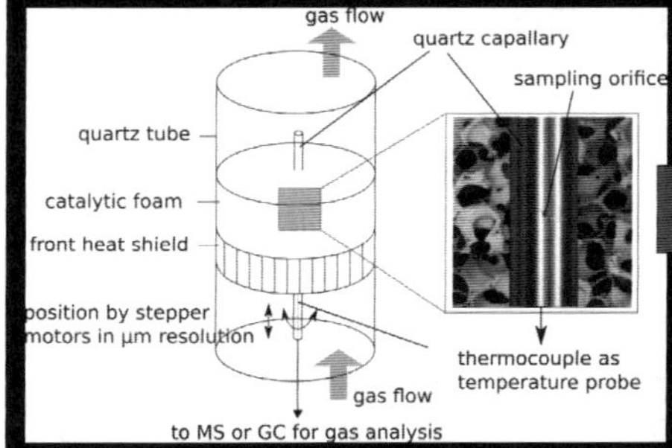

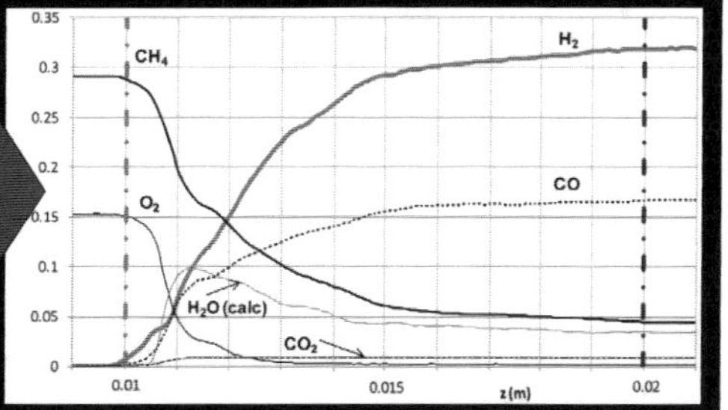

Catalytic partial oxidation of methane
$CH_4 + ½ O_2 \rightarrow 2 H_2 + CO$

α-Al_2O_3 foam monoliths
10 mm in length and 16.5 mm in diameter

$CH_4/O_2/Ar$ = 0.291/0.153/0.556
$\dot{V}_{total}$= 5000 mln/min^{-1} → $v_{in} \approx$ 2 m/s
S_V = 4650 m²/m³
ε = 0,7
$F_{cat/geo}$ = 2

Dalle Nogare et al. (2008) *Journal of Catalysis, 258*, 131-142

catFM and CFD setup

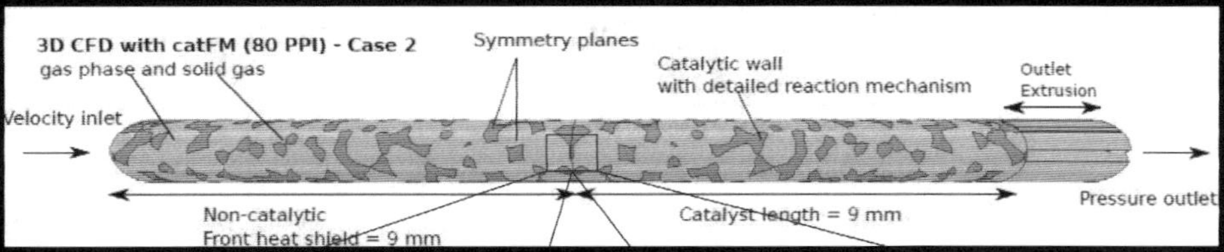

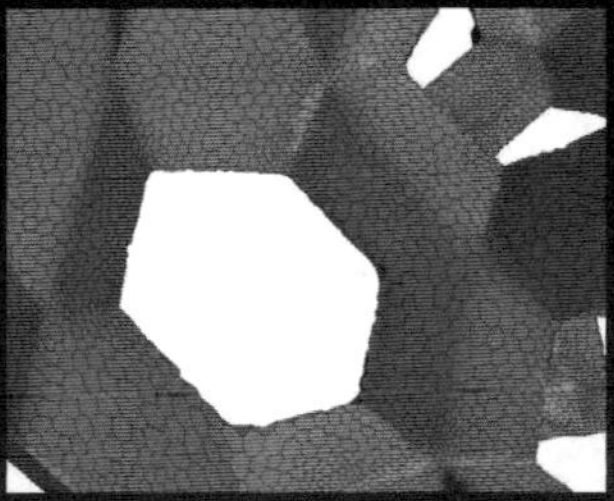

strut geometry

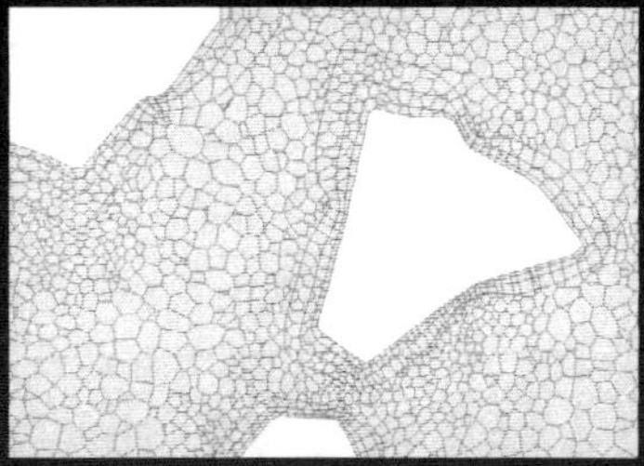

mesh resolution

Transport phenomena

CD-adapco

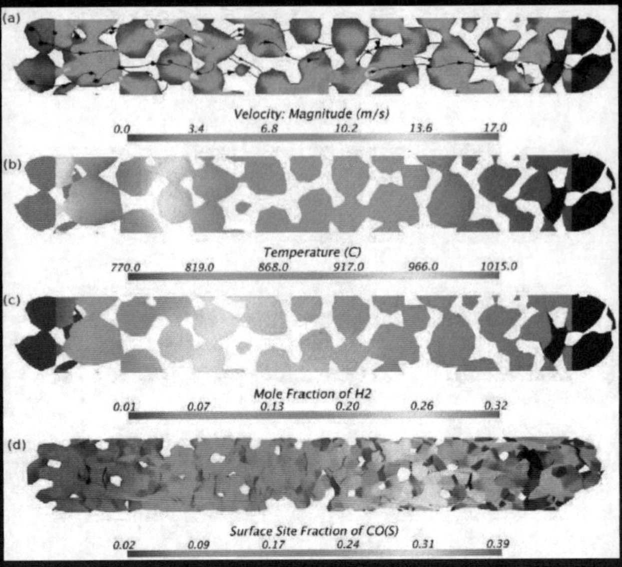

Axial temperature and mole fractions of CPOX

CD-adapco

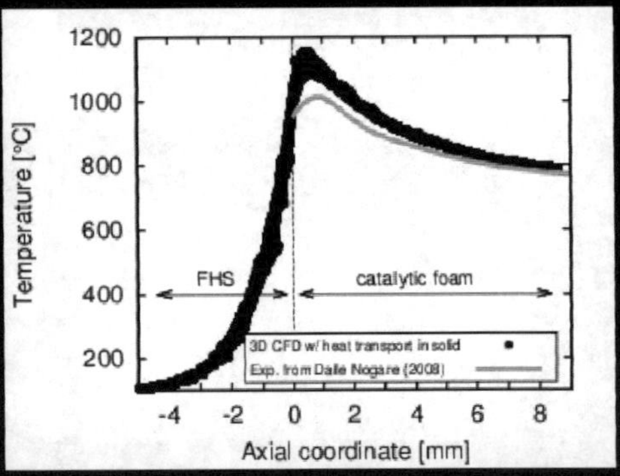

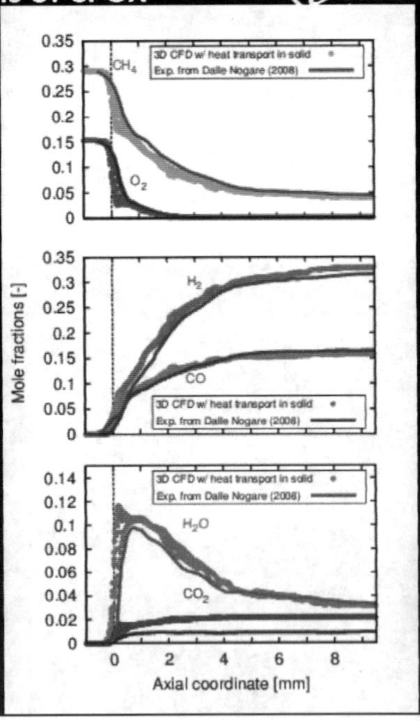

Conclusion

CD-adapco

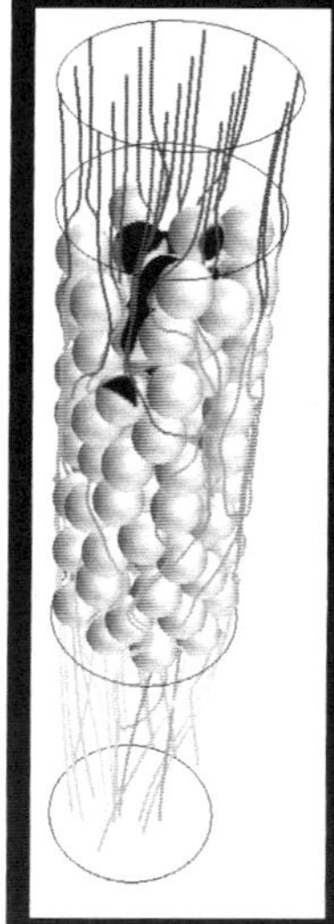

- Conventional plug flow and pseudo-homogeneous kinetic models fail in small *N* fixed-bed reactors.
- Adequate modeling with detailed CFD in combination with detailed kinetics
- Interactions between chemical kinetics and transport of momentum, heat and mass
- Quantitative comparison of catalyst performance between different particle shapes
- catFM able to model artificially foam structures

➢ Better understanding of multi-scale chemical reactors

CD-adapco

Carbon Dioxide Conversion Through Tri-Reforming: Reactor/Process Design and Optimization

Benson, T. J., Lamar University
Zhang, Y., Lamar University
Raihan, M. E., Lamar University
Originally presented on: 10/1/2015 11:00:00 - 11:30:00

Conversion of CO_2 into valuable products is developing as a promising technology for reduction of exogenous CO_2 contributions in the atmosphere. According to the International Energy Agency, the total worldwide CO_2 produced in 2007 was approximately 3.2×10^{10} tons. Although capture and storage technologies have been well studied in recent years, they are costly due to separation of CO_2 and steam using amine-type absorbers. Since governmental regulations are likely to require fewer CO_2 emissions by industry, it is necessary to explore economical solutions to the CO_2 problem. One avenue to combat the growing release of CO_2 into the atmosphere is to convert it to useful chemicals via tri-reforming. Tri-reforming is the conversion of carbon dioxide, methane, and water to produce syngas without the separation of CO_2 and steam. Through careful catalyst design, a novel catalyst has been synthesized using reverse micelles to create highly active nano-sized NiSn particles onto metal-oxide supports. These catalysts have been characterized and tested in our laboratory. Subsequently, a process design using ASPEN simulation has been developed to efficiently convert CO_2.

HOW NEW REACTOR INTERNALS CAN BE USED TO ENHANCE PROFITABILITY

Shell Global Solutions

Pankaj Desai
Licensing Sales Manager

DISCLAIMER

Shell Global Solutions is a network of independent technology companies in the Shell Group. In this document, the expressions "Shell" or "Shell Global Solutions" are sometimes used for convenience where reference is made to these companies in general, or where no useful purpose is served by identifying a particular company.

The information contained in this material is intended to be general in nature and must not be relied on as specific advice in connection with any decisions you may make. Shell and Shell Global Solutions are not liable for any action you may take as a result of you relying on such material or for any loss or damage suffered by you as a result of you taking this action. Furthermore, these materials do not in any way constitute an offer to provide specific services.

Some services may not be available in certain countries or political subdivisions thereof.

The information contained in this presentation is provided for general information purposes only. No representations or warranties, express or implied, are made by Shell Global Solutions International B.V. or its affiliates concerning the applicability, suitability, accuracy or completeness of the information contained herein and neither Shell Global Solutions International B.V. nor its affiliates accept any responsibility whatsoever for the use of this information. Neither Shell Global Solutions International B.V. nor any of its affiliates shall be liable in any way for any liability, loss, cost or damage incurred as a result of the receipt or use of the information contained herein.

AGENDA

- How new reactor internals can unlock additional value
- Shell reactor internals
- Increased catalyst volume
- Maintenance and safety advantages
- Case study
- Summary

COULD NEW REACTOR INTERNALS UNLOCK ADDITIONAL VALUE FOR YOU?

If you have a need to…

- Meet tighter product specifications
- Process more difficult feeds
- Increase capacity
- Enhance profitability
- Improve safety
- Extend cycle length
- Reduce turnaround times

Or if you have experienced any of the following in your reactor…

- High pressure drop
- Bed fouling
- Thermal maldistribution
- Short cycle length
- High weighted average bed temperature (WABT)
- Reliability issues

…then installing state-of-the-art reactor internals could help to unlock additional value.

WHAT COULD NEW REACTOR INTERNALS BE WORTH TO YOU?

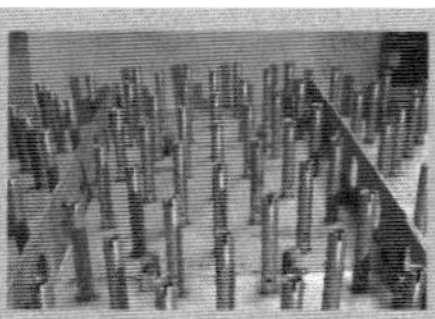

ENHANCE SAFETY
Power tools, welding equipment and hot-work permits are not required

PROCESS HIGHER-MARGIN FEEDS
A 5% increase in ICO feed in a 100-t/ h unit when the ICO upgrade margin is $50/ t more than the gas oil upgrade margin: **$2.2 million**

IMPROVE HYDROCRACKER PRODUCT YIELDS
Installing a higher selectivity catalyst in a 200-t/ h unit to produce more middle distillates, when the naphtha diesel spread is $100/ t: **$3.5 million**

INCREASE CYCLE LENGTH
Fewer interim shutdowns in a 100-t/ h hydrotreater with an upgrade margin of $20/ t: **$670,000**

CUT TURNAROUND TIMES
Cutting turnarounds in a 100-t/ h hydrotreater with an upgrade margin of $20/ t by two days: **$96,000**

SHELL REACTOR INTERNALS

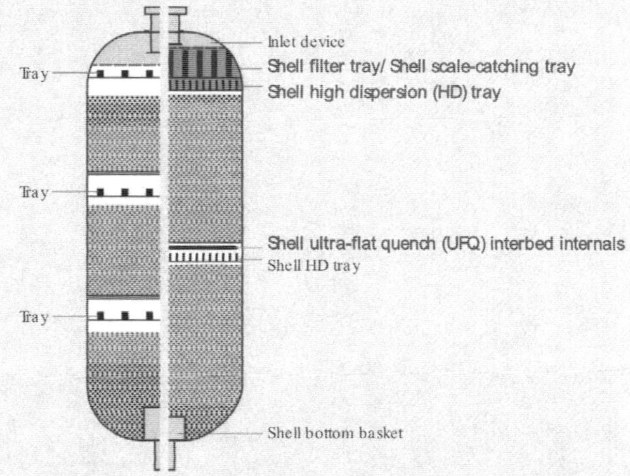

FOULING ABATEMENT

- Fouling can restrict unit throughput and affect performance
- Shell filter trays and Shell scale-catching trays have a strong track record of reducing bed fouling

06 October 2015

SHELL FILTER TRAYS AND SHELL SCALE-CATCHING TRAYS

FEATURES OF SHELL'S FOULING ABATEMENT TECHNOLOGIES

- Offer additional filtering capacity by trapping contaminants and sediments
- Minimize the height of the top-bed grading layers

CUSTOMER VALUE

- Reduced pressure drop
- Enhanced reliability and reduced shutdown frequency
- Higher catalyst loading capacity

Shell filter tray: Chemical and physical fouling abatement

Shell scale-catching tray: Physical fouling abatement and scale trapping

 7th AIChE Southwest Process Technology Conference CONFIDENTIAL 06 October 2015 8

FILTER AT WORK

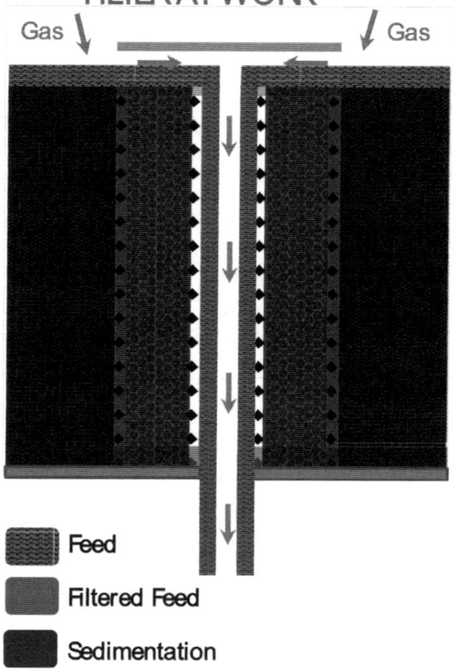

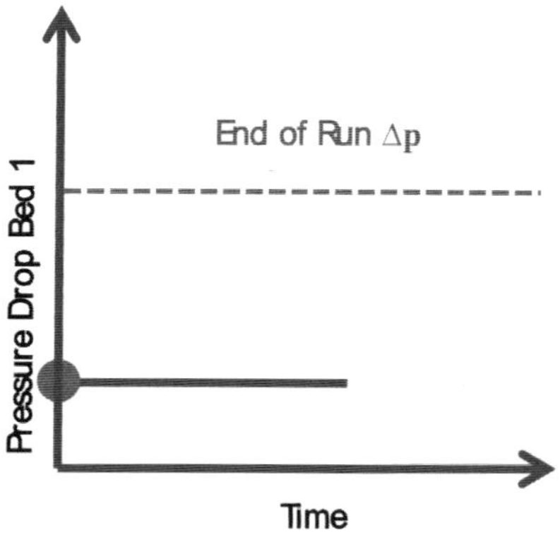

- Feed filtered by filter basket without increasing pressure drop over the 1st bed
- Filter basket is no longer effective pressure drop over the 1st bed increases
- End Of Run is reached later than without tray

HOW SHELL FILTER TRAYS HELPED TO REDUCE PRESSURE DROP IN A HYDROCRACKER

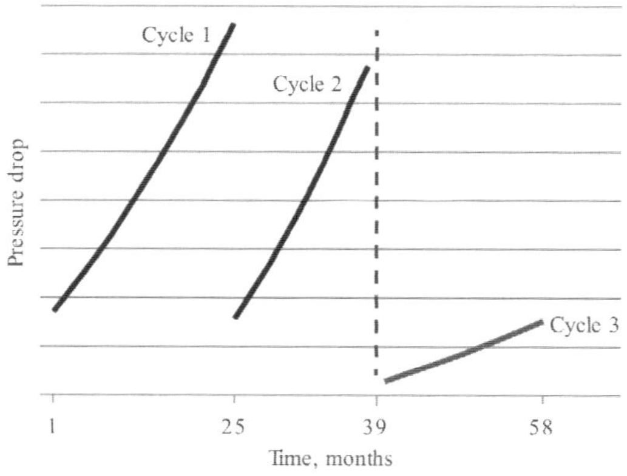

- Cycles 1 and 2: the reactor suffered rapid increases in pressure drop
- Cycle 3 (the first after Shell filter trays were installed): the pressure drop increased much more slowly

An integrated solution that featured catalyst analysis, catalyst loading, Shell filter trays and a grading layer helped a North American refiner to reduce the pressure drop in its hydrocracking reactor.

SHELL HD TRAYS

FEATURES OF SHELL HD TRAYS

- Create an almost perfect uniform liquid distribution with close to 100% catalyst utilization
- Help to increase catalyst volume (active catalyst can be loaded to within 6 in. of the bottom of the Shell HD tray)
- Boltless design

CUSTOMER VALUE

- Increased cycle lengths owing to
  - Lower WABT
  - Lower inlet temperatures
  - Improved thermal distribution
- Maximized catalyst loading
- Easy installation and maintenance

HOW SHELL HD TRAYS INCREASE CATALYST UTILIZATION

Shell HD trays

- Enable nearly 100% of the catalyst inventory to be utilized
- Offer high feed rate flexibility

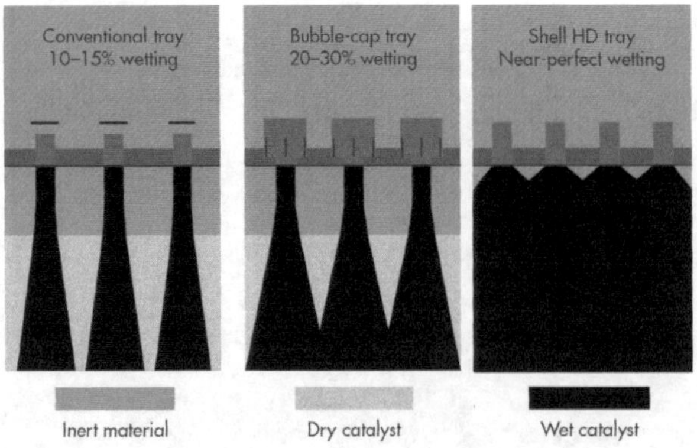

HOW SHELL HD TRAYS IMPROVED THERMAL DISTRIBUTION

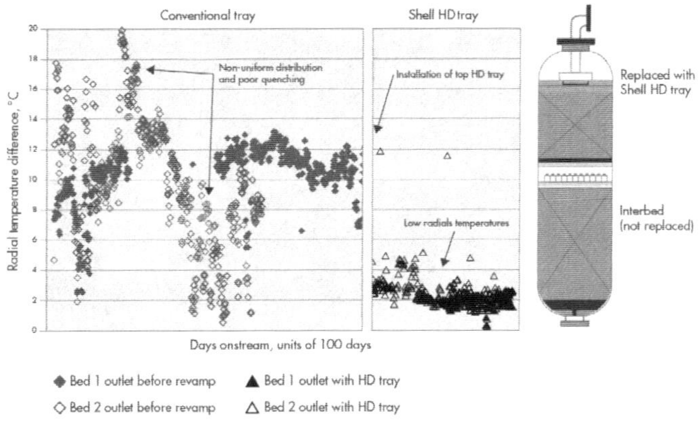

When a bubble-cap distribution tray at the top of Bed 1 was replaced with a Shell HD tray, the radial thermal maldistribution in both beds was virtually eliminated.

HOW A SHELL HD TRAY INCREASES CYCLE LENGTH

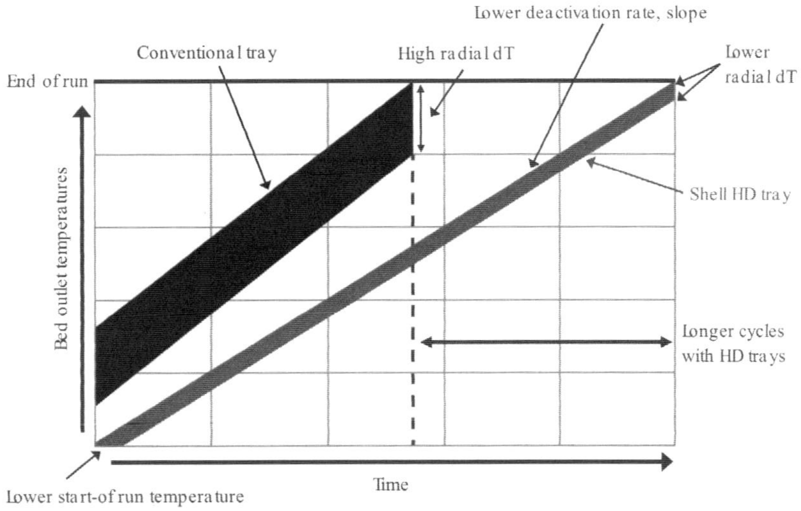

ACHIEVING PERFECT QUENCH MIXING

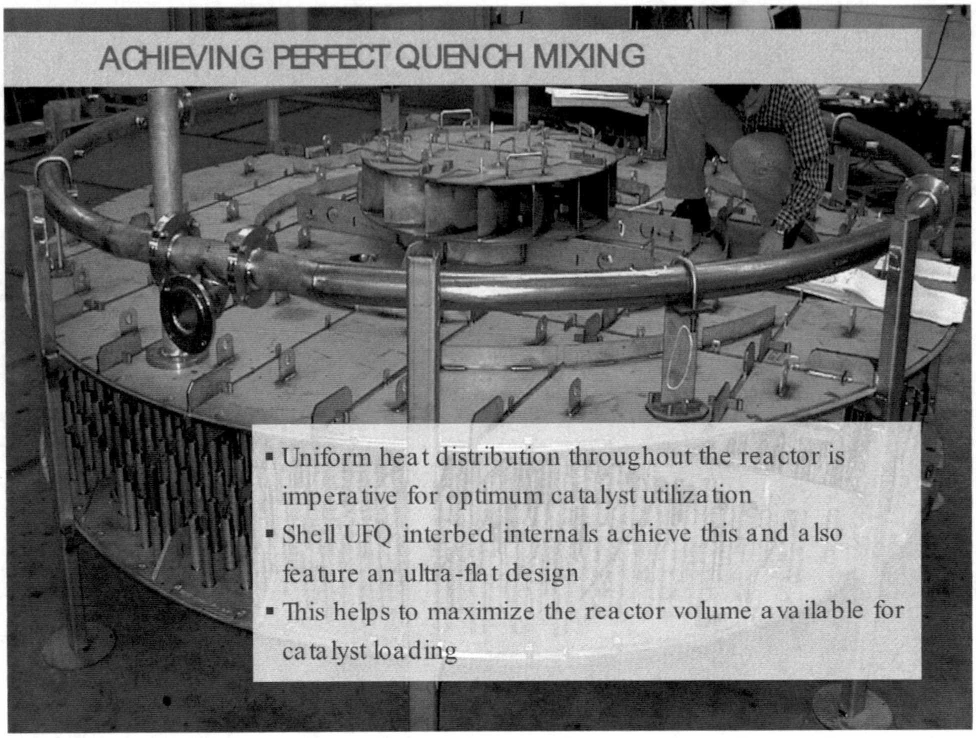

- Uniform heat distribution throughout the reactor is imperative for optimum catalyst utilization
- Shell UFQ interbed internals achieve this and also feature an ultra-flat design
- This helps to maximize the reactor volume available for catalyst loading

SHELL UFQ INTERBED INTERNALS

FEATURES OF SHELL UFQ INTERBED INTERNALS

- Proven technology: over 50,000 t/d processed
- Radial thermal maldistribution essentially eliminated
- Wide operating window (±50%)
- Quench load independent performance
- Integrated with the Shell HD tray

CUSTOMER VALUE

Because more catalyst can be loaded into the reactor, refiners can

- Increase cycle lengths, throughput and conversion rate
- Achieve tighter product quality specifications for sulfur, boiling point, density, cetane and cold flow properties

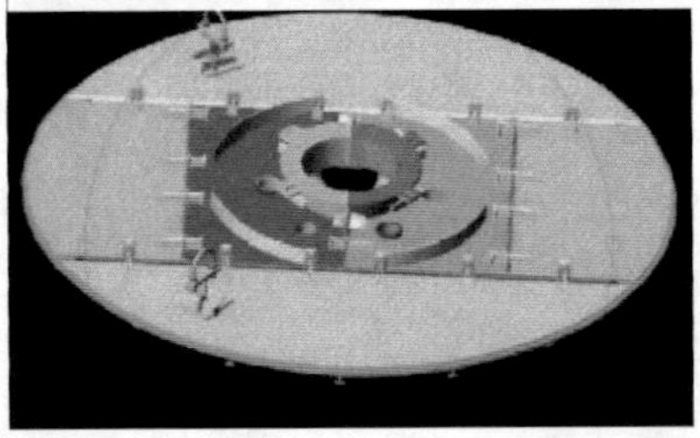

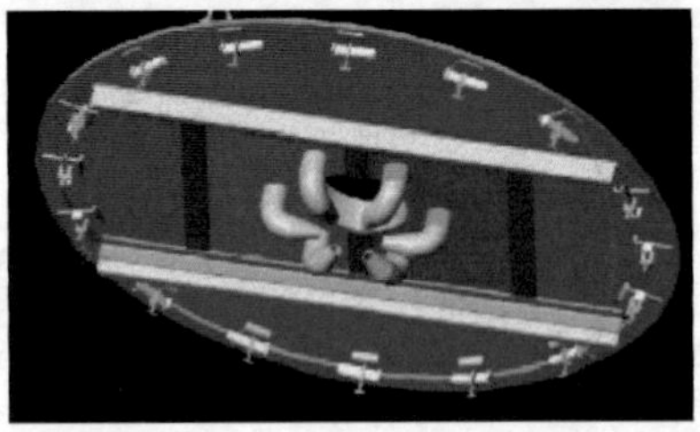

HOW SHELL UFQ INTERBED INTERNALS HELPED TO REDUCE RADIAL TEMPERATURE MALDISTRIBUTION

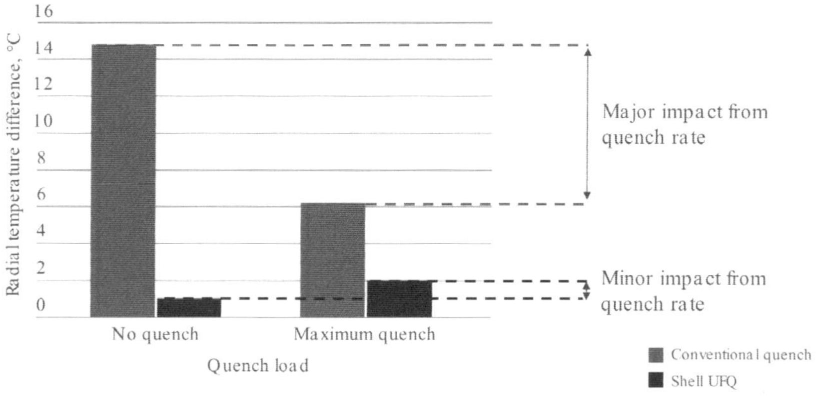

Shell UFQ interbed internals provide high quench efficiency and significantly reduce radial temperature maldistribution, regardless of the quench flow rate.

INCREASING CATALYST VOLUME AND UTILIZATION

THE COMBINATION OF NEW-GENERATION REACTOR INTERNALS AND CATALYSTS CAN HAVE A MAJOR IMPACT

- The amount of catalyst loaded in the reactor can be increased by up to 30%
- In addition, Shell reactor internals utilize the loaded catalyst better
  - Conventional designs typically use about 80% of the catalyst
  - Shell reactor internals enable close to 100% utilization
- These two factors provide refiners with options to generate more margin:
  - Cycle life extension
  - Processing heavier, less-expensive feeds

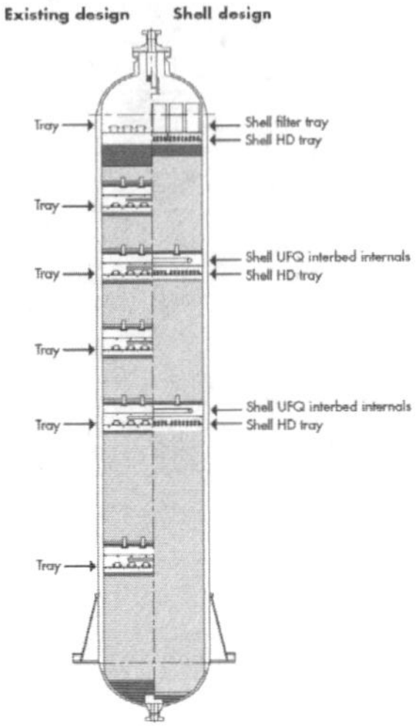

EASE OF INSTALLATION, MAINTENANCE ADVANTAGES, AND ENHANCED STAFF SAFETY

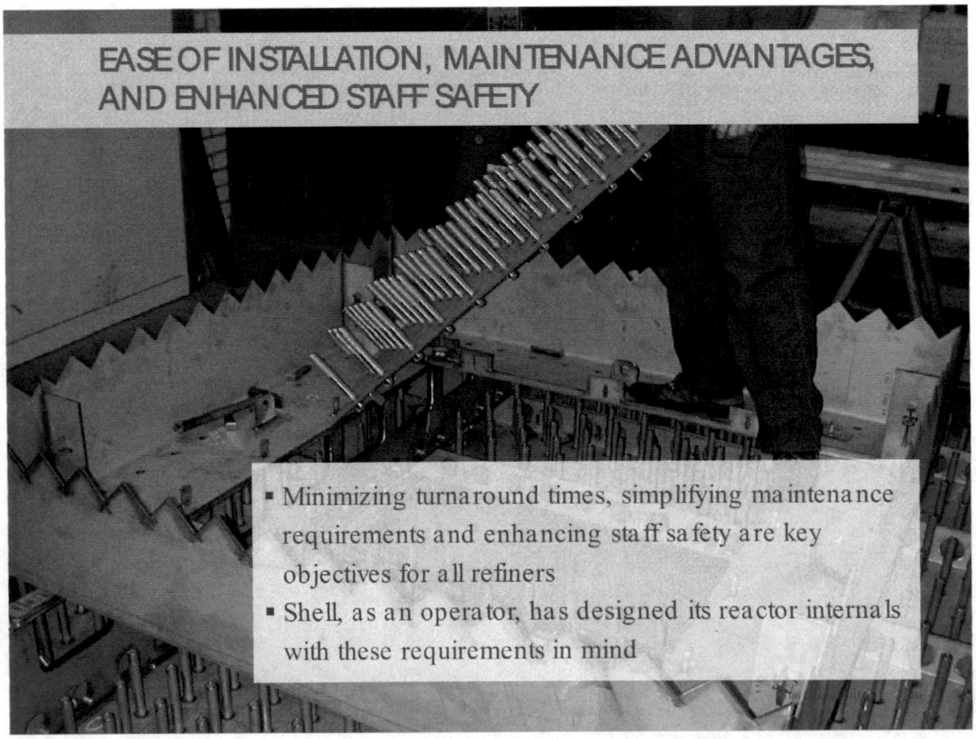

- Minimizing turnaround times, simplifying maintenance requirements and enhancing staff safety are key objectives for all refiners
- Shell, as an operator, has designed its reactor internals with these requirements in mind

SAFE AND EASY INSTALLATION AND MAINTENANCE

- Wedge pins and split keys **(no nuts and bolts)**
- Easy to install and maintain
- Significantly reduces time for inert entry and turnarounds
- No welding
- One tray per shift can be installed
- Manways: ½ hour to open, ½ hour to close

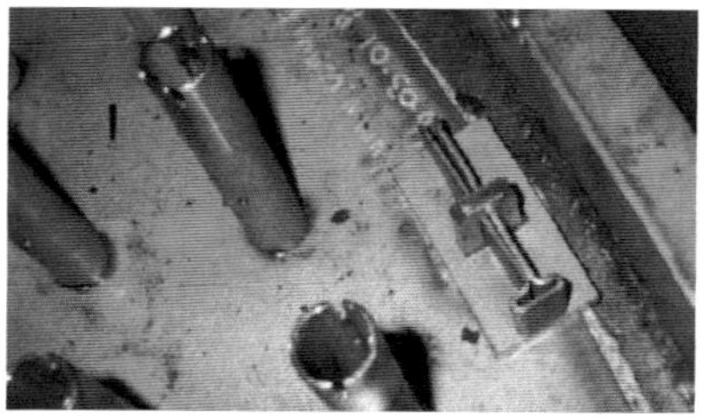

Shell reactor internals have a boltless, weldless and ergonomic design that facilitates fast installation, easy maintenance and a shorter turnaround.

TRACK RECORD

- Shell reactor internals are helping more than 300 customers around the world to stay competitive
  - 500+ designed
  - 250+ online
- 20 years' operational experience
- Installed in more than 1,500 reactor vessels across the world, of which about 80% are in non-Shell-operated refineries

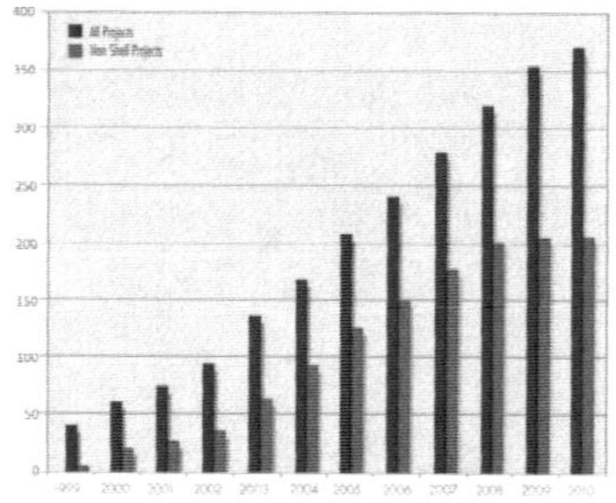

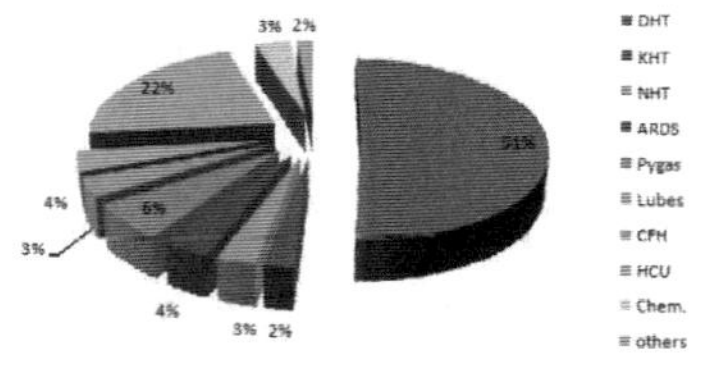

Shell Global Solutions is the market leading licensor of reactor internals and has the largest installed base.

CASE STUDY: PETROBRAS' CLEAN FUELS JOURNEY

DRIVERS

Petrobras needed to adapt its units to make 10-ppm-sulfur diesel and to increase its refining capacity.

The refiner intended to revamp an existing, moderate-pressure diesel hydrotreater at its Paulínia refinery for ultra-low-sulfur-diesel service. The unit was originally designed to produce 500-ppm-sulfur diesel at a feed rate of 5,000 m^3/d.

SOLUTION

Technologists from Petrobras, Shell Global Solutions and Criterion investigated solutions. The team found a way to achieve the objectives through a drop-in solution that involved Shell reactor internals in combination with a new state-of-the-art Criterion catalyst. In less than six months, the reactor internals were designed and manufactured.

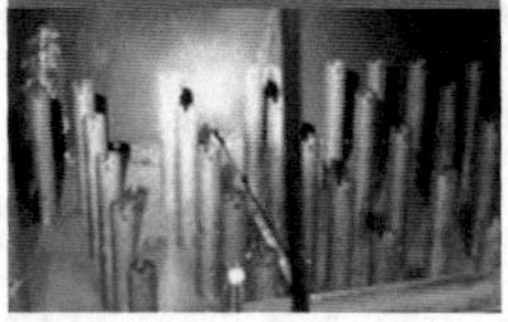

VALUE DELIVERED

The loaded catalyst volume increased by 19% without increasing the total reactor volume, which helped to increase the unit's capacity to 6,000 m^3/d. Not having to change the reactor was a significant capital saving.

All the modifications were also completed in the normal turnaround period of just over a month.

 7th AIChE Southwest Process Technology Conference CONFIDENTIAL 06 October 2015 24

INCREASED CATALYST VOLUME AND UTILIZATION

Increased catalyst volume and utilization	Existing reactor	Revamped reactor
Total reactor volume (m^3)	347	347
Usable catalyst volume (%)	67.1	79.5
Loaded catalyst volume (m^3)	233	276
Distribution effectiveness (%)	85	100
Catalyst in use (m^3)	198	276
Catalyst activity (RVA) (%)	100	140
Effective volume gain (m^3)	198	386 (+94% gain)
Predicted cycle life (months)	<3	30

Diesel S10 production	Existing reactor	Revamped reactor
Density 20/4°C	0.851	0.835
Sulfur (wtppm)	4,000	4
Nitrogen (wtppm)	1,200	40
Cetane index ASTM D4737	46	51

Petrobras was able to use the same reactor to produce ultra-low-sulfur diesel.

CONCLUSION: HOW NEW REACTOR INTERNALS CAN BE USED TO ENHANCE PROFITABILITY

If you answer "yes" to any of these questions, then new reactor internals could provide an opportunity to unlock additional value from your operations.

- Has your reactor shown signs of fouling, high pressure drop or thermal maldistribution?
- Has it been a challenge to achieve any particular product specs? Density? Cetane? Cloud point?
- Is reactor internals maintenance a critical path activity during your turnarounds?
- Would it be beneficial to increase the severity of your feeds?
- Have you had any difficulties in achieving your desired cycle length?
- Are you under pressure to increase product yields?

Shell reactor internals have helped refiners worldwide and have been subject to ongoing research and development, informed by real-world operational challenges, for over 20 years.

Q&A

 7th AIChE Southwest Process Technology Conference CONFIDENTIAL 06 October 2015 27

Sustainable Energy Reductions by Optimization of Flare Systems

Shah, K.,
Lazzaroni, M.,
Nelson, D.,

Originally presented on: 10/1/2015 9:30:00 - 10:00:00

As part of Albemarle's commitment as an Energy Star Partner to reduce energy usage and adopt more energy-efficient operations, we recently made several improvements to the plant's flare systems to result in significant cost savings and sustainable energy reductions.

Excess hydrocarbons that cannot be reused or recycled are safely and efficiently combined with excess air and burned in flares at our site. Natural gas assist is used in the flares to ensure that environmental Btu limits are met while destruction and removal efficiency (DRE) of at least 98% is achieved. A cross-functional team evaluated the flare operation and identified several optimization opportunities. This presentation will cover how improvements in process flow measurement and a new controller technology based on a neural network algorithm was able to achieve significant reduction in natural gas usage at the flare.

LyondellBasell At A Glance

- One of the world's largest plastics, chemical and refining companies
- Annual revenues of $45.6 billion
- 13,100 employees worldwide

U.S. $ IN MILLIONS	2012	2013	2014
Sales and other operating revenues	$45,352	$44,062	$45,608
Operating income	$4,676	$5,102	$5,736
Income from equity investments	$143	$203	$257
Income from continuing operations	$2,858	$3,860	$4,172
Diluted EPS from continuing operations	$4.96	$6.76	$8.00
Diluted share count (millions)	577	570	521
Cash flow from operations	$4,787	$4,835	$6,048
Capital expenditures	$1,060	$1,561	$1,499
EBITDA	$5,808	$6,311	$7,050
LCM adjustment	-	-	$760
EBITDA excluding LCM	$5,808	$6,311	$7,810
Diluted EPS from continuing operations excluding LCM	-	-	$8.92

lyondellbasell.com

lyondellbasell 2

Global Capacity Positions

- Polypropylene
- Polypropylene Compounds
- Oxygenated Fuels

1st

- Propylene Oxide

2nd

- Polyolefins Licensing

3rd

- Polyethylene
- Ethylene
- Propylene

5th

Refining Capacity
268,000
barrels per day

Sources: Third party consultants data.

Note: Capacities and worldwide capacity positions are as of December 31, 2012, and include our pro rata share of joint ventures.

lyondellbasell.com lyondellbasell 3

Global Presence

LyondellBasell's global presence spans **18 countries** and includes **55 manufacturing sites** and **13,300 employees**.

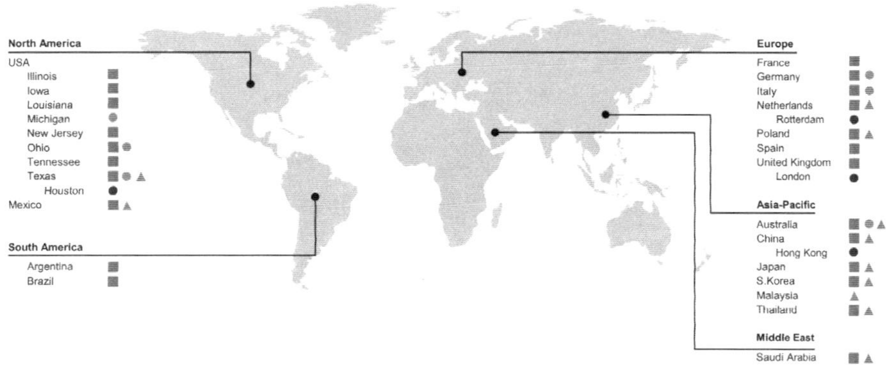

● Administrative Offices/Headquarters ■ Manufacturing ● Technology Centers ▲ Joint Ventures

Owned and operated by LyondellBasell subsidiaries and/or joint ventures.

lyondellbasell.com lyondellbasell 4

Olefin Manufacturing Locations

- USA
  - Channelview, TX
  - La Porte, TX
  - Corpus Christi, TX
  - Morris, IL
  - Clinton, IA
- Europe
  - Munchmunster, Germany
  - Wesseling, Germany
  - Berre, France

La Porte Site

- Gas Cracker, 1.5 Billion lbs/year Ethylene nameplate capacity.
- M.W. Kellogg Design
- Started up in 1991
- Feed Flexibility from Ethane, Propane, Butane to Natural Gasoline.
- Ability to produce polymer grade or chemical grade Propylene.
- Debottlenecked in 1996 with addition of a KTI furnace to 1.8 Billion lbs/yr Ethylene.

La Porte Complex
QE-1 Olefins Unit

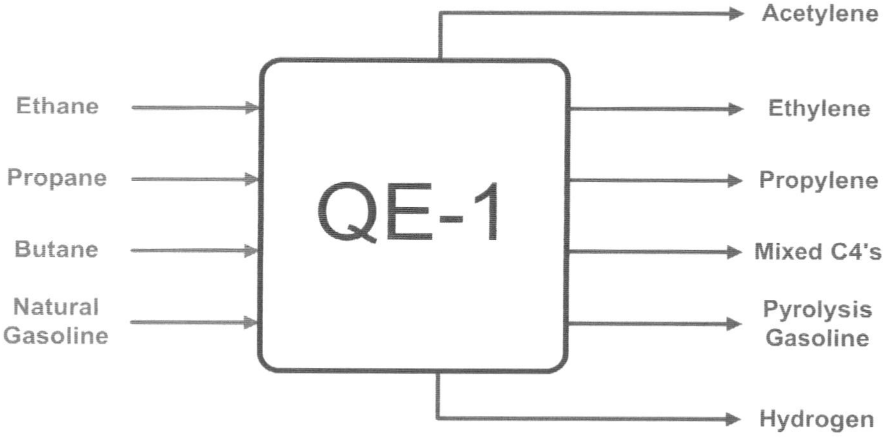

lyondellbasell.com

lyondellbasell

Shale Gas Developments

- Abundant recovery of Ethane from shale gas has made it a highly desirable feedstock.
- Demand for Ethylene in the US has remained robust since 2009
- Movement across the industry to use maximum amount of Ethane as feedstock.
- Recognized a significantly attractive opportunity to debottleneck La Porte plant and increase Ethylene production from Ethane.
- Similar effort in progress at the Channelview, Corpus and Morris sites.

lyondellbasell.com

lyondellbasell 8

La Porte Expansion

- Project initiated to expand capacity to the limits of the C2 Splitter.
- Retained a single train concept.
- Project executed in 2014. New nameplate capacity 2.6 Billion lbs./year Ethylene.
- Economic criteria was to maintain cents per pound project cost below a green field grassroots plant.
- An energy efficient design was developed within the constraints of this criteria.

Energy Efficient Strategy

- New efficient trays were employed to achieve maximum capacity.
- Aluminum core in kettle condensers were installed to improve approach temperatures in the C2 Splitter. Helped improve refrigeration efficiency.
- Employing higher feedstock conversions with new furnaces.
- New furnaces were designed to produce more high pressure steam and operate and with longer run lengths.

Energy Efficient Strategy

- Substituted waste heat for steam in the de-ethanizer side reboiler.
- All the major compressors were re-wheeled or replaced to restore high efficiency.
- Demin water circuit re-designed to recover waste heat efficiently.
- Maximized hydrogen sales.

Results

- Monthly average energy consumption per pound of product mix has reduced by about 5 %. Actual energy saving is in the order of billions of Btus per month.

Questions?

Energy Efficiency in Plant Equipment
7th AIChE Southwest Process Technology Conference
October 1, 2015

Enhancing heat exchanger performance

Thomas Lestina
Vice President, Engineering Services

HTRI

Steps to improve performance

- ✓Avoid design mistakes
- ✓Mitigate fouling
- ✓Optimize temperature profile
- ✓Define complete operating profile
- ✓Minimize excess area
- ✓Use small flow passages
- ✓Use enhancement technologies

Example: Production heater

- Four stacked TEMA E shells, with two tubepasses
- 3°C approach
  Water inlet minus feed outlet

E shell with multipass tubes not suitable with 3 °C approach temperature

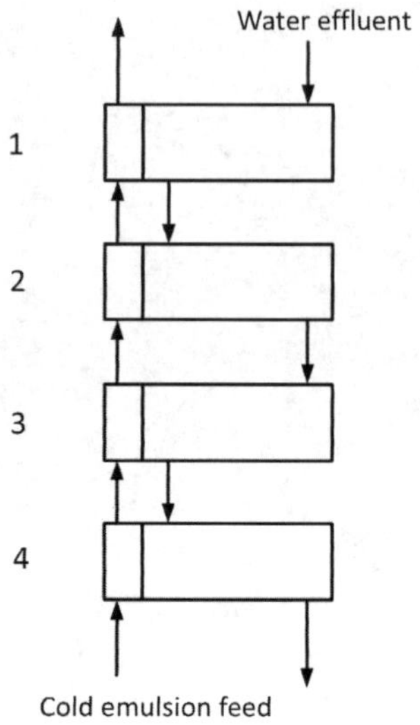

HTRI

Tubeside limited design

Thermal resistance, %

Shell:	4.73
Tube:	72.2
Fouling:	22.43
Metal:	0.64

Calculated ΔP: 4.914 kPa
Allowable ΔP: 50 kPa

Higher velocities and ΔP are desired

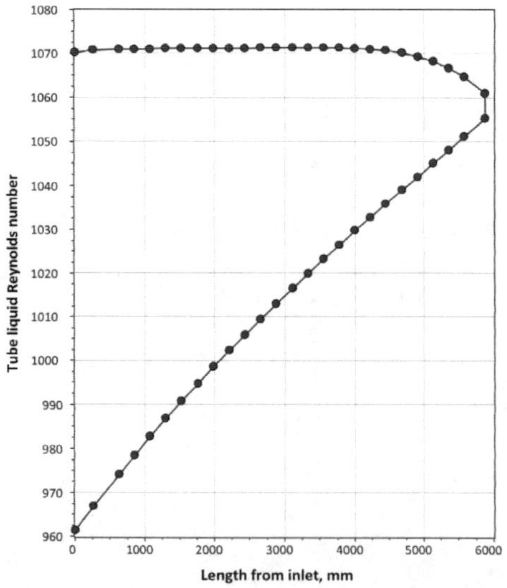

HTRI

Design mistakes increased fouling and reduced process efficiency

- 3 °C design approach
- 11 °C operating approach

Fired heater duty increased to heat feed

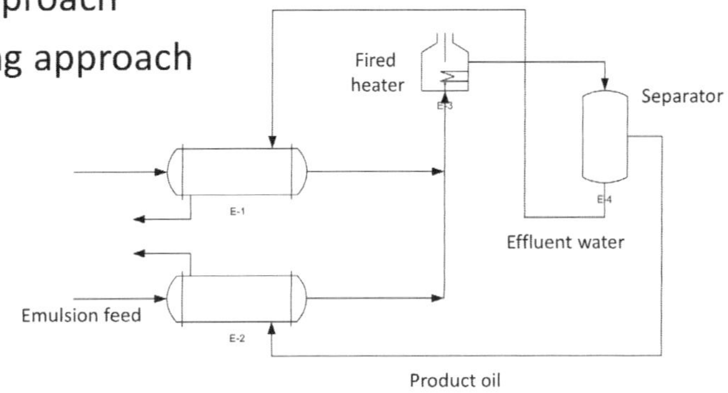

HTRI

Process conditions and exchanger design determine temperature profile

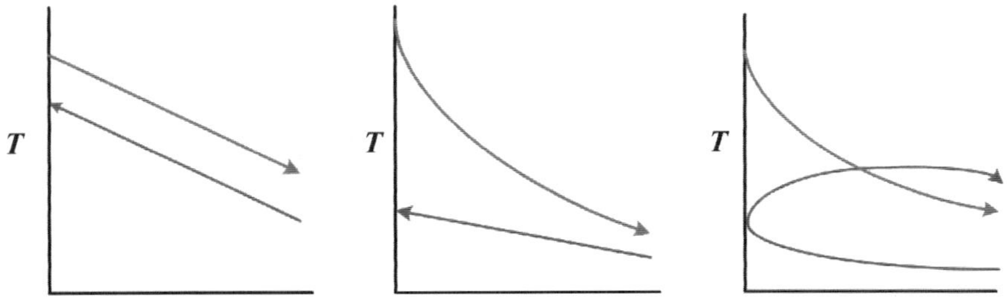

Length
Ideal counterflow

Length
Different hot and cold stream capacities

Length
Temperature reversal

HTRI

Complete operating profile has many conditions

- Clean and fouled performance: BOR, MOR, EOR
- Turndown conditions
- Bypass operation

How is performance affected by large variations in flow rate?

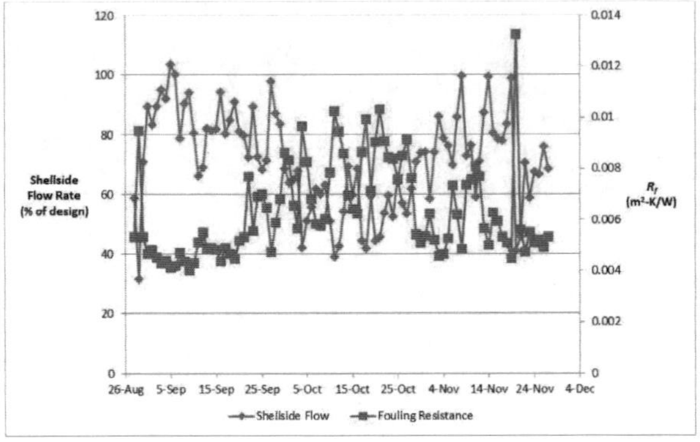

HTRI

7

Typical excess areas for current process heat exchangers

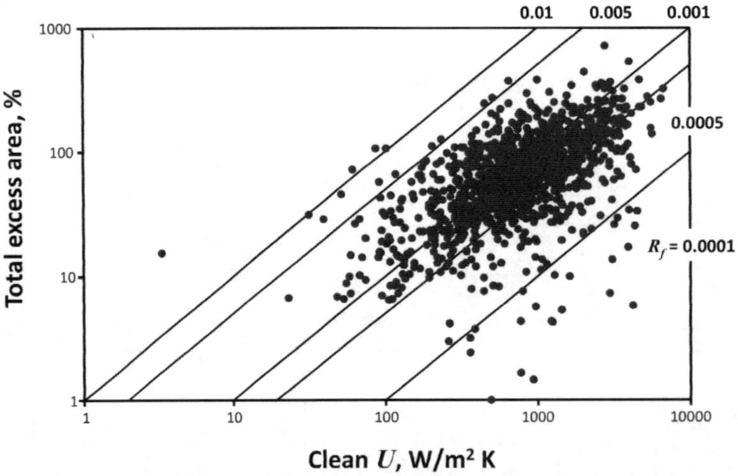

Excess areas less than 20% are needed for well designed exchangers

HTRI

8

Use small flow passages

Typical tube diameter: 19.05 mm or 25.4 mm

Smaller hydraulic diameters enhance performance in same volume

- For plate-frame channels, 4 – 10 mm
- For plate-fin passages, 1 – 6 mm

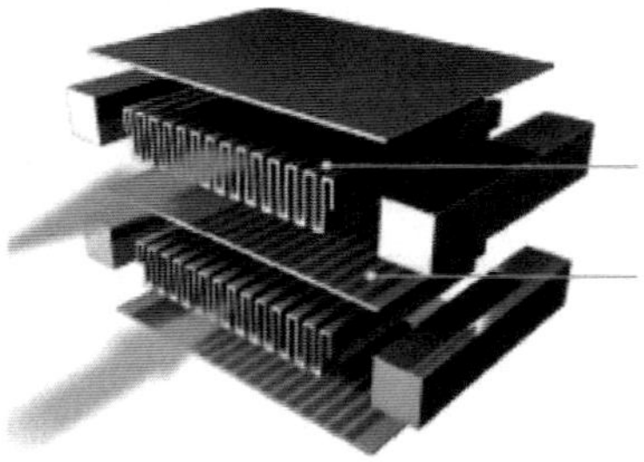

Typical plate-fin configuration

HTRI

When should enhanced technologies be considered?

- When temperature difference between hot and cold stream is low (less than 5 °C)
- When capacity upgrades are planned but no room is available for larger exchangers
- When exchanger performance limits production (debottlenecking)

HTRI

Surface enhancements

- Single phase
  - Disrupt boundary layer
  - Roughen surface
- Boiling
  Promote bubble formation
- Condensation
  - Encourage dropwise vs. film condensate
  - Promote condensate removal

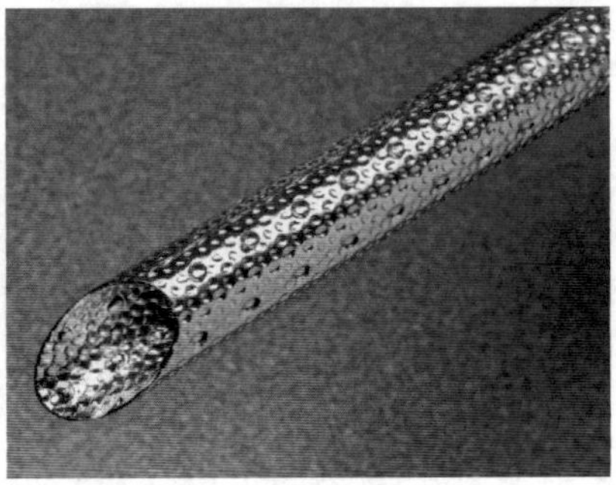

Vipertex® dimpled tube

Image source: *Energy Management and Efficiency for the Process Industries*, eds. A. P. Rossiter and B. P. Jones, Figure 10.3, Wiley & Sons, New York (2015).

HTRI 11

Extended surfaces increase surface area

Surface type	Integral low fin	High fin	Continuous plate fin
Fin height	1 – 3 mm	—	—
Outside/inside area ratio	3 – 5	20 – 25	10 – 25

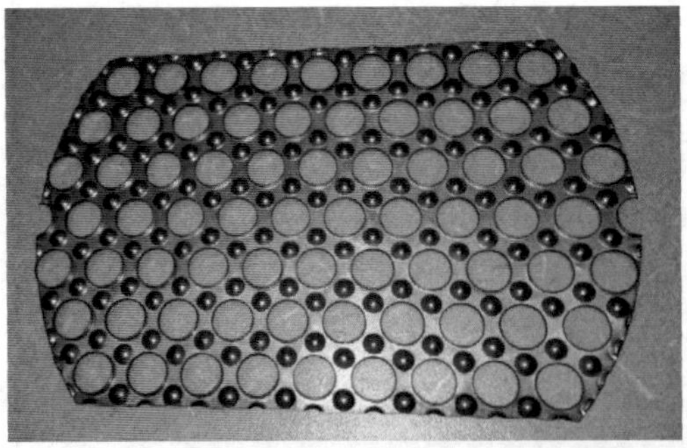

Continuous plate fin with dimple texture

Image source: *Energy Management and Efficiency for the Process Industries*, eds. A. P. Rossiter and B. P. Jones, Figure 10.6, Wiley & Sons, New York (2015).

HTRI 12

Inserts improve flow distribution

Tube

Static mixer

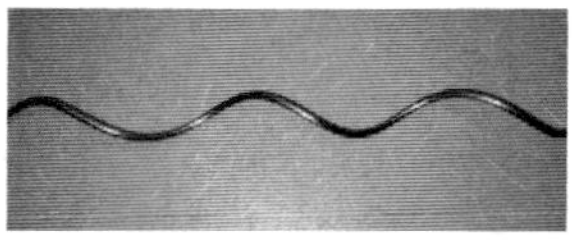

Wire coil boundary interrupter

Twisted tape

Wire wrapped core insert

Shell

Non-TEMA baffles

- Helical
- EMbaffle™
- Strip
- Rod
- Disc-and-donut

Image source: *Energy Management and Efficiency for the Process Industries*, eds. A. P. Rossiter and B. P. Jones, Figure 10.8, Wiley & Sons, New York (2015).

Example: C3 splitter reboiler

C3 splitter

Column pressure, bar a	20
Liquid static head, m	7.3
Steam pressure, bar a	1.2
Boiling range, °C	0.18
Tube diameter, mm	19.05
Tubecount	2200
y_o	0.23

Thermal resistance, %

Shell: 12.94
Tube: 8.90
Fouling: 73.17
Metal: 4.99

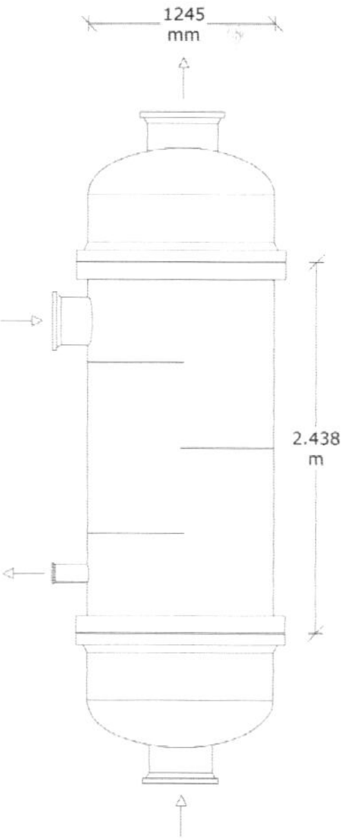

How will this operate when clean?

Thermal resistance, %

Shell: 11.72
Tube: 83.75
Fouling: 0.00
Metal: 4.53

Film boiling in all increments

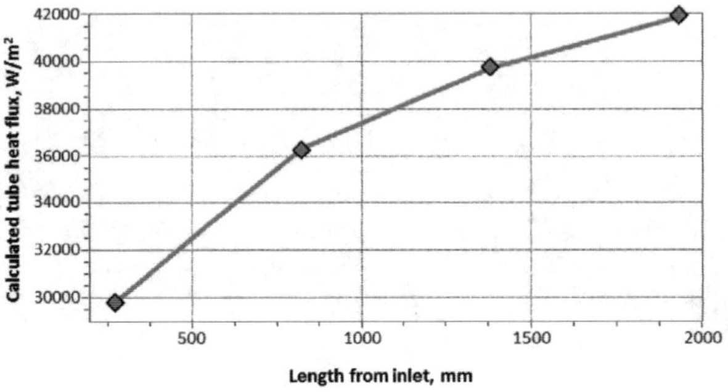

HTRI

How much will inserts improve performance?

Approximately 3-fold increase in heat flux

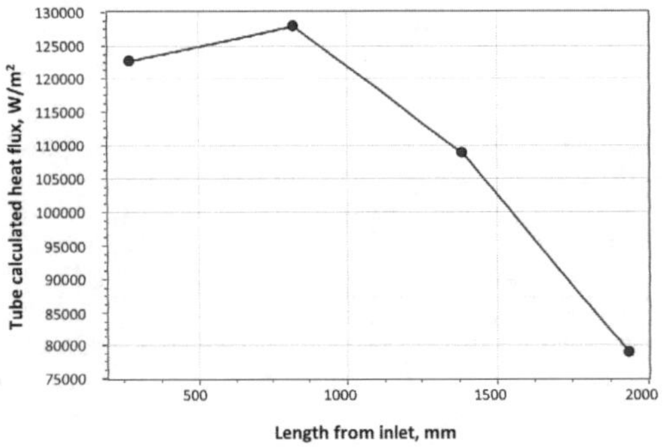

Twisted tapes in reboiler at HTRI Research & Technology Center

HTRI

Industry trends

- Optimize maintenance schedule with operating profile and thermal design technologies
- Use operating data to mitigate risk

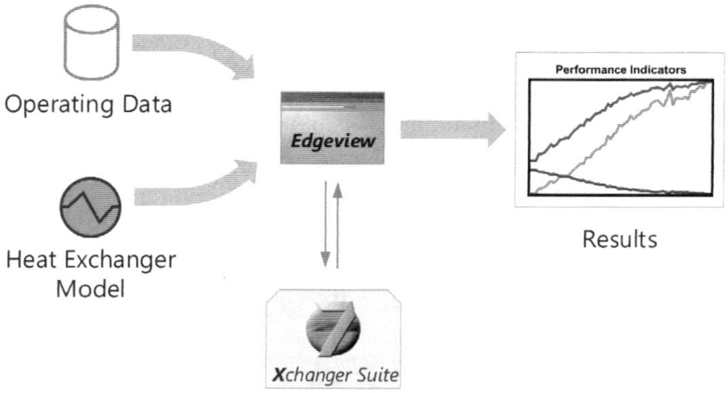

HTRI

17

ExxonMobil

1st Oct 2015, 11:30AM-12:00PM
7th AIChE Southwest Process Technology Conference

Distillation Tower Energy Efficiency Improved by Advanced Control Techniques

Srinivas Karra
ExxonMobil Chemical Company

Energy lives here

Energy Efficiency of Distillation Towers

- Overview
  - A combined total of 44 MMBTU/H energy savings were obtained which contribute to ~36% of total energy consumed in co-product recovery unit distillation towers via improvements to advanced control applications.
- Tasks
  - Minimization of unit recycle stream
  - Employing prediction-error-feed-forward techniques for tighter spec control
  - Preferential loading on towers in series
  - Deployment of Inferential Estimators for Tower optimization
- Benefits

Total energy savings = 44 MMBTU/H
Annualized energy saved in 2014 = 382,374 MMBtu/Yr
Annualized Net CO_2 Emissions Reduction = 4,432 (Tons CO_2 or CO_2 Equiv)

Task 1

Minimization of Recycle Stream

Recycle is used to be ensure the tower A loading

- ✓ Identified the right set of indicators for tower weeping
  - ✓ DP low threshold
  - ✓ Minimum reboiler duty
- ✓ Implemented recycle minimization strategy against thresholds on tower minimum loading indicators
- ✓ Resulted in cutting recycle stream by more than 50% while ensuring good Tower A operations
- ✓ Reduced Tower C loading by 25%, which offered debottlenecking of throughput limitation.

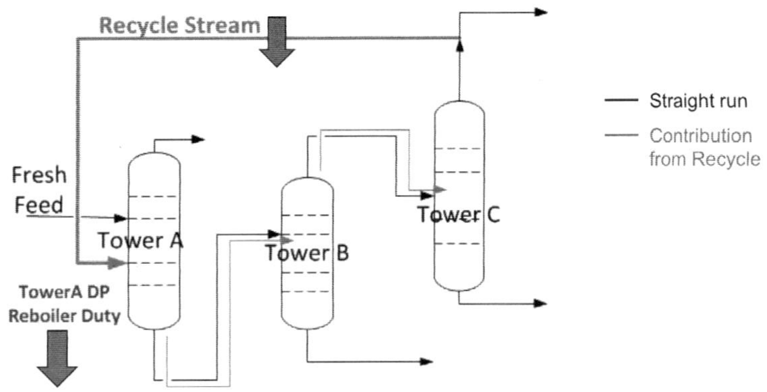

ExxonMobil

3

Task 2

Prediction Error Feed Forward - Methodology

Tower with Long Time-to-steady-state (C3Splitter)

- ✓ Response time for a change in OH analyzer as a result of change in reflux – 6 to12 hours
- ✓ Inclusion of a tray analyzer as a controlled variable for generating prediction error to identify the presence of feed composition disturbances
- ✓ Making moves on reflux flow in response to projected effect on OH analyzer
- ✓ Resulted in tighter OH spec control and reduced reboiler duty via reduced reflux

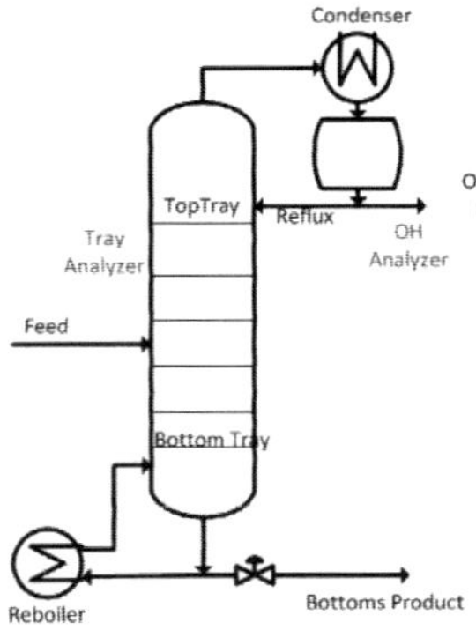

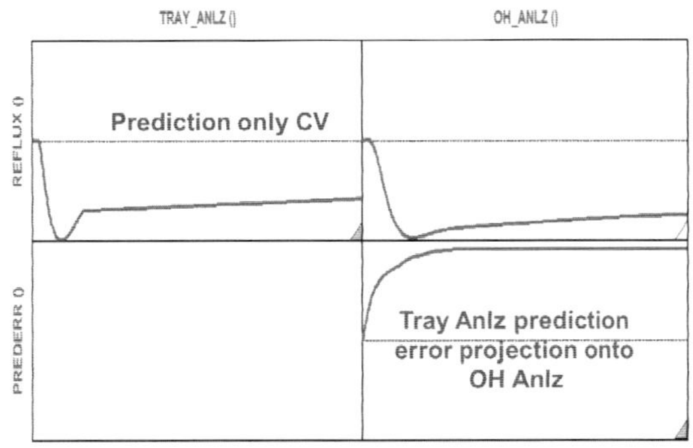

ExxonMobil

4

Task 2

Prediction Error Feed Forward - Results

Resulted in

- ✓ Tighter OH spec control
- ✓ Reduced reflux on the tower
- ✓ Reduced reboiler duty on tower by >35%

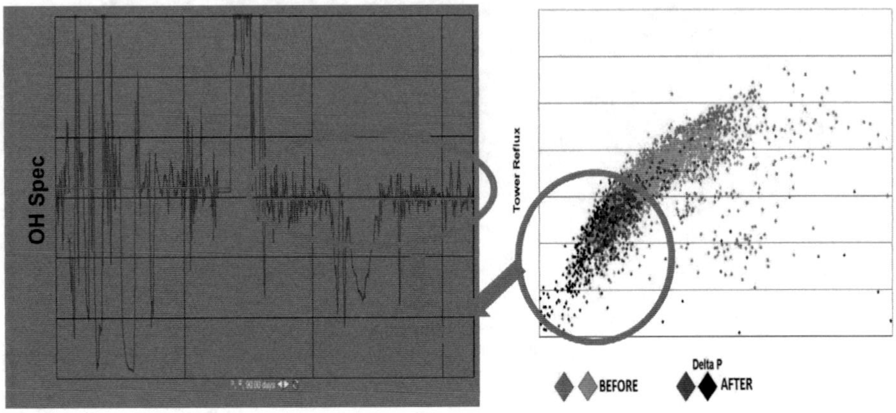

ExxonMobil

Task 3

Preferential Loading on Towers in Series

Tower A is made to take additional load

- ✓ Refrigerant is more expensive than CW medium for OH condenser
- ✓ Maximization of reboiler duty against minimum threshold on TowerB loading

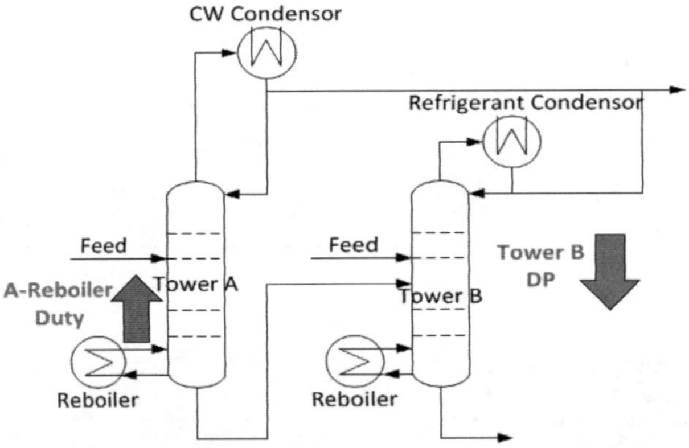

ExxonMobil

Inferential Estimators for Tower Optimization

Maximize the high value product recovery from tower bottoms via

- ✓ Lowering tower reflux against high limit on **inferred* overheads 90% point**
- ✓ Raising the cut-point temperature against low limit on bottoms flow (to ensure no-plug operation)
- ✓ Lower tower pressure to achieve same separation with low reboiler duty

Resulted in 25% reduction in fired-heater reboiler duty

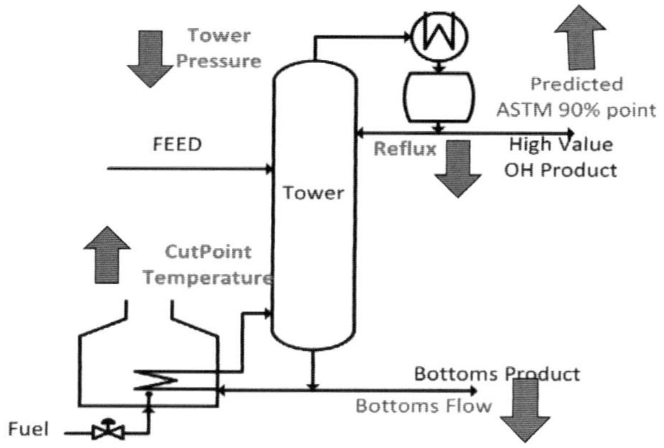

ExxonMobil

Combined Energy Savings

Total energy savings

- ✓ Total energy savings (all towers combined) = 44 MMBTU/H
- ✓ Annualized energy saved in 2014 = 382,374 MMBTU/Yr
- ✓ Annualized Net CO2 Emissions Reduction = 4,432 Tons CO2 or CO2 Equiv.

ExxonMobil

Thank you!

ExxonMobil

9

The Intriguing Demise of Jurassic Control Strategies

Brackin, I., Flint Hills Resources

Originally presented on: 10/1/2015 9:30:00 - 10:00:00

The petro-chemical industry has been around for decades. The industry has made advances in many facets of the business, but there are some parts that are reliable and have not changed in many years.

There is a term in economics known as "Creative Destruction". Creative Destruction is the process of replacing outdated entities and stagnation with new innovation. Regulatory process control in many petro-chemical plants continue to use strategies developed thirty years ago. Process control consists of monitoring inputs and manipulating outputs using software, hardware, and human machine interfaces.

When the industry was young control valves opened and closed with air pressure that traveled through ¼" tubing from the control room to the valve actuator. In some cases, the copper tubing was up to 300 feet long. This technology made it quite expensive to manipulate valves from the control room. Control engineers developed split range control schemes that reduced the installation cost. This allowed for moving two valves with one piece of ¼" tubing.

The downside of the split range is having two valves but only one knob. Today the valve is still actuated with air but there is a pair of small wires attached to the transducer converting a digital signal to a pneumatic signal that changes the air pressure to the valve actuator. The pair of wires is inexpensive compared to ¼" copper tubing. This approach is called twin integrated control, and each valve has a dedicated set of wires. Two valves will have two knobs. The digital age is here.

Graphic interface to control systems is another area that needs innovation. Digitally we can do anything but often people who design graphic interfaces have very little knowledge of human factors engineering. Most graphics are designed using the P&ID approach instead of task oriented design.

The tough part of "Creative Destruction" is it may be painful to realize tremendous gain. Humans are creatures of habit and change stings.

Steam System Management: Energy, Environmental and Operational Issues

White, D. C., Emerson Process Management

Originally presented on: 10/1/2015 10:00:00 - 10:30:00

Control of the overall steam system in a refinery or a large petrochemical plant can be one of the most difficult automation challenges, yet one of the most important. In many plants the steam usage will be 40% or more of the total energy use at the site. There will normally be multiple steam headers at different pressure levels, each with multiple users and suppliers. Demand at each level is changing and changes in one header can induce disturbances in the others. The many users and producers distributed widely around the site with local controls further complicating coordination. There are also multiple control objectives for the steam system with differing priorities. Older plants frequently have steam systems that have been enlarged and modified frequently with limited overall planning leading to many opportunities for improvement. Some required measurements and controllers to implement improvements are commonly either missing or underperforming. Significant steam can be wasted through uncoordinated operation and the response to large disturbances such as equipment trips is often unsatisfactory. Complicating the operation further are the strong interactions among the steam, power, fuel and hydrogen systems – disturbances in one often propagate to the others. Current and pending environmental emission regulations place additional constraints on the flexibility required to respond to disturbances. Improved automation and optimization can provide significant benefits and an improved capability to balance the competing demands of energy reduction, meeting environmental constraints, and satisfying operational needs.

metso ExperTune

Leveraging Control Systems to Achieve Operational Reliability Excellence

AIChE
Southwest Process Technology Conference

7th AIChE Southwest Process Technology Conference | October 1-2, 2015 | Galveston, TX

Agenda

- Motivations – Why Bother?
- Instrument Performance
- Control Valve Performance
- Process Upsets and Interactions
- Practical Matters - Implementation

AIChE
Southwest Process Technology Conference

7th AIChE Southwest Process Technology Conference | October 1-2, 2015 | Galveston, TX

We've Been Working Hard at the Wrong Things

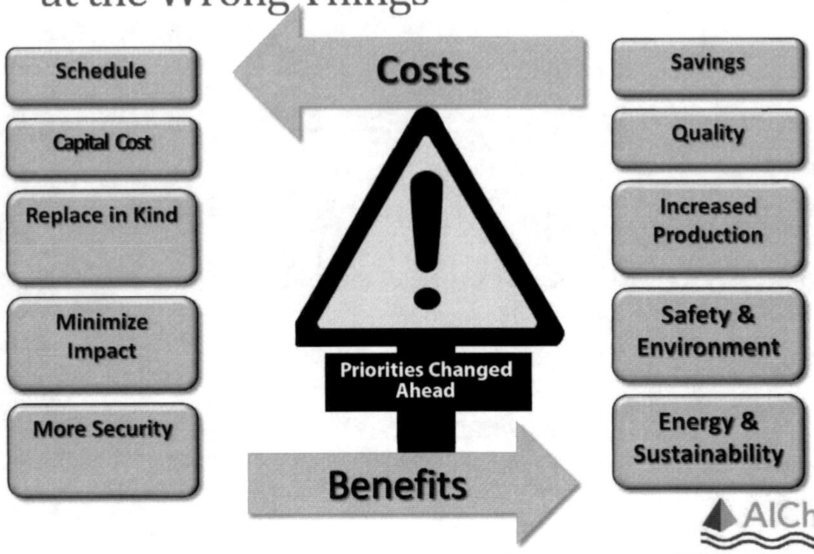

AIChE Southwest Process Technology Conference

7th AIChE Southwest Process Technology Conference | October 1-2, 2015 | Galveston, TX

Finding the Business Value of Basic Control Performance

- Is There Any Opportunity?
- What Should We Do About It?
- Can We Fix It?
- What Are the Results?

Measure | Decide | Action

AIChE Southwest Process Technology Conference

7th AIChE Southwest Process Technology Conference | October 1-2, 2015 | Galveston, TX

Controls Can Fail in Many Ways

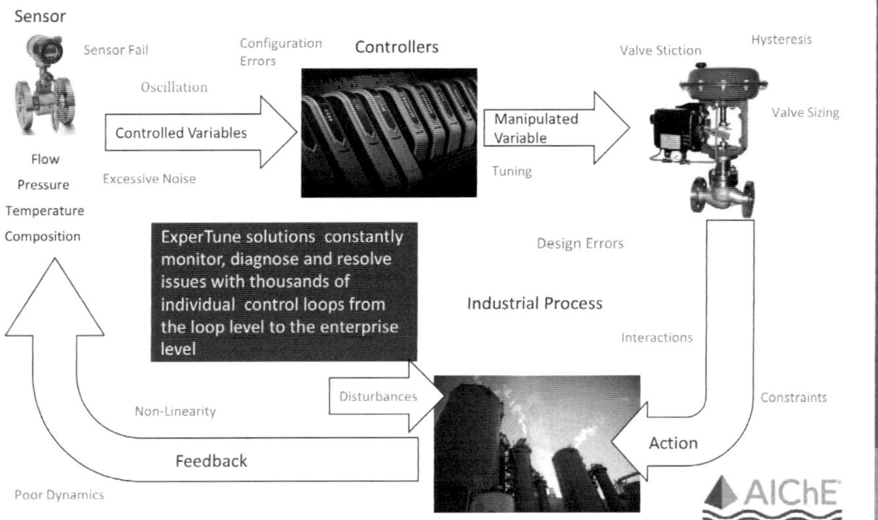

AIChE Southwest Process Technology Conference

7th AIChE Southwest Process Technology Conference | October 1-2, 2015 | Galveston, TX

Motivations:

Typical Control System Performance

Typical Statistics

- 10% - 90% Control Loops in Manual
- 30% Loops-Tuning is <u>completely</u> wrong
- 40% Loops Oscillating
- Improper PID Loop Configuration
  - 30% DCS Systems
  - 95% PID on PLC's
- 85% Sub-Optimal Tuning
- 75% Control Loops <u>increase</u> variability

AIChE Southwest Process Technology Conference

7th AIChE Southwest Process Technology Conference | October 1-2, 2015 | Galveston, TX

Why Low Variability Counts
Part 1

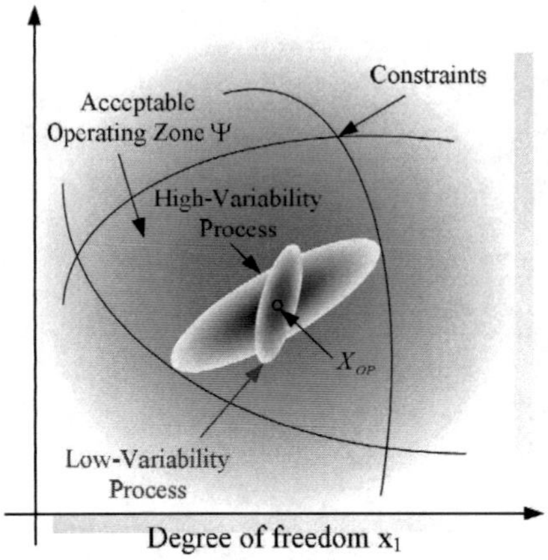

Shahriari & Ding, IEEE 2011 Conference on Control Applications

AIChE
Southwest Process Technology Conference

Why Low Variability Counts
Part 2

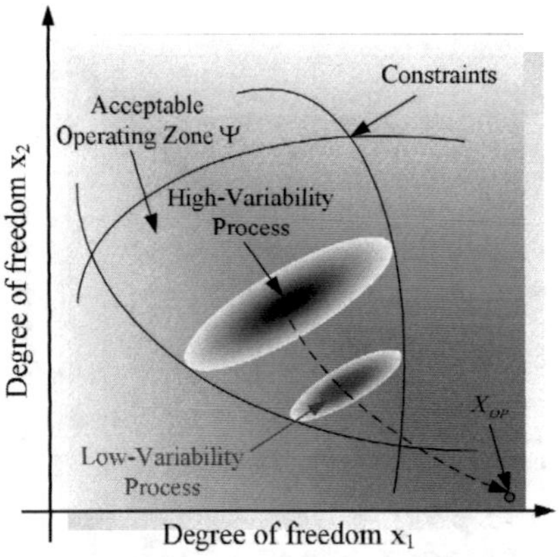

Shahriari & Ding, IEEE 2011 Conference on Control Applications

AIChE
Southwest Process Technology Conference

7th AIChE Southwest Process Technology Conference | October 1-2, 2015 | Galveston, TX

Control Performance Can Be Managed

How do you know if your control system is living up to it's purpose?

"You can't manage what you can't control, and you can't control what you don't measure."

Measure → Decide → Action

AIChE Southwest Process Technology Conference

7th AIChE Southwest Process Technology Conference | October 1-2, 2015 | Galveston, TX

Measuring Control Performance

- Criteria for Good Real-Time Performance Measures:

Meaningful	• Affects the Bottom Line • Easily Understood
Measurable	• Can be Measured • In Real Time, 24 x 7
Actionable	• Direct Actions • Immediate Results

AIChE Southwest Process Technology Conference

7th AIChE Southwest Process Technology Conference | October 1-2, 2015 | Galveston, TX

What to Measure?

Opportunity Gap	Harris Index	
SP Changes	Oscillation	Period
Stiction	Valve Sizing	Noise Band
Time at Limits	Service Factor	
Std Dev	Variance	Noise Band
Min of CO	Max of CO	Avg. of CO
Min of PV	Max of PV	Avg. of PV

AIChE
Southwest Process Technology Conference

Conclusions

- Basic Controls are a Profit Center, not a Cost Center!
- Reduced Variability Drives Efficiency
- 24 x 7 Monitoring Identifies the Issues
- Many Mechanisms to Achieve Savings:
  - Valve Repairs
  - Instrumentation
  - Tuning
  - Mechanical Issues
  - Process Interaction

AIChE
Southwest Process Technology Conference

7th AIChE Southwest Process Technology Conference | October 1-2, 2015 | Galveston, TX

Agenda

- Motivations – Why Bother?
- Instrument Performance
- Control Valve Performance
- Process Upsets and Interactions
- Practical Matters - Implementation

Southwest Process Technology Conference

Types of Instrument Problems

- Noisy
- Flat-Lining
- Scale / Range
- Calibration
- Over-Filtered
- Other Failure Modes

Southwest Process Technology Conference

7th AIChE Southwest Process Technology Conference | October 1-2, 2015 | Galveston, TX

Basic Measures of Instrument Performance

Time at Limits

Std Dev | Variance | Noise Band

Min of PV | Max of PV | Avg. of PV

Equipment Diagnostics : Instrumentation

He's dead, Jim.

- Dead Instrument

PV Max = PV Min

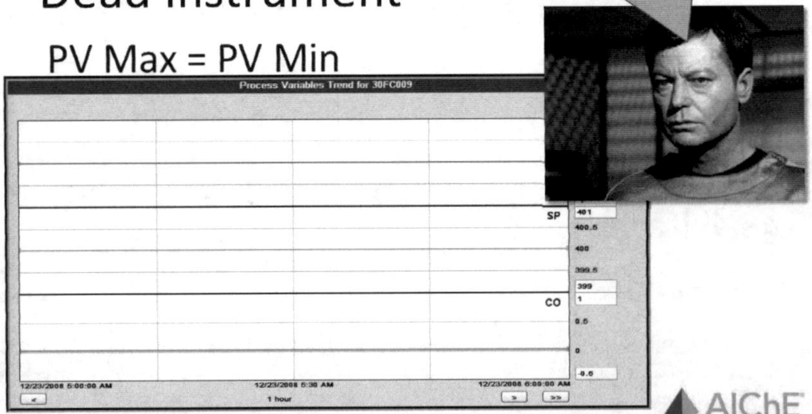

7th AIChE Southwest Process Technology Conference | October 1-2, 2015 | Galveston, TX

Equipment Diagnostics: Instrumentation

- Spiking

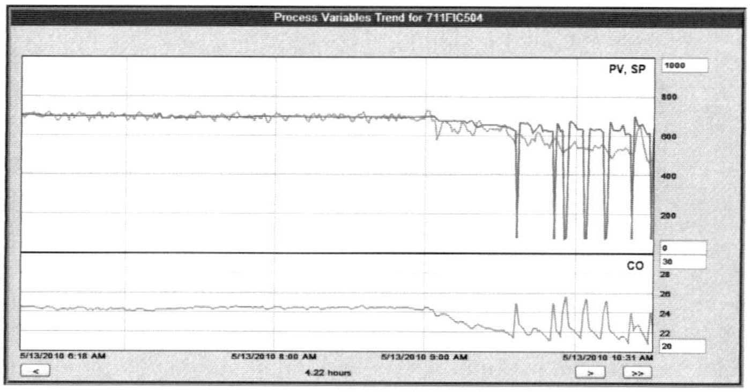

Southwest Process Technology Conference

7th AIChE Southwest Process Technology Conference | October 1-2, 2015 | Galveston, TX

Noisy Instrument?

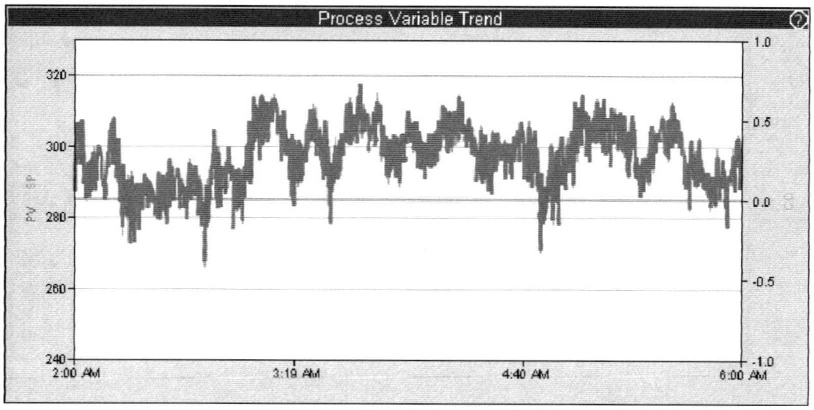

Southwest Process Technology Conference

7th AIChE Southwest Process Technology Conference | October 1-2, 2015 | Galveston, TX

Agenda

- Motivations – Why Bother?
- Instrument Performance
- Control Valve Performance
- Process Upsets and Interactions
- Practical Matters - Implementation

AIChE Southwest Process Technology Conference

7th AIChE Southwest Process Technology Conference | October 1-2, 2015 | Galveston, TX

Root Cause of Valve problems

- 30% of Control Valves have problems:
  - Valve Under-Sized
  - Valve Over-Sized
  - Stiction
  - Hysteresis

AIChE Southwest Process Technology Conference

7th AIChE Southwest Process Technology Conference | October 1-2, 2015 | Galveston, TX

Do you know how bad the problem is?

- How do you know if a valve is bad?
- How to avoid repairing good valves?

Answer:
You must be able to accurately measure valve performance

AIChE Southwest Process Technology Conference

7th AIChE Southwest Process Technology Conference | October 1-2, 2015 | Galveston, TX

What to Measure?

Oscillation
Period
Stiction
Valve Travel
Time at Limits
Std Dev
Variance
Noise Band
Min of CO
Max of CO
Avg. of CO

AIChE Southwest Process Technology Conference

7th AIChE Southwest Process Technology Conference | October 1-2, 2015 | Galveston, TX

Equipment Diagnostics: Valve Sizing

- Why does this happen?
- Symptoms?
- Detection: Time at limits,

  3.0 < Process Gain < 0.33

Loop	Description	Min of CO	Max of CO	Output at limit (%)	Process gain
Sort	Sort	Sort	Sort	Sort	Sort
120TC088	MP Superheat In A	-2*	8.204*	82.67*	-
PC061	HP to IP Letdown Big	-2*	3.486*	66.29*	-
PC064	IP to LP letdown	-2*	2.747*	75.09*	-
15FC002	HC out to Tank	0*	1.729*	5.082	-
10FC029	Flash Drum Btms to mix tee	0.02076†*	0.02076†*	0†	0.37†
14FC088	NH3 injection	0.2708*	2.764*	0	-
C05FIC045	Ovhd to BL	1.431	2.047	0	2.1†
C04PDIC013	DeC4 hot vap. Bypass	10.05	12.05	0	-
PC039	Decarbonator	10.83	16.13	0	-
C04FIC031	Finishing Col. Btms	11	11	0	1.3†
C04FIC024	2nd stage contactor liquid	11.64	12.49	0	-
110LC048	Blowdown Level	15.33	16.91	0	-
14FC027	LC Cut to Exch 67	17.41	19.73	0	-
14FC028	LC Cut to Exch 68	17.69	19.72	0	-
Stats	**Description**	Min of CO	Max of CO	Output at limit (%)	Process gain
Minimum	N/A	-2	0.02076	0	0.37
Average	N/A	6.405	9.215	16.37	1.257
Median	N/A	10.05	11	0	1.3
Maximum	N/A	17.69	19.73	82.67	2.1

AIChE
Southwest Process Technology Conference

Hysteresis

- What is it?
- Why does it happen?
- Symptoms?
- How to Fix It?

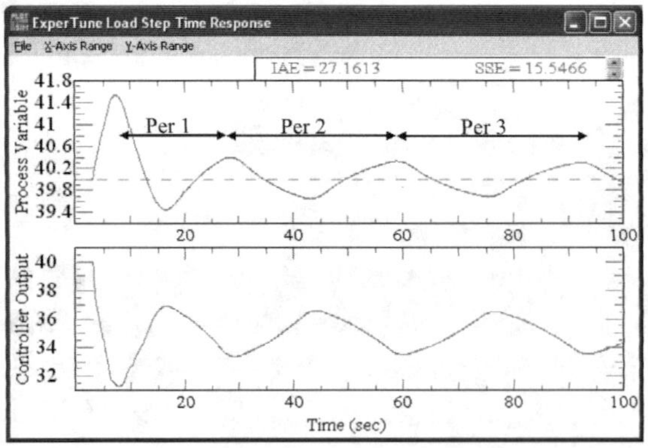

AIChE
Southwest Process Technology Conference

Stiction

- What is it?
- Why does it happen?
- Symptoms?
- How to fix it?

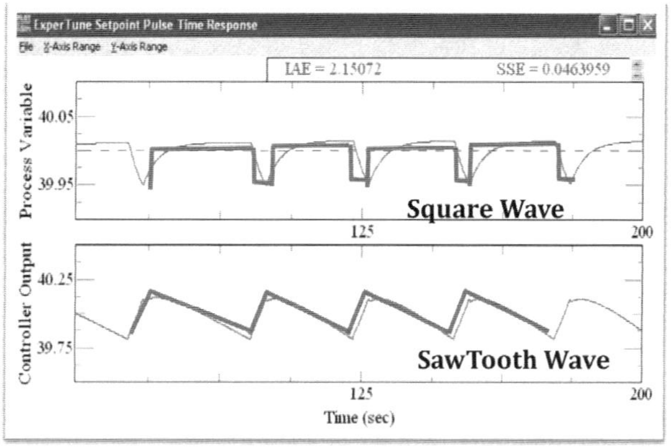

AIChE
Southwest Process Technology Conference

Variability caused by Stiction

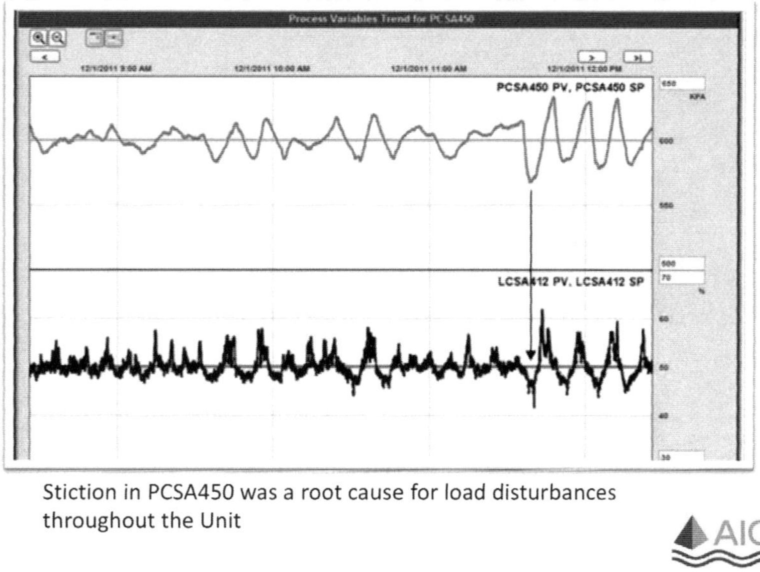

Stiction in PCSA450 was a root cause for load disturbances throughout the Unit

AIChE
Southwest Process Technology Conference

Address the Root Cause

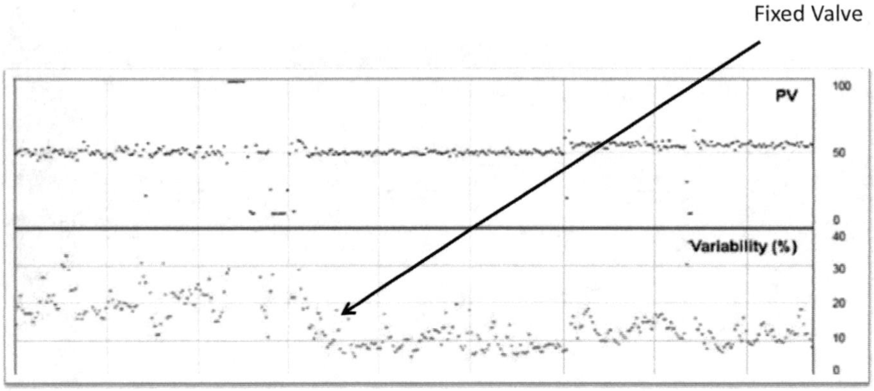

- Is there still room for improvement?

Retune the Control Loop

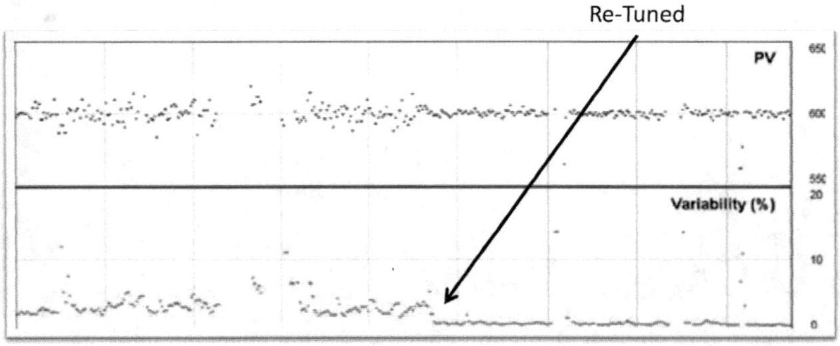

- There is still room for more improvement!

Decrease in ClO2 Consumption in Final Reaction Tower

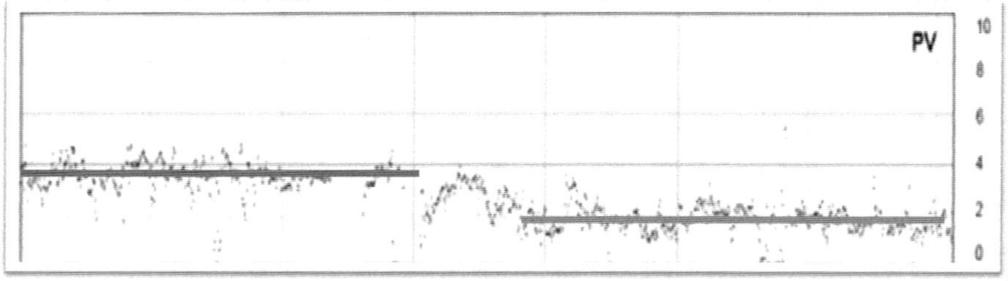

Savings = $170,000 / year

AIChE
Southwest Process Technology Conference

ClO2 flow rate has been reduced by approximately 50%

7th AIChE Southwest Process Technology Conference | October 1-2, 2015 | Galveston, TX

Monitoring – Easier than Ever to Find the Issues

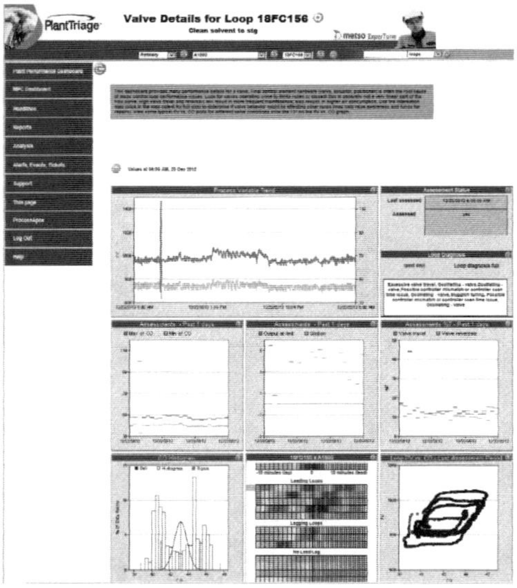

- Identify Valves with problems
- Shutdown Planning
  - Do not repair valves that are OK
  - Know which parts you will need
  - Prioritize the work
- Valve problems, cause process problems, cause money problems

AIChE
Southwest Process Technology Conference

7th AIChE Southwest Process Technology Conference | October 1-2, 2015 | Galveston, TX

Agenda

- Motivations – Why Bother?
- Instrument Performance
- Control Valve Performance
- Process Upsets and Interactions
- Practical Matters - Implementation

AIChE Southwest Process Technology Conference

Metrics can reveal process problems

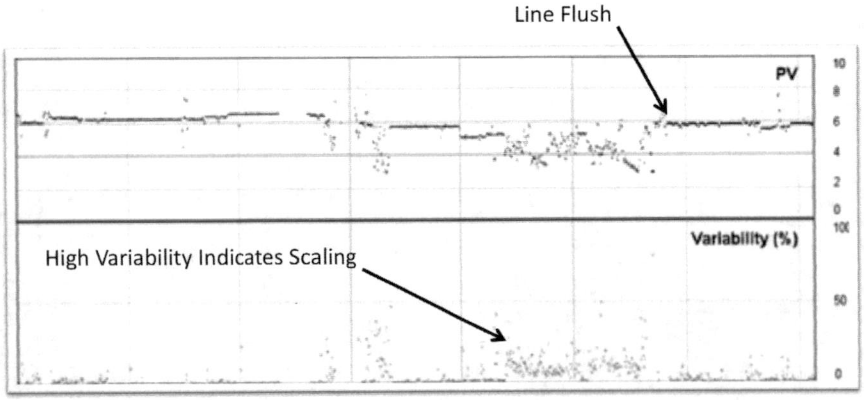

- The flow rate returns to normal after flushed with fresh water.

AIChE Southwest Process Technology Conference

High variability indicates that the line is scaling and can no longer maintain the proper flow rate.

What if Variability Comes From Someplace Else?

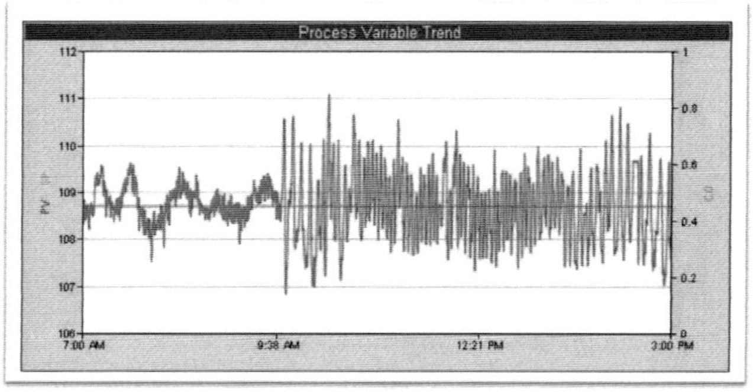

AIChE
Southwest Process Technology Conference

7th AIChE Southwest Process Technology Conference | October 1-2, 2015 | Galveston, TX

Identify Oscillations

- Oscillation Period
  - Fourier Transform
- Oscillation Shape
- Comparisons can lead to root cause.

AIChE
Southwest Process Technology Conference

7th AIChE Southwest Process Technology Conference | October 1-2, 2015 | Galveston, TX

Find interactions within our measurements

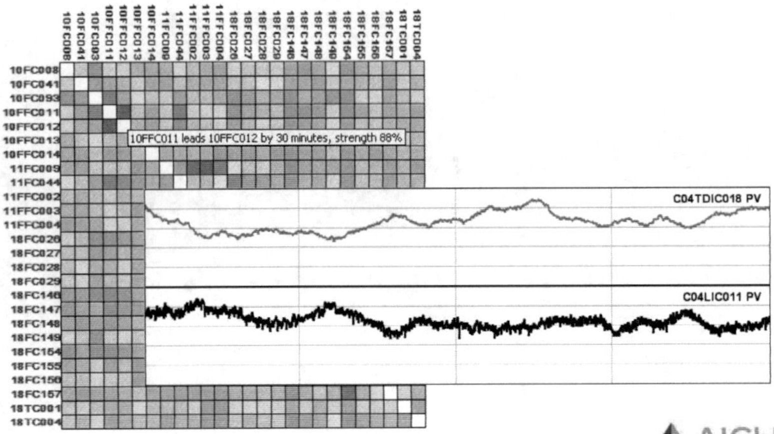

AIChE
Southwest Process Technology Conference

7th AIChE Southwest Process Technology Conference | October 1-2, 2015 | Galveston, TX

Where is it coming from?

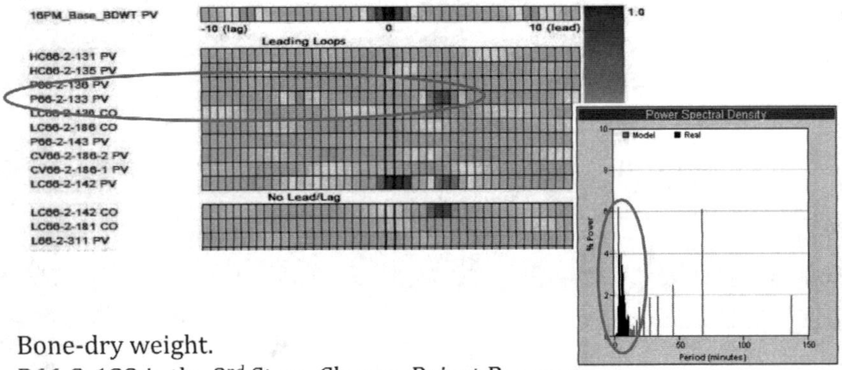

Bone-dry weight.
P66-2-133 is the 2nd Stage Cleaner Reject Pressure.

Consistency was too heavy and the cleaner rejects were plugging up and releasing at a regular cycle of 6 minutes.

AIChE
Southwest Process Technology Conference

7th AIChE Southwest Process Technology Conference | October 1-2, 2015 | Galveston, TX

Agenda

- Motivations – Why Bother?
- Instrument Performance
- Control Valve Performance
- Process Upsets and Interactions
- **Practical Matters - Implementation**

Southwest Process Technology Conference

37

Evaluate where you are

To point out the improvement potential and current control environment situation

- Survey of the current control environment
- Recommended improvements and solutions

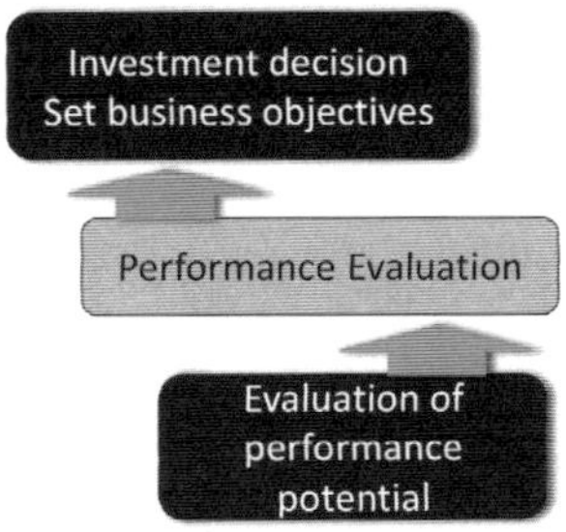

Southwest Process Technology Conference

38

Capturing Control System Performance Data

- Capture Data in Real-Time
- Standardize on Metrics (KPI's):
  - Measurable
  - Meaningful
  - Actionable
- Diagnostics built on these metrics
  - Instrument & Valve Failures
  - Loss of Control
  - Cause of Oscillation
- Historize, Track, Drill Down

AIChE Southwest Process Technology Conference

7th AIChE Southwest Process Technology Conference | October 1-2, 2015 | Galveston, TX

Best Ways to Get Measurable Results

- Focus on 1 or 2 objectives
- Clear Scope Boundaries
- Documented Baselines (Tech. & Econ.)
- Dedicated Resources
- Dedicated Time Frame
- Go after low hanging fruit
- Document, Document, Document

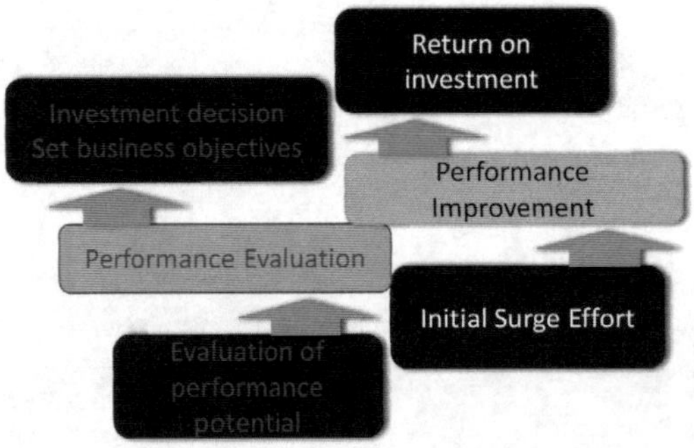

AIChE Southwest Process Technology Conference

7th AIChE Southwest Process Technology Conference | October 1-2, 2015 | Galveston, TX

What to Report? To Whom?

- Management Wants to Know:
  - Are we making progress?
  - What is it worth?
  - Can you sustain it?

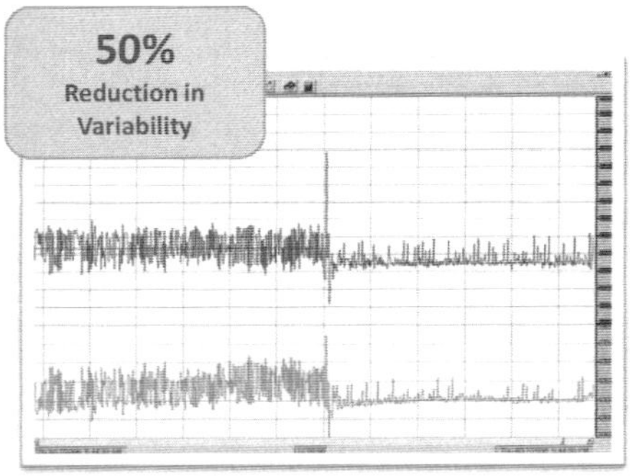

- Reduced Variability
- Reduced Off-Spec Product
- Increased High-Value Product
- Reduced Recycle

Make it Part of Daily Routines

- What is Your **<u>Existing</u>** Work Flow?
  - Who?
  - Decides What?
  - When?
- Bring the Right Info to the Right Person:
  - E-Mail Reports
  - Condition Alerts

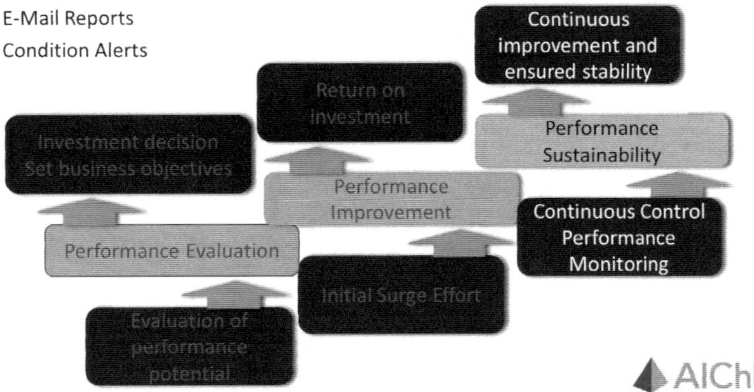

7th AIChE Southwest Process Technology Conference | October 1-2, 2015 | Galveston, TX

Conclusions

Control system performance analysis can provide significant business benefits

Opportunities exist for:
- Costs: Energy, Material, Waste, Production Rate
- Safety & Environmental
- Quality

To be successful:

- Match Business Objectives with Metrics
- Cover each production process fully
  - Most Upsets Come from Unexpected Places
- Integrate the Information Into Business Processes
- Follow-Up
- Document the benefits

AIChE Southwest Process Technology Conference

7th AIChE Southwest Process Technology Conference | October 1-2, 2015 | Galveston, TX

metso ExperTune

Thank You!

AIChE Southwest Process Technology Conference

44

7th AIChE Southwest Process Technology Conference | October 1-2, 2015 | Galveston, TX

INGENERO

Excellence Through Insight

Big Process Data Analysis
improves operational efficiency in Ethylene plants

Dr. Pratap Nair

7th AIChE Southwest Process Technology Conference

Oct 1, 2015

INGENERO

Brief background on Ingenero

INGENERO

2

Ingenero – Overview

A High End Technical Services Company Supporting Operating and Engineering Companies in the Process Industry

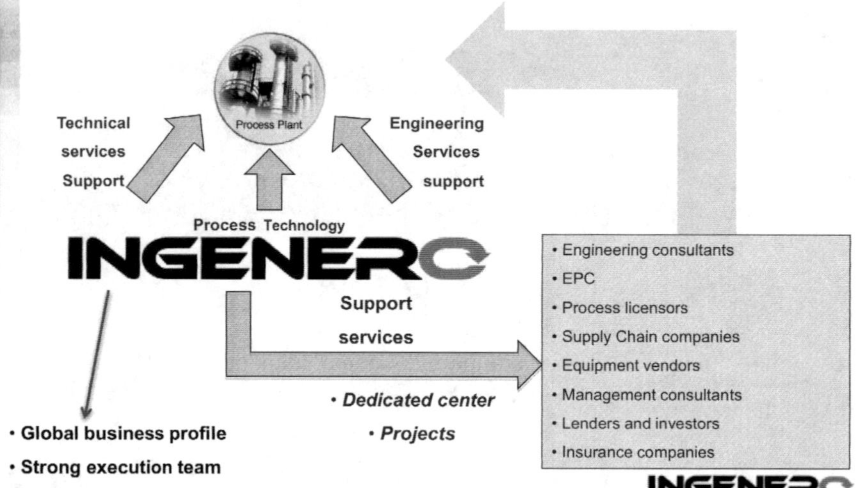

INGENERO

3

Global Business Profile

We serve customers in multiple continents with key offices in USA, India and KSA

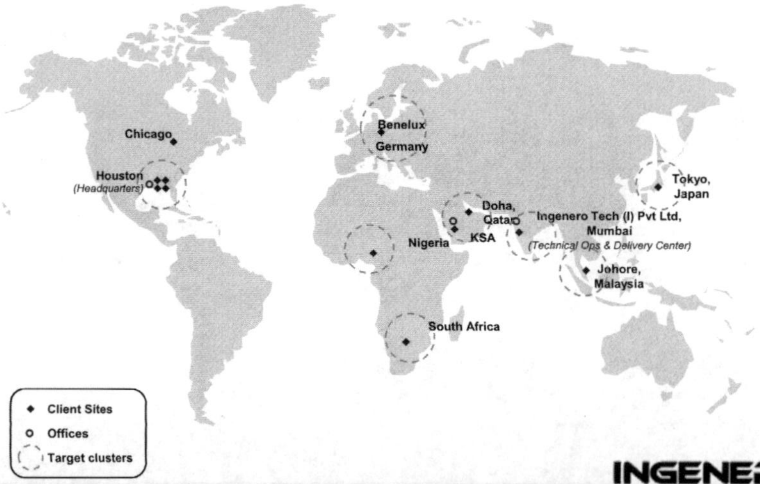

INGENERO

4

Broad Ingenero Service Areas

Ingenero combines extensive knowledge with various process expertise and operations experience to deliver highly ~~specialized 360⁰~~ engineering solutions

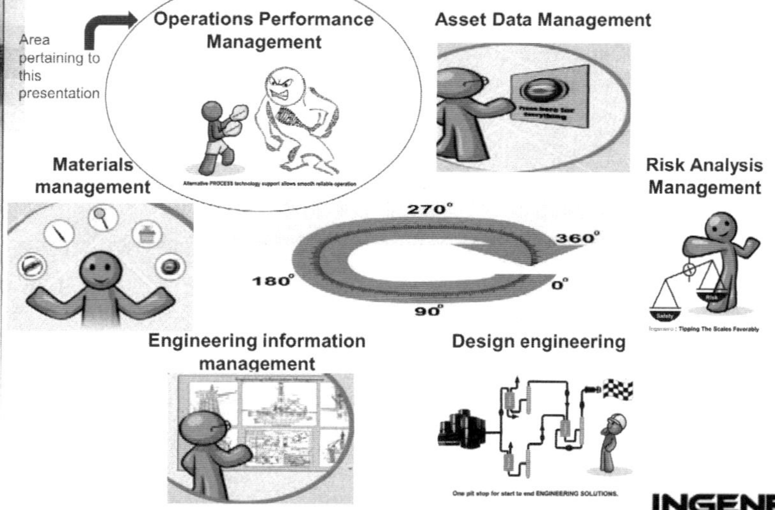

5

The Industry challenge that is being addressed

6

The Motivation: Exposure and Risk

Lack of an in depth understanding of process behavior leads to exposure to tangible & intangible threats & losses

- Safety risks (i.e. Explosions, fires and releases)
- Lack of environmental compliance
- Decreased efficiency – Energy, Productivity, Yield
- Diminished efficacy and reduced reliability – targets are missed
- Unplanned shutdowns, extended turnarounds, poor maintenance planning, and problematic start-ups
- Lost opportunity for strategic change/modification
- Poor project execution and management
- Uncontrollable swings in plant operation
- Fire fighting mode of operation
- Declining workforce moral

INGENERO

7

Typical Challenges Running Ethylene Plants

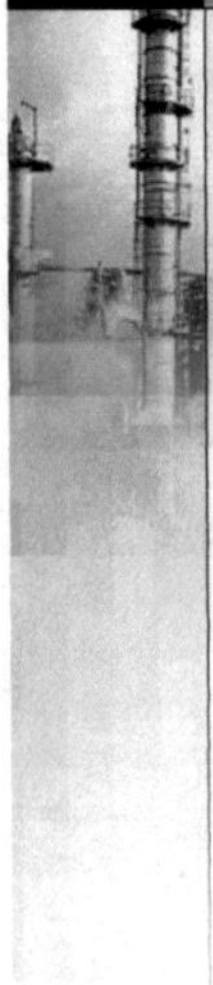

Starting up and sustaining production levels consistently from an ethylene plant does not always match expectations

- Furnace run lengths below expectation
- Unexpected tube failures and lower than expected tube life
- Rapid coking
- Inefficient furnace decokes
- Equipment reliability and lower than expected uptime
- Below capacity operation
- Recurring fouling in TLEs, chillers, CGC circuit
- Utility limitations
- Refrigerant and caustic losses
- Lower than expected product recovery
- Higher than expected specific energy consumption
- Off spec product
- Excessive flaring

INGENERO

8

The Source of Operating Challenges

Traditional plant operating paradigms contribute to some of the plant operating challenges observed

- Plants usually operated based on personal experience of the operator/engineer, guided by SOPs provided by the licensor
- Early detection of anomalies often missed due to inability to uncover truths and insights that aren' t readily obvious or don' t align with conventional process wisdom or intuition
- Sometimes even lack of sufficient expertise contributes to inadequate plant operating decisions
- Our experience has shown that each process has its own idiosyncrasies often varying by 30% or more from a plant employing the same process technology
  - ✓ Differences arise from design execution discrepancies, raw material variations from that expected in the design, outdated SOPs not changed to reflect the specific plant, operators running using experience from another similar process that may not work for the given plant
- Relying on intuition based on prior experience alone, to make plant operating decisions, can pose challenges on account of the plant idiosyncrasies
- While there is sufficient data to make informed decisions, it is not readily usable

9

Data Availability

Data availability is not a constraint but utilization is

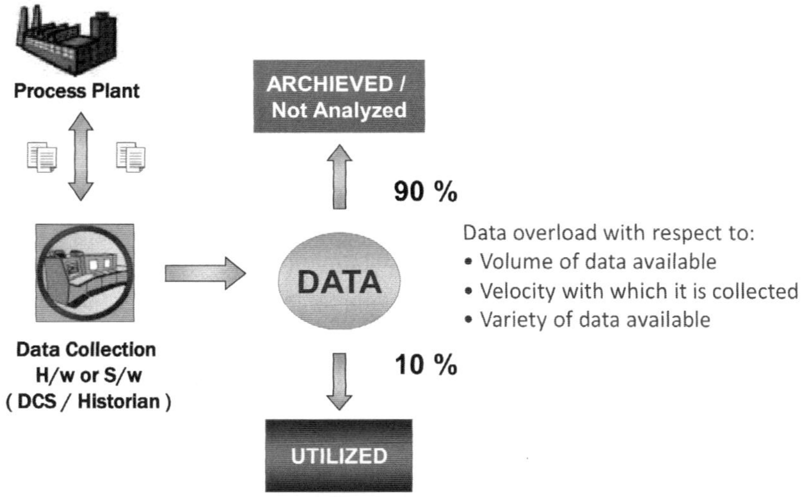

10

The Basic Challenge

The availability and utilization of time of the onsite plant engineers to utilize available data aided by technology tools for analysis

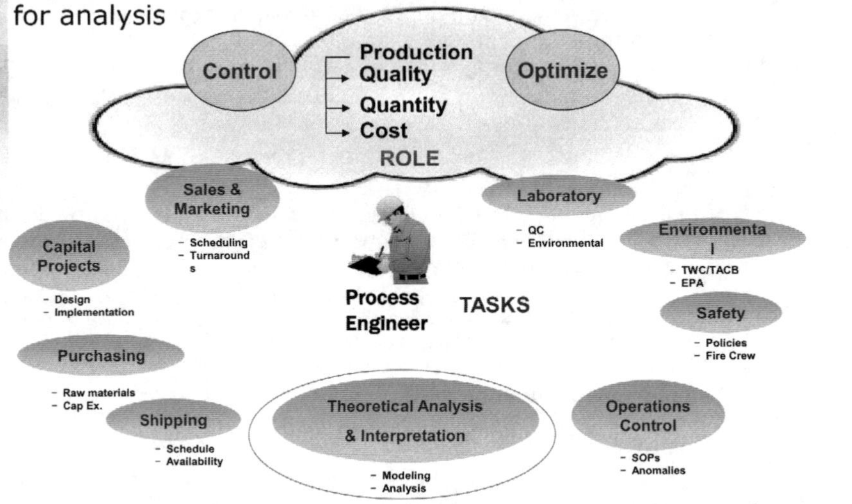

INGENERO

11

The approach and solution to address the challenge

INGENERO

12

Intuition and Judgment Enhancement

Utilization of data available from the process plant can address the challenges arising from lack of process expertise or variations in the process

- Analysis of larger data sets available from plant and business systems and its transformation into actionable information can
  - ✓ Reveal more powerful **descriptive**, historical insights into what actually happened
  - ✓ Be used to develop **predictive** models to gain insights into what is likely to happen
  - ✓ Be synthesized to provide better **prescriptive** insights to better determine the course of action
- The data analysis does not replace but enhances judgment and intuition
- It can back up and validate an intuition based position
  - ✓ Can agree with intuition
  - ✓ Further review, analysis and refinement, if it does not agree with intuition in a "sniff test"
  - ✓ Provides a solid back up to a decision instead of purely basing it on experience based intuition alone

INGENERO

The Process Data Analysis Solution

The solution provides actionable inputs to the operating team while enhancing their knowledge and intuition at a faster pace

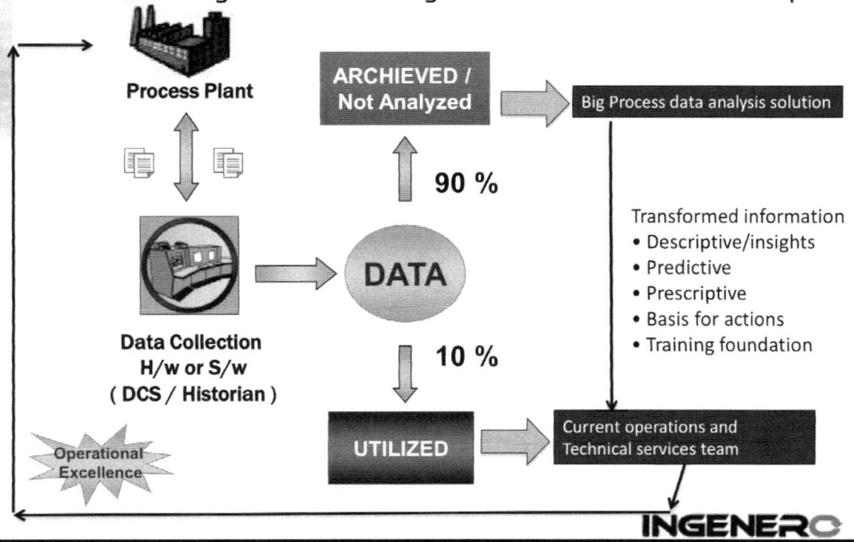

INGENERO

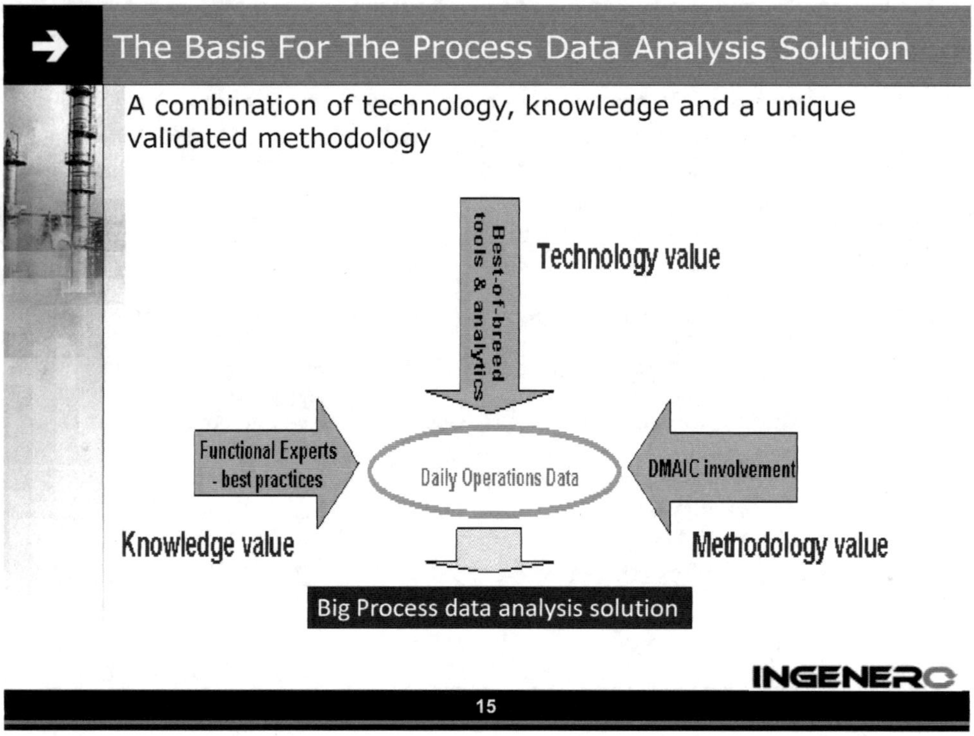
The Basis For The Process Data Analysis Solution
A combination of technology, knowledge and a unique validated methodology
Best-of-breed tools & analytics
Technology value
Functional Experts - best practices
Daily Operations Data
DMAIC involvement
Knowledge value
Methodology value
Big Process data analysis solution
INGENERO
15

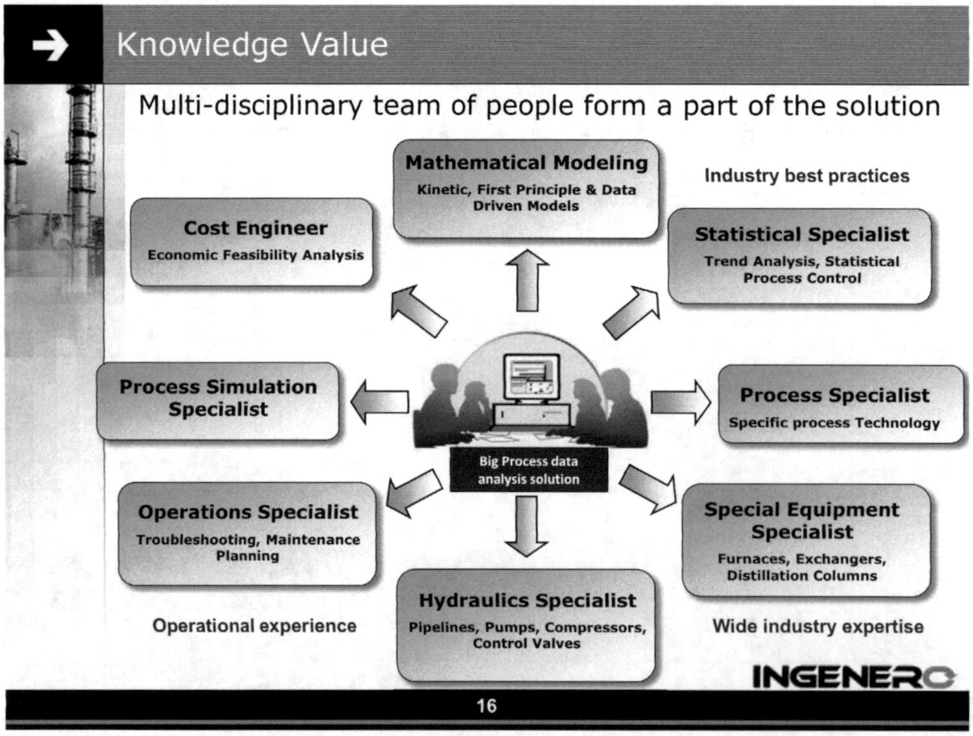
Knowledge Value
Multi-disciplinary team of people form a part of the solution
Mathematical Modeling
Kinetic, First Principle & Data Driven Models
Industry best practices
Cost Engineer
Economic Feasibility Analysis
Statistical Specialist
Trend Analysis, Statistical Process Control
Process Simulation Specialist
Process Specialist
Specific process Technology
Big Process data analysis solution
Operations Specialist
Troubleshooting, Maintenance Planning
Special Equipment Specialist
Furnaces, Exchangers, Distillation Columns
Hydraulics Specialist
Pipelines, Pumps, Compressors, Control Valves
Operational experience
Wide industry expertise
INGENERO
16

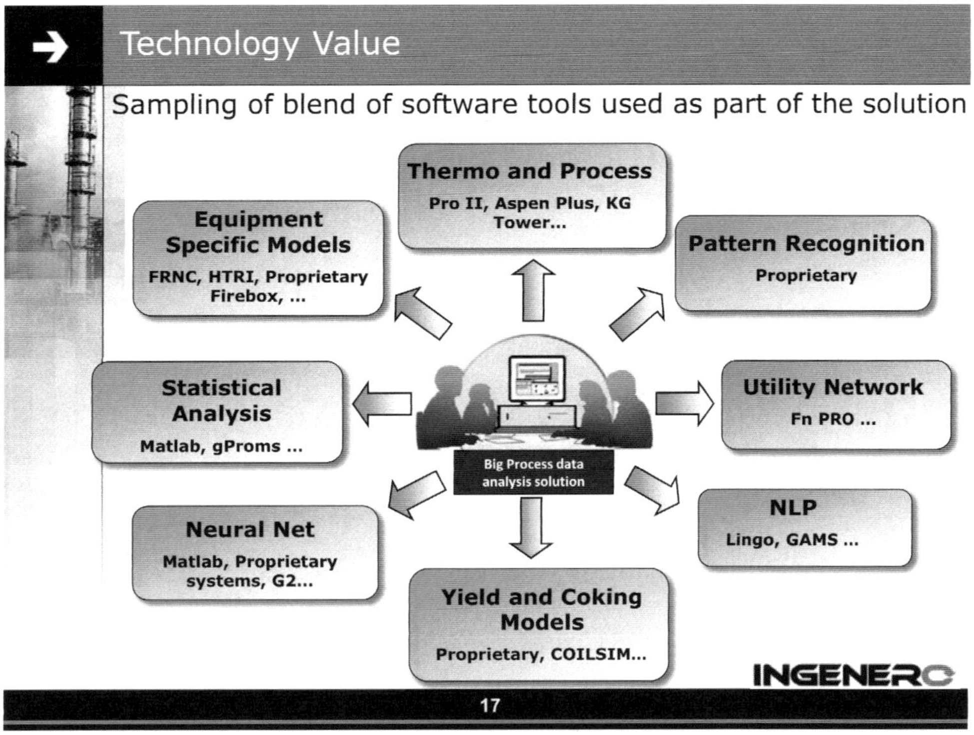
Technology Value
Sampling of blend of software tools used as part of the solution
Thermo and Process
Pro II, Aspen Plus, KG Tower...
Equipment Specific Models
FRNC, HTRI, Proprietary Firebox, ...
Pattern Recognition
Proprietary
Statistical Analysis
Matlab, gProms ...
Utility Network
Fn PRO ...
Big Process data analysis solution
Neural Net
Matlab, Proprietary systems, G2...
NLP
Lingo, GAMS ...
Yield and Coking Models
Proprietary, COILSIM...
INGENERO
17

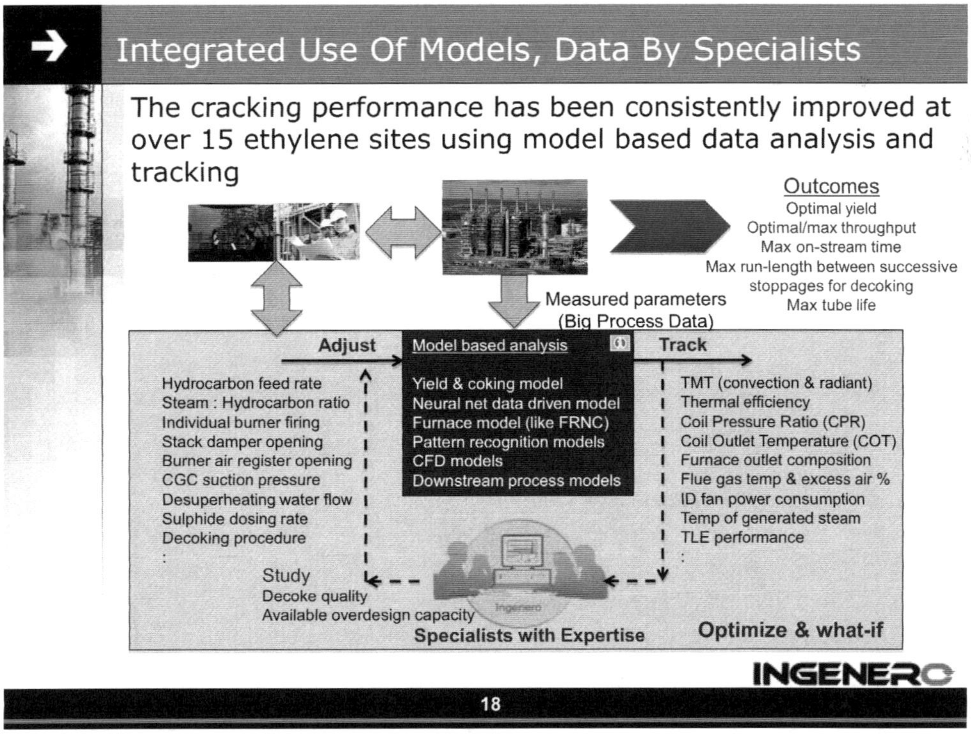
Integrated Use Of Models, Data By Specialists
The cracking performance has been consistently improved at over 15 ethylene sites using model based data analysis and tracking
Outcomes
Optimal yield
Optimal/max throughput
Max on-stream time
Max run-length between successive stoppages for decoking
Max tube life
Measured parameters (Big Process Data)
Adjust
Model based analysis
Track
Hydrocarbon feed rate
Steam : Hydrocarbon ratio
Individual burner firing
Stack damper opening
Burner air register opening
CGC suction pressure
Desuperheating water flow
Sulphide dosing rate
Decoking procedure
Yield & coking model
Neural net data driven model
Furnace model (like FRNC)
Pattern recognition models
CFD models
Downstream process models
TMT (convection & radiant)
Thermal efficiency
Coil Pressure Ratio (CPR)
Coil Outlet Temperature (COT)
Furnace outlet composition
Flue gas temp & excess air %
ID fan power consumption
Temp of generated steam
TLE performance
Study
Decoke quality
Available overdesign capacity
Specialists with Expertise
Optimize & what-if
INGENERO
18

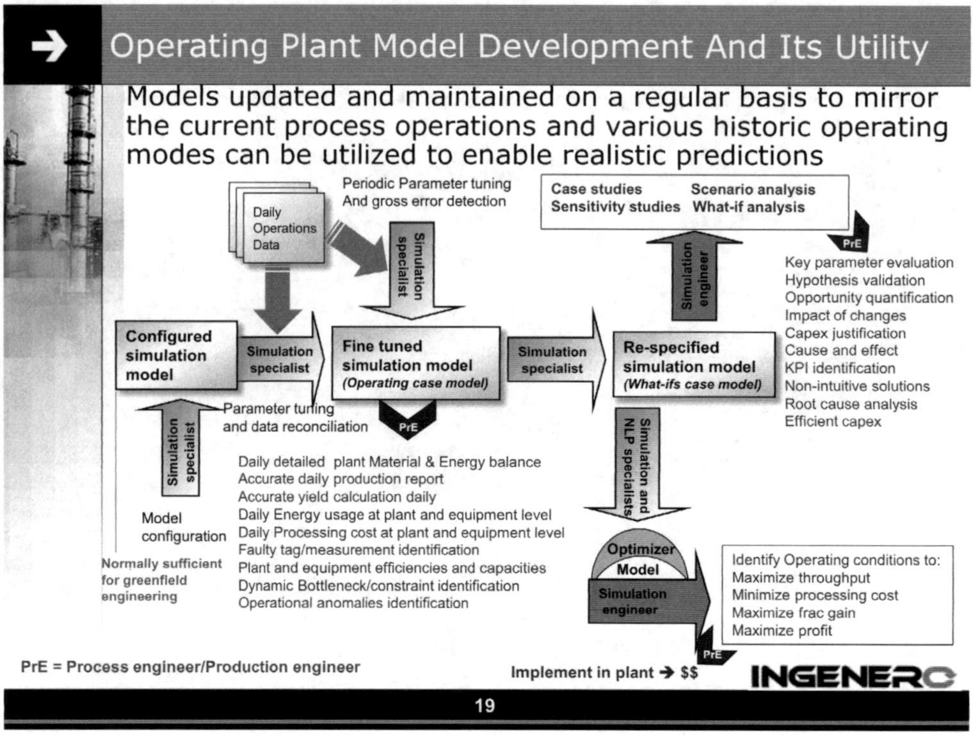
Operating Plant Model Development And Its Utility
Models updated and maintained on a regular basis to mirror the current process operations and various historic operating modes can be utilized to enable realistic predictions
Daily Operations Data
Periodic Parameter tuning And gross error detection
Simulation specialist
Case studies
Scenario analysis
Sensitivity studies
What-if analysis
PrE
Simulation engineer
Key parameter evaluation
Hypothesis validation
Opportunity quantification
Impact of changes
Capex justification
Cause and effect
KPI identification
Non-intuitive solutions
Root cause analysis
Efficient capex
Configured simulation model
Simulation specialist
Fine tuned simulation model
(Operating case model)
Simulation specialist
Re-specified simulation model
(What-ifs case model)
Parameter tuning and data reconciliation
PrE
Simulation specialist
Simulation and NLP specialists
Daily detailed plant Material & Energy balance
Accurate daily production report
Accurate yield calculation daily
Daily Energy usage at plant and equipment level
Daily Processing cost at plant and equipment level
Faulty tag/measurement identification
Plant and equipment efficiencies and capacities
Dynamic Bottleneck/constraint identification
Operational anomalies identification
Model configuration
Normally sufficient for greenfield engineering
Optimizer Model
Simulation engineer
Identify Operating conditions to:
Maximize throughput
Minimize processing cost
Maximize frac gain
Maximize profit
PrE
PrE = Process engineer/Production engineer
Implement in plant → $$
INGENERO
19

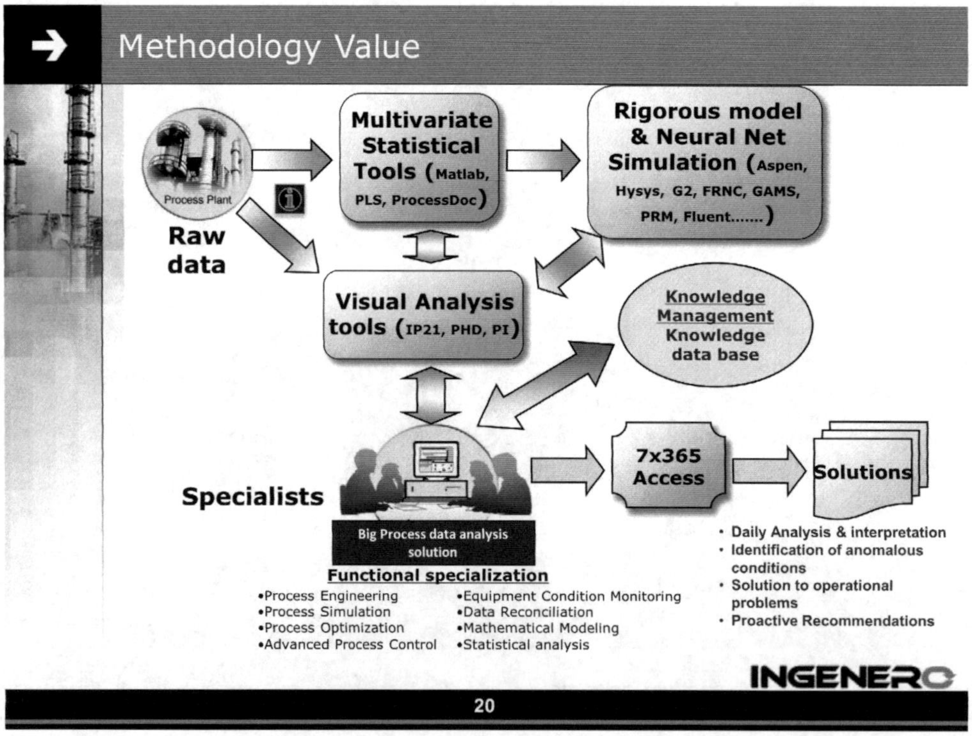
Methodology Value
Process Plant
Raw data
Multivariate Statistical Tools (Matlab, PLS, ProcessDoc)
Rigorous model & Neural Net Simulation (Aspen, Hysys, G2, FRNC, GAMS, PRM, Fluent.......)
Visual Analysis tools (IP21, PHD, PI)
Knowledge Management
Knowledge data base
Specialists
7x365 Access
Solutions
Big Process data analysis solution
Functional specialization
•Process Engineering
•Process Simulation
•Process Optimization
•Advanced Process Control
•Equipment Condition Monitoring
•Data Reconciliation
•Mathematical Modeling
•Statistical analysis
• Daily Analysis & interpretation
• Identification of anomalous conditions
• Solution to operational problems
• Proactive Recommendations
INGENERO
20

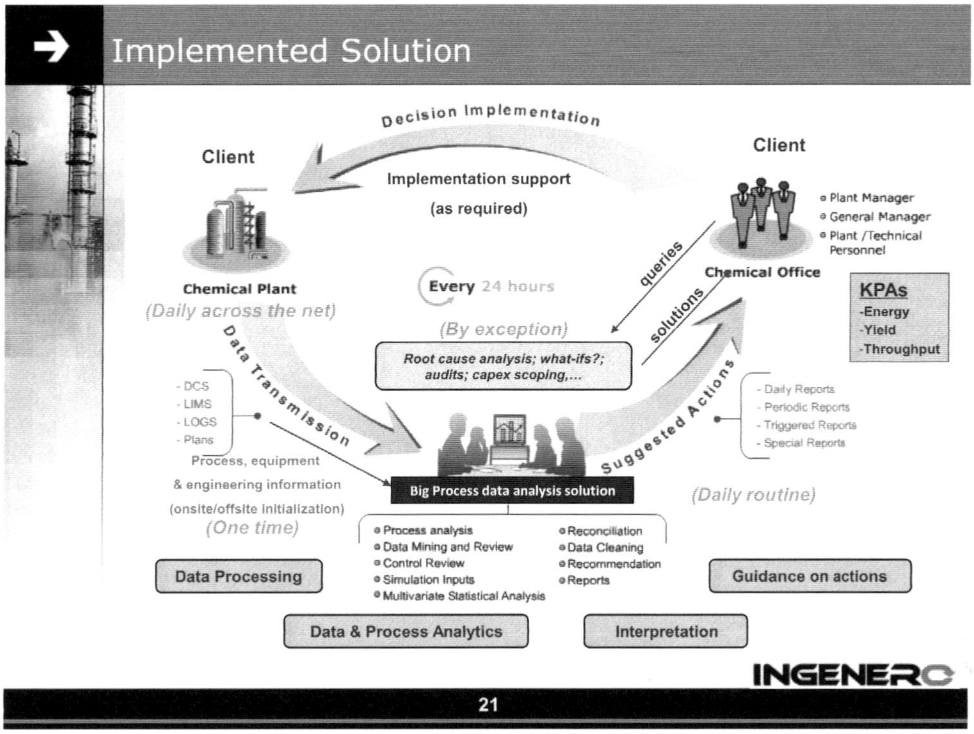
Implemented Solution
Decision Implementation
Client
Implementation support
(as required)
Client
Plant Manager
General Manager
Plant /Technical Personnel
Chemical Office
KPAs
-Energy
-Yield
-Throughput
queries
solutions
Chemical Plant
(Daily across the net)
Every 24 hours
(By exception)
Root cause analysis; what-ifs?; audits; capex scoping,...
Data Transmission
- DCS
- LIMS
- LOGS
- Plans
Process, equipment
& engineering information
(onsite/offsite initialization)
(One time)
Suggested Actions
- Daily Reports
- Periodic Reports
- Triggered Reports
- Special Reports
Big Process data analysis solution
(Daily routine)
Process analysis
Data Mining and Review
Control Review
Simulation Inputs
Multivariate Statistical Analysis
Reconciliation
Data Cleaning
Recommendation
Reports
Data Processing
Guidance on actions
Data & Process Analytics
Interpretation
INGENERO
21

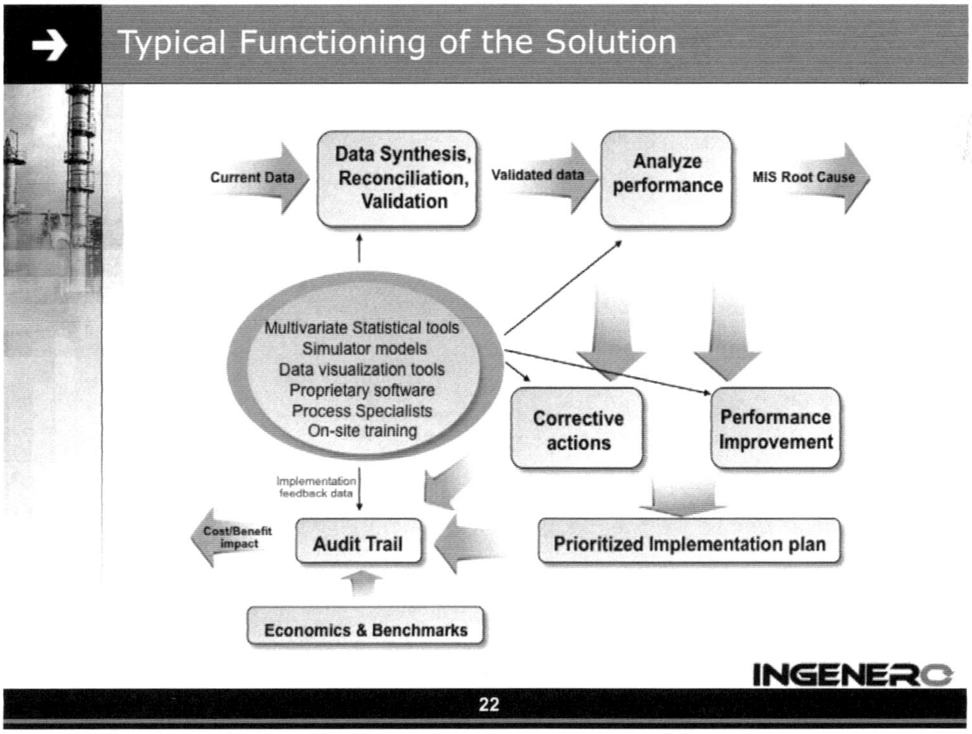
Typical Functioning of the Solution
Current Data
Data Synthesis, Reconciliation, Validation
Validated data
Analyze performance
MIS Root Cause
Multivariate Statistical tools
Simulator models
Data visualization tools
Proprietary software
Process Specialists
On-site training
Corrective actions
Performance Improvement
Implementation feedback data
Cost/Benefit impact
Audit Trail
Prioritized Implementation plan
Economics & Benchmarks
INGENERO
22

Results and Benefits

INGENERO

23

Typical Value From Process Data Analysis Solution

Typical Benefits Derived from the Process Data Solution Application

Proactive vs Reactive

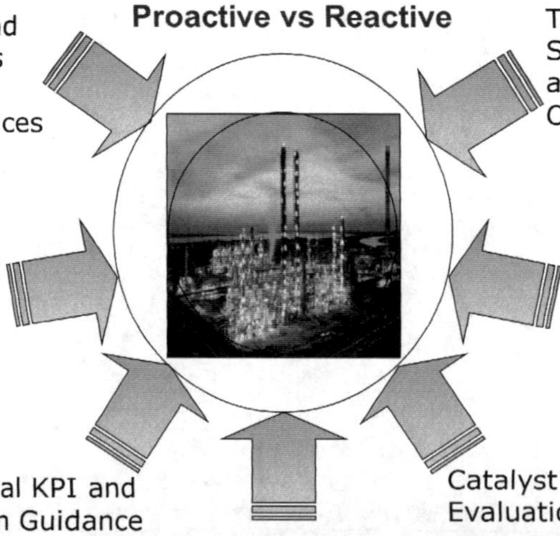

INGENERO

24

Improved Furnace Run-length

Case example of outcome from the Big Data Solution at an Ethylene plant

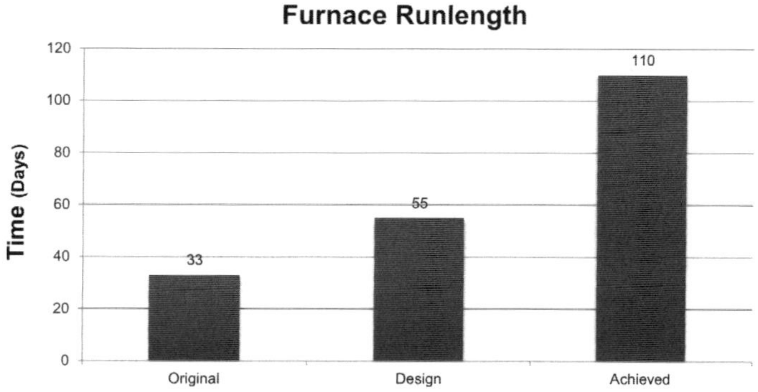

Furnace run lengths between decoking doubled over design expectation

INGENERO

25

Improved Production

Case example of outcome from the Process Data Analysis Solution at an Ethylene plant

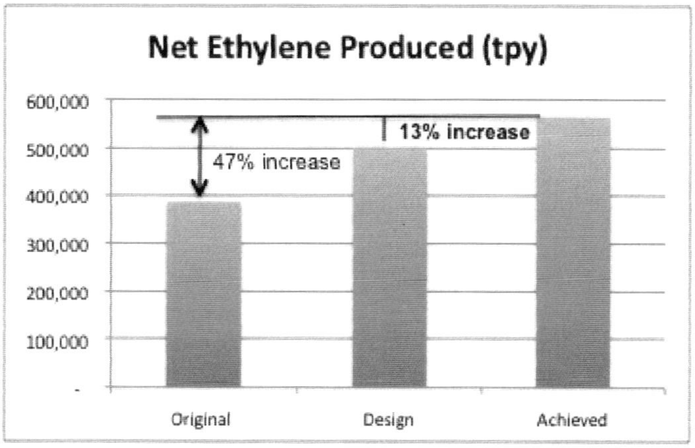

Increased capacity without capex

INGENERO

26

Improved Ethylene yield

Case example of outcome from the Process Data Analysis Solution at an Ethylene plant

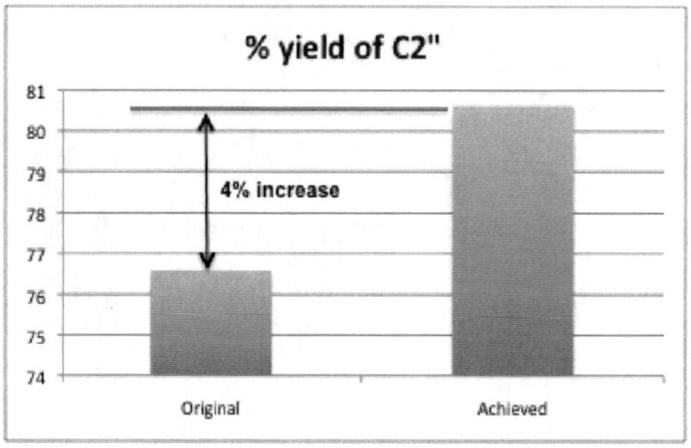

Increased raw material conversion efficiency

INGENERO

27

Improved Turnaround Efficiency

This Solution has enabled enhanced turnaround efficiency for several manufacturers globally

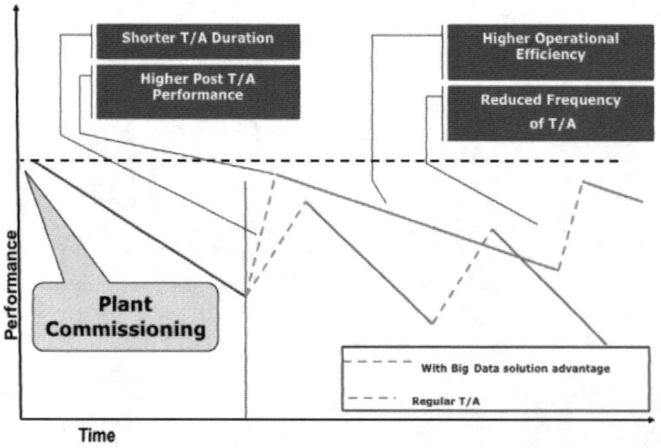

INGENERO

28

Sampling of results delivered by the Solution

Other results the implementation of the Process Data Analysis Solution has offered its implementers

- Helped run the plant an extra year before a turnaround, optimally
- Early detection of furnace tube leak to avert tube failure
- Helped process manufacturer team ascertain design flaws to win against licensor
- Failure predictions and residual life analysis prevented multiple safety and maintenance events
- Debottlenecking study at 50% cost and time for 25% capacity increase
- Increased record production levels
- Multiple cases of early detection of anomalous conditions prevents a plant upset
- Increased furnace reliability
- Optimal furnace feed selection
- Reduced specific energy consumption

INGENERO

29

In Conclusion

The Process Data Analysis Solution has allowed process manufacturers sustained and improved performance on various parameters and achievement of Operations Excellence

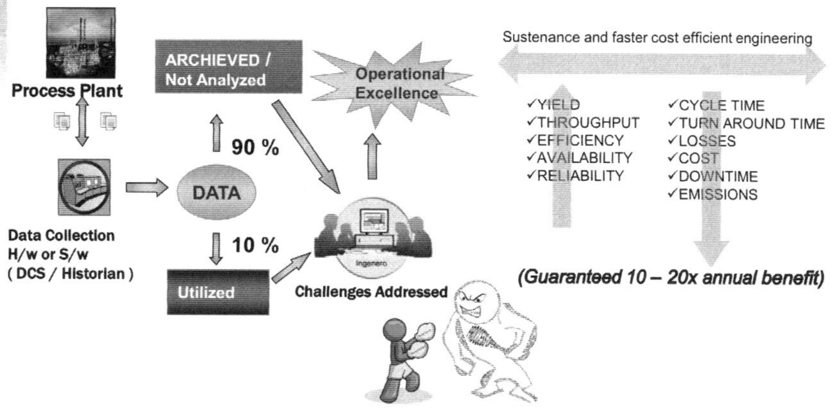

INGENERO

30

INGENERO

Excellence Through Insight

4615 Southwest Freeway, Suite 320,
Houston, Texas, 77002, USA

www.ingenero.com info@ingenero.com

INGENERO

31

Addendum slides

INGENERO

32

Data Collection Process

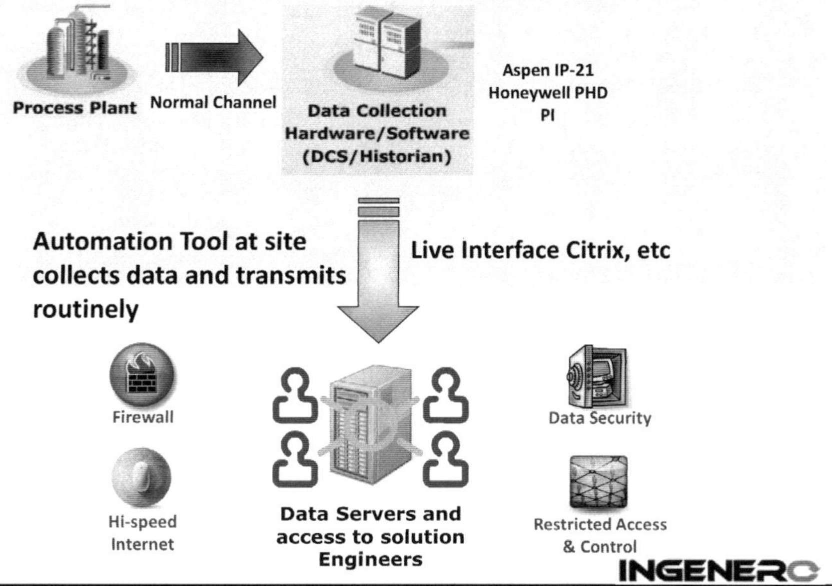

33

Technology Utilization Examples

34

Integrated Process Monitoring Platform

An integrated platform that ties in diverse software tools enables efficient analysis

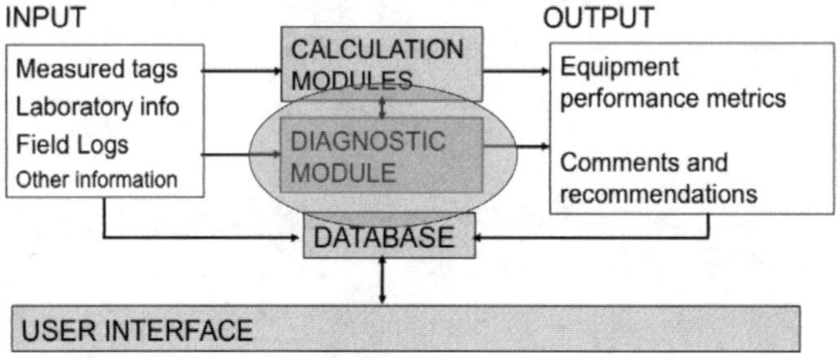

Tag Management
Long term trends
Customized Reports
Knowledge Repository
Create rule based recommendation

INGENERO

Utilization Of Models In Diagnostic Module

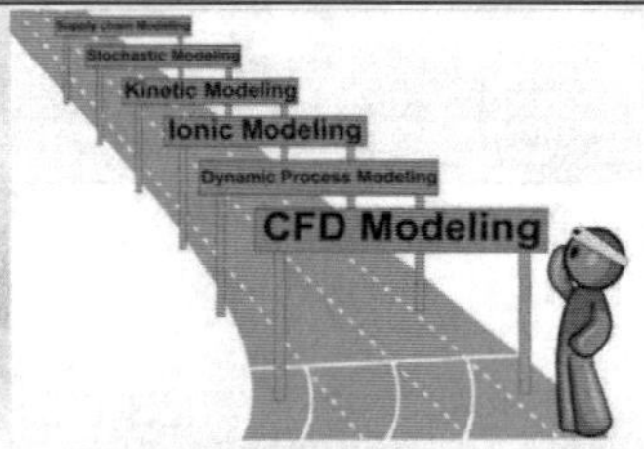

- Steady state process simulation & optimization (MINLP based)
  - Aspen suite, Hysys, FRNC, HTRI.....
    - Petrochemicals, Ionic, Specialty chemicals, Pharmaceutical intermediates, Refining, Oleochemicals
- Dynamic models
  - Operator training, model predictive control
- Fundamental first principle models
  - Reaction kinetics, mass & heat transfer, CFD, Thermodynamics
- Data driven models
  - Statistical, neural net, pattern recognition, Winter's model, Weibull analysis, PCA, time series
- Supply chain models
  - Integrated supply chain model (supply, production, inventory, distribution), Distribution network design & analysis, logistics optimization, strategic planning models, scheduling models, Material flow analysis, Optimal feed distribution

INGENERO

Technology Utilization To Analyze Data

The appropriate technology is utilized as part of the solution to help the plant achieve the greatest production and extend furnace run lengths

- Convection section modeling (e.g. FRNC 5) to achieve maximum throughput / cracking safely
- CFD to address flue gas channeling, flame impingement and hot spots on tubes
- Fire box modeling (geometry / heat flux) to control rate of fouling at the individual coil level
- Yield models allow appropriate adjustments for incoming feed
- ANN modeling to predict run lengths and adjust run lengths to plan decoking cycle
- LP Model to maximize production options appropriately

INGENERO

Model Utilization Example - FRNC

Convection section modeling allows maximum through-put without exceeding design tube skin temperature

Capacity limitation: DSS design skin temperature

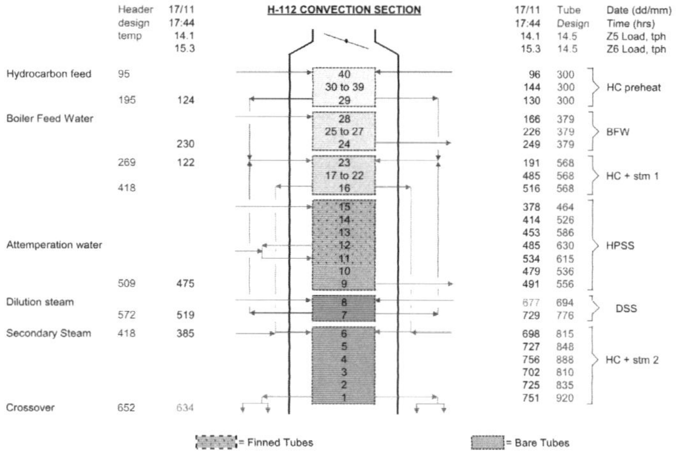

INGENERO

Tool Utilization Example: Pattern Recognition

Pattern recognition and clustering techniques often utilized as part of the analysis

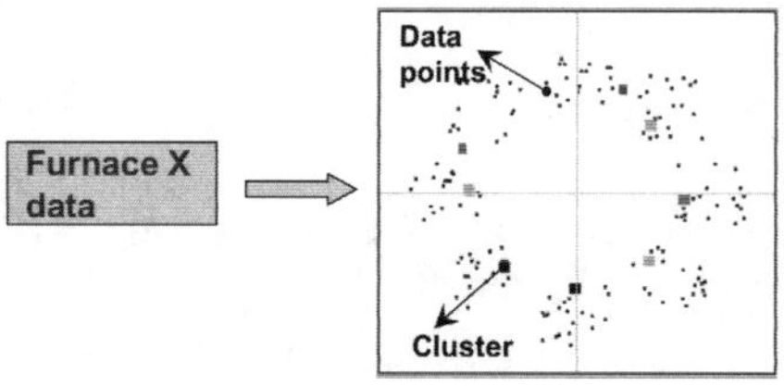

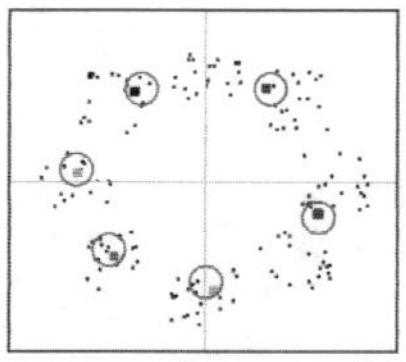

Clusters for complete data

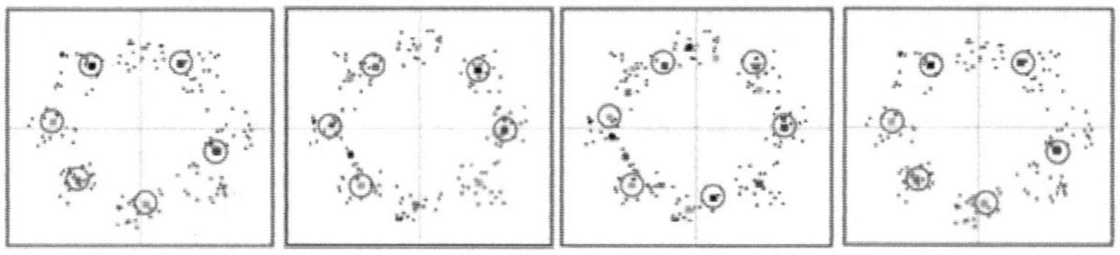

Clusters per zone

INGENERO

39

Technique Utilization Example - Visualization

Sample MIS – Modular View Of Energy Flow

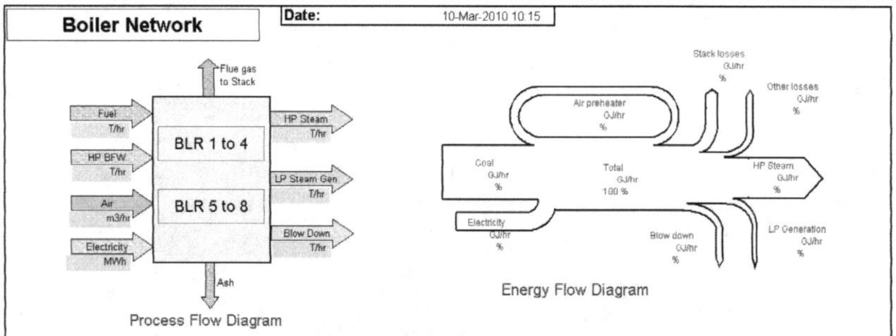

Metrics	Unit	Std Value	Present Value
Sp. Energy Consumption	GJ/T		
Sp. Coal Consumption	T/T		
Sp. Electrical consumption	MWh/T		
Sp. LP steam consumption	T/T		
Air to fuel ratio	m3/T		
Blow down %	%		
Evaporation ratio	T/T		
Air preheater energy	GJ/hr		
HP steam pressure	kg/cm2 g		
HP steam superheated temp	deg C		

Back to Main Page

Turbines Unit

INGENERO

40

Model Detail Utilization Example

Fire box (geometry / heat flux) models allows individual burner management

Burner Adjustment for Controlling TMT

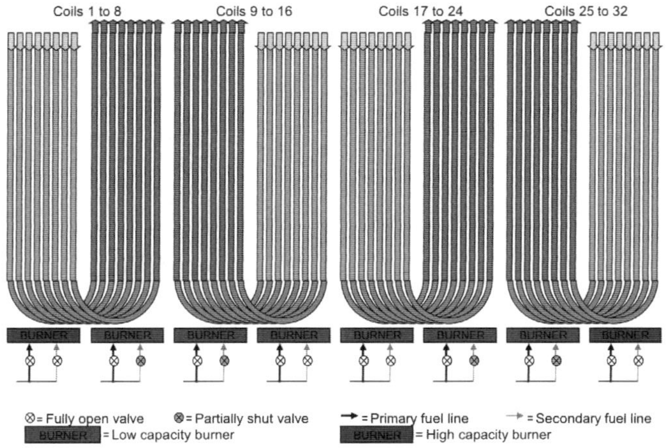

INGENERO

Model Utilization Example - Yield

Yield model provides insights on key furnace parameters and their interaction

Furnace outlet stream composition as a function of:

- Coil outlet temperature (COT)
- Coil pressure
- Dilution steam to hydrocarbon ratio

INGENERO

Model Utilization Example – LP

LP modeling sets optimal operating parameters for furnace feed distribution

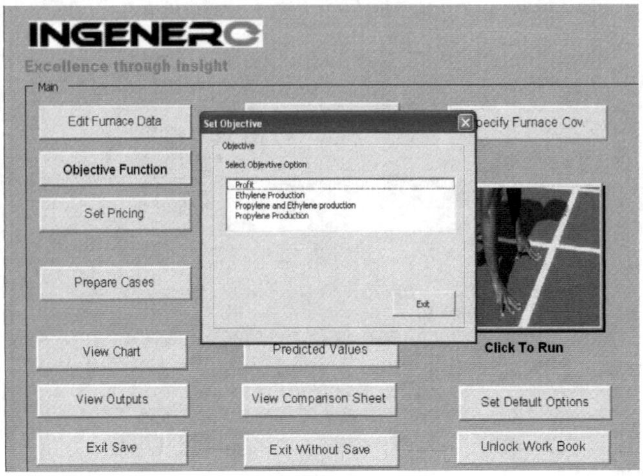

Model Utilization Example – Process Simulator

The recovery section of ethylene plants is about operating effectively and efficiently – process simulation helps

- Minimize chemical usage
- Minimize energy
- Debottleneck the operations

Steady state simulation provides the foundation for improvement

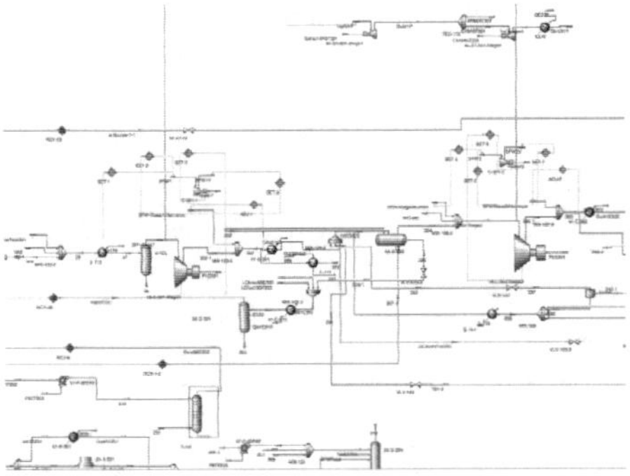

Technique Utilization Example – Energy analysis

Energy Audits and Pinch Analysis are extremely effective in stabilizing energy usage and most efficient heat exchanger network

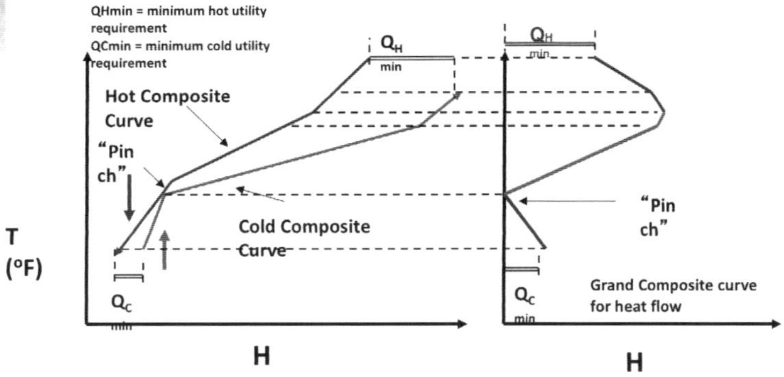

INGENERO

45

Model Utilization Example – CW Network

Cooling water network models can identify problems and find solutions

Example using fn-Pro

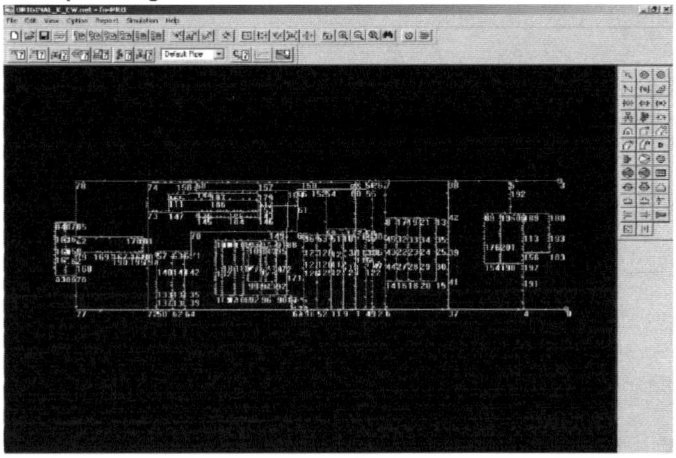

INGENERO

46

Improving Life-Cycle Costs of Large Integrated Capital Projects in the Chemical Industry

Mike Strobel, The Fidelis Group LLC
Partner

Houston, Texas
Since 2001

FIDELIS GROUP

Agenda

- Simulation Modeling Overview
- Large Integrated Project Example
- Conclusion

providing confidence, quantifying risk FIDELIS GROUP 2

Management of Capital Assets

- Good Resource Allocation ***Requires*** Good Decision Making Ability
- More Profitable Decisions ***Require*** Quantitative Answers

providing confidence, quantifying risk   FIDELIS GROUP   3

Modeling Methods

Gut Feel
Trend Models
Analytical

Limited by complexity and change (aging, design, maintenance, operations)

Simulation →

Unconstrained by complexity
More accurate
More profitable

providing confidence, quantifying risk   FIDELIS GROUP   4

Monte Carlo Method

Stochastic simulation or discrete event simulation

- Runs many life cycles, each with a different stochastic behavior, and aggregates the results
- Predicts expected system behavior and variation
- Provides basis for "what-if" analysis

providing confidence, quantifying risk FIDELIS GROUP 5

Inputs

- Configuration (Index Flow Sheet, Process Flow Diagram, Piping and Instrumentation Diagram)
- Failure/Repair Data (client, industry, Fidelis)
- Operation & Maintenance Logic (demand, alternate supplies, scheduled activities, tank level management, logistics)

providing confidence, quantifying risk FIDELIS GROUP 6

Level of Detail

Sub-Systems
Process Units
Process Site
Entire Supply Chain

- Mechanical equipment
  - fixed, rotating
- Pipes
- Tanks
- Reactors
- Instruments
- Electrical systems
  - switches, breakers, bus bars
- Human behavior
- Failure modes
- Scheduled activities

providing confidence, quantifying risk FIDELIS GROUP 7

Additional Output Quantification

- Availability
- Throughput
- System Sensitivities
- Capacity Utilization/Bottlenecks
- Downtime
- Spares Convergence/Optimization
- Undesired Events (number, duration, magnitude)
- Logistic Events
- Tank Events
- Life Cycle Costs

providing confidence, quantifying risk FIDELIS GROUP 8

Large Project Example

Issues/Challenges:

- Multiple ETC Teams located on multiple continents
- JV Partner relationships and confidentiality
- Complete a very complex unit level model with inputs that were not fixed
- Optimize Power/Utility configuration & usage
- Frequently update model structure, product mix and results
- Long model run times due to model complexity
- Optimize unit capacity versus capital cost

providing confidence, quantifying risk FIDELIS GROUP 9

Complete a Very Complex Unit Level Model

- Due to the size and complexity of the site, it was initially created as 4 distinct models
  - Raw Material / Hydrocarbon Envelope
  - Derivative 1 Envelope
  - Derivative 2 Envelope
  - Power/Utilities

- After model validation of the 4 envelopes, the models were merged together

providing confidence, quantifying risk FIDELIS GROUP 10

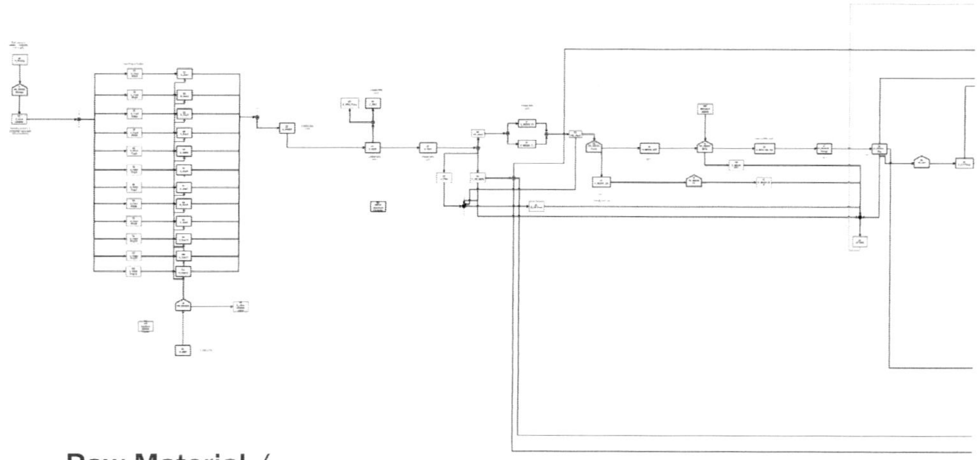

Raw Material / Hydrocarbon

FIDELISGROUP

11

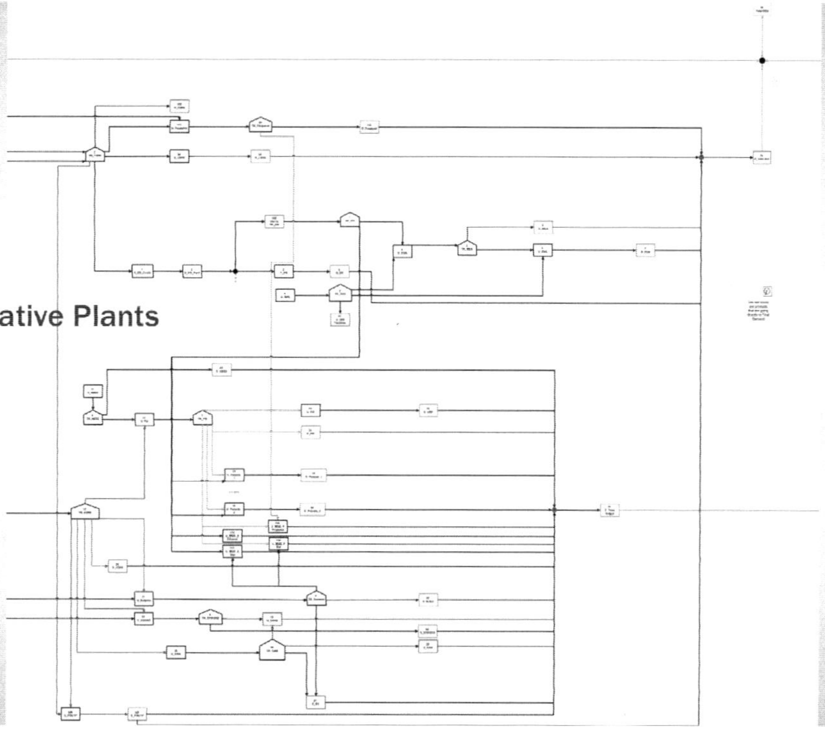

Derivative Plants

12

Frequently Update Model Structure & Product Mix

- Auto-Name Pipes
- Prefix Unit and Tank Names
- Built an Excel Macro to complete data checks prior to importing data into Titan
  - Pipe and Tank Balances
  - Name Matches
  - Check for Missing/New Pipes
- Import Model Updates to Pipes, Tanks and components directly into Titan
- Use of Sim Customize Sub in VBA

providing confidence, quantifying risk

FIDELIS GROUP

13

Update Results and Report

- Export Titan Table results directly out to Excel
- Titan configured to generate custom results in excel format
  - Bander
  - Result Groups
  - Bottleneck Identification and Ranking through Pipe Limit Feature
- Created Excel Macro to read these formats and automatically update all results
- Results automatically updated to Report format

providing confidence, quantifying risk

FIDELIS GROUP

14

Bander

	A	B	C	D	E	F	G	H	I	J	K	L	M	N	O	P	Q	R	S	T	U	V	W
1		0%	1%	2%	3%	4%	5%	6%	7%	8%	9%	10%	11%	12%	13%	14%	15%	16%	17%	18%	19%	20%	21%
2	Alpha	2.38%	0.00%	0.00%	0.00%	0.00%	0.00%	0.00%	0.00%	0.00%	0.00%	0.00%	0.00%	0.00%	0.00%	0.00%	0.00%	0.00%	0.00%	0.00%	0.00%	0.00%	0.00%
3	Bravo	2.38%	0.00%	0.00%	0.00%	0.00%	0.00%	0.00%	0.00%	0.00%	0.00%	0.00%	0.00%	0.00%	0.00%	0.00%	0.00%	0.00%	0.00%	0.00%	0.00%	0.00%	0.00%
4	Charlie	2.38%	0.00%	0.00%	0.00%	0.00%	0.00%	0.00%	0.00%	0.00%	0.00%	0.00%	0.00%	0.00%	0.00%	0.00%	0.00%	0.00%	0.00%	0.00%	0.00%	0.00%	0.00%
5	Delta	4.21%	0.00%	0.00%	0.00%	0.00%	0.00%	0.00%	0.00%	0.00%	0.00%	0.00%	0.00%	0.00%	0.54%	0.00%	0.00%	0.00%	0.00%	0.00%	0.25%	0.01%	0.00%
6	Echo	2.82%	0.00%	0.00%	0.00%	0.00%	0.00%	0.00%	0.00%	0.00%	0.00%	0.00%	0.00%	0.00%	0.00%	0.00%	0.00%	0.00%	0.00%	0.00%	0.00%	0.00%	0.00%
7	Foxtrot	2.82%	0.00%	0.00%	0.00%	0.00%	0.00%	0.00%	0.00%	0.00%	0.00%	0.00%	0.00%	0.00%	0.00%	0.00%	0.00%	0.00%	0.00%	0.00%	0.00%	0.00%	0.00%
8	Golf	2.82%	0.00%	0.00%	0.00%	0.00%	0.00%	0.00%	0.00%	0.00%	0.00%	0.00%	0.00%	0.00%	0.00%	0.00%	0.00%	0.00%	0.00%	0.00%	0.00%	0.00%	0.00%
9	Hotel	2.82%	0.00%	0.00%	0.00%	0.00%	0.00%	0.00%	0.00%	0.00%	0.00%	0.00%	0.00%	0.00%	0.00%	0.00%	0.00%	0.00%	0.00%	0.00%	0.00%	0.00%	0.00%
10	India	2.82%	0.00%	0.00%	0.00%	0.00%	0.00%	0.00%	0.00%	0.00%	0.00%	0.00%	0.00%	0.00%	0.00%	0.00%	0.00%	0.00%	0.00%	0.00%	0.00%	0.00%	0.00%
11	Juliett	2.82%	0.00%	0.00%	0.00%	0.00%	0.00%	0.00%	0.00%	0.00%	0.00%	0.00%	0.00%	0.00%	0.00%	0.00%	0.00%	0.00%	0.00%	0.00%	0.00%	0.00%	0.00%
12	Kilo	4.94%	0.00%	0.00%	0.00%	0.00%	0.00%	0.00%	0.00%	0.00%	0.00%	0.00%	0.00%	0.00%	0.00%	0.00%	0.00%	0.00%	0.00%	0.00%	0.00%	0.00%	0.00%
13	Lima	4.99%	0.00%	0.00%	0.00%	0.00%	0.00%	0.00%	0.00%	0.00%	0.00%	0.00%	0.00%	0.00%	0.00%	0.00%	0.00%	0.00%	0.00%	0.00%	0.00%	0.00%	0.00%
14	Mike	5.02%	0.00%	0.00%	0.00%	0.00%	0.00%	0.00%	0.00%	0.00%	0.00%	0.00%	0.00%	0.00%	0.00%	0.00%	0.00%	0.00%	0.00%	0.00%	0.00%	0.00%	0.00%
15	November	5.01%	0.00%	0.00%	0.00%	0.00%	0.00%	0.00%	0.00%	0.00%	0.00%	0.00%	0.00%	0.00%	0.00%	0.00%	0.00%	0.00%	0.00%	0.00%	0.00%	0.00%	0.00%
16	Oscar	5.02%	0.00%	0.00%	0.00%	0.00%	0.00%	0.00%	0.00%	0.00%	0.00%	0.00%	0.00%	0.00%	0.00%	0.00%	0.00%	0.00%	0.00%	0.00%	0.00%	0.00%	0.00%
17	Papa	5.02%	0.00%	0.00%	0.00%	0.00%	0.00%	0.00%	0.00%	0.00%	0.00%	0.00%	0.00%	0.00%	0.00%	0.00%	0.00%	0.00%	0.00%	0.00%	0.00%	0.00%	0.00%
18	Quebec	4.97%																					
19	Romeo	4.97%																					
20	Sierra	4.99%																					
21	Tango	5.00%																					
22	Uniform	4.98%																					
23	Victor	5.02%																					
24	Whiskey	4.96%																					
25	Xray	4.93%																					
26	Yankee	4.97%																					
27	Zulu	5.00%																					
28	One	4.98%																					
29	Two	4.94%																					
30	Three	4.94%																					
31	Four	5.01%																					

Turn up Track

Ready

Focus Unit Utilization Data

Focus Unit	
Ranges	**Data**
0%	2.82%
1% to 45%	1.91%
46% to 54%	0.04%
55% to 64%	0.92%
65% to 74%	1.32%
75% to 83%	4.41%
84% to 85% (Demand)	67.21%
86% to 100% (Annual)	5.51%
95% of Design Rates	15.84%
100% of Design Rates	0.00%
Total	**100.0%**

Focus Unit Turn - Up Histogram

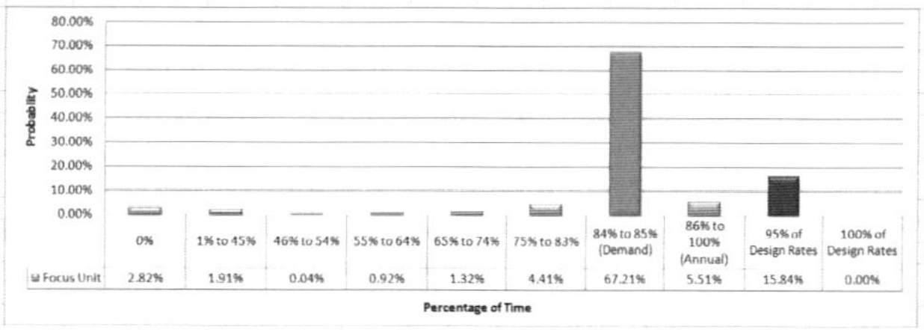

- ❖ Can be used to band tanks and pipes also

Bottleneck Identification and Ranking

Summary Report

	Unconstrained Throughput	Potential Product	Constrained Throughput
P 10	13113390.94	1705545.819	11407845.12
P 50	14570434.37	2469034.888	12101399.49
P 90	16027477.81	3308527.232	12718950.58

Mission Time	1825	Days

providing confidence, quantifying risk FIDELIS GROUP

Bottleneck Identification and Ranking

P50 Bottleneck Analysis

Main Area	Sub Area	Common Name	Titan Pipe Name	Capacity	Time Stamp	Weight (bbls)
Gasification	Gasifiers	Gasifier Feed	Gasifiers Gasifier Feed	200	1/1/2014	0
Gasification	Cracking	Gasifier Product	Cracking Gasifier Product	10000	3/1/2016	0
C3	Polyol 1	Batch 1	Polyol 1 Batch 1	175	1/1/2014	0
C3	Polyol 2	Batch 2	Polyol 2 Batch 2	10000	3/1/2016	0
Gasification	Methanol	Methanol Generation	Methanol Methanol Generation	250	1/1/2014	0
C3	Butanol	Butanol	Butanol Butanol	250	1/1/2014	0
C2	EDC	EDC	EDC EDC	300	1/1/2014	986007.6518
Gasification	Waste	Waste	Waste Waste	10000	3/1/2016	0
C3	PG	PG	PG PG	10000	3/1/2016	0

- Identify Bottlenecks
- Quantify the weight of each bottleneck
- De-bottleneck single or groups of bottlenecks to create project rankings and time

providing confidence, quantifying risk FIDELIS GROUP

Long Run Times

- The model had long run times due to model complexity
- During this project, we started development of what has now become Hydra.

Hydra is a parallel processing system that enables hundreds of lifecycles (virtual histories) to run concurrently.

- This development cut run time down from over 24 hours to as short as 2 hours.

providing confidence, quantifying risk FIDELIS GROUP 18

Optimization Unit Capacity versus Capital Cost

- Brute force through alternate cases (known cases)
- Op-Tite – Full Optimization Engine
  - Price per unit capacity increase
  - Min and Max unit capacity ranges
  - Smart Optimization – only run until the maximum is reach; not exhaustive
- Conclusion = Optimization yielded similar results as alternate case studies in much less time and effort.

providing confidence, quantifying risk FIDELIS GROUP 19

Conclusion

- Use of Simulation Modeling helps to improve efficiency and reduce project timelines
- Integrate Model with Process Simulator to directly feed inputs
- Millions of dollars in reduced capital and/or increased utilization is the pay off for up front simulation work
- Continued Use of Model through all phases of project will optimize results
- Hand-Off Model to plant personnel for ongoing optimization

providing confidence, quantifying risk FIDELIS GROUP 20

Guidelines for Dynamic Modeling of Column Startup

Chemstations, Inc.

Authors: Wendy Young Reed &
Stuart Schwab

Chemstations™ www.chemstations.com

Agenda

- Background of case study
- Determining scope of model
- Simulation Guidelines
- Case Study
- Summary

Chemstations™ 2

Background of Case Study

- Customer – Calumet Montana Refining, LLC
- Had existing steady-state model of entire facility
- Shared P&IDs, PFDs, procedures, & process data from previous start-up
- New crude unit being built & would like accurate simulation of startup conditions

Chemstations™ 3

Original Steady State Simulation

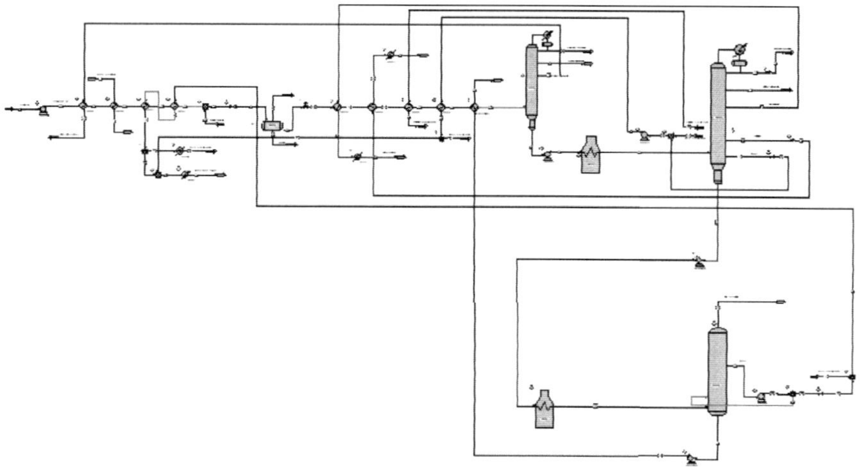

Chemstations™ 4

Goals of Case Study

- Simulate crude unit only
- Model existing tower startup
- Compare to historical startup data
- Determine if dynamic simulation could be used as tool to predict startup performance of new crude unit

Chemstations 5

Determining Scope of Model

Use operating procedure to determine sequential steps & model targets:

1) Crude charge introduced to tower at increasing temp & flow rate
2) Target level reached in overhead accumulator
3) Reflux system placed in service
4) Stripping steam introduced to tower
5) Product streams reach specification

Chemstations 6

Simulation Guidelines

- Start with a simple model
- Get model to converge
- Validate results
- Add complexity as required to reach desired engineering solution

Chemstations

7

Start with a Simple Model

- Crude tower only
- Change to dynamic simulation mode
- Pumparounds/side strippers not in service
- Crude Heater modeled as temp & flow ramps
- Use actual data & equipment parameters
  - Feed temps & flows from actual data
  - Accumulator dimensions & condenser setpoints
  - # trays, feed stage, etc.

Chemstations

8

Simplified Flowsheet

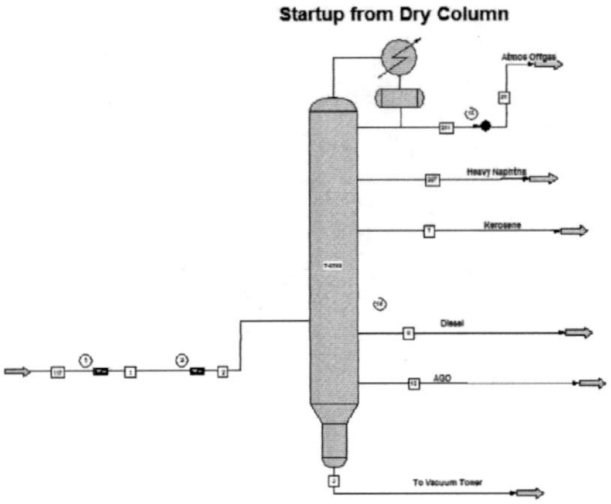

Chemstations™

9

Change to Dynamic Simulation Mode

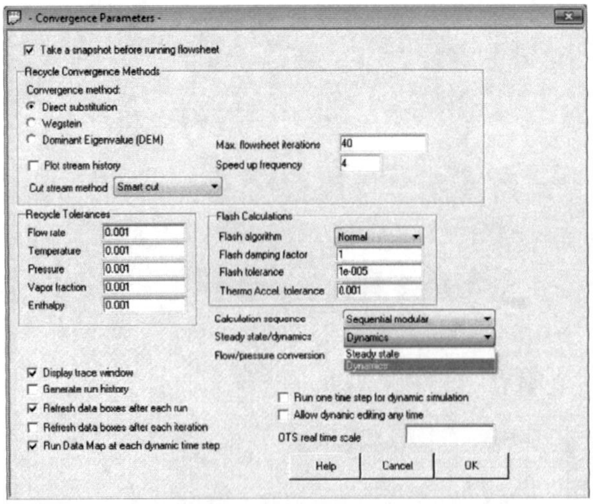

Chemstations™

10

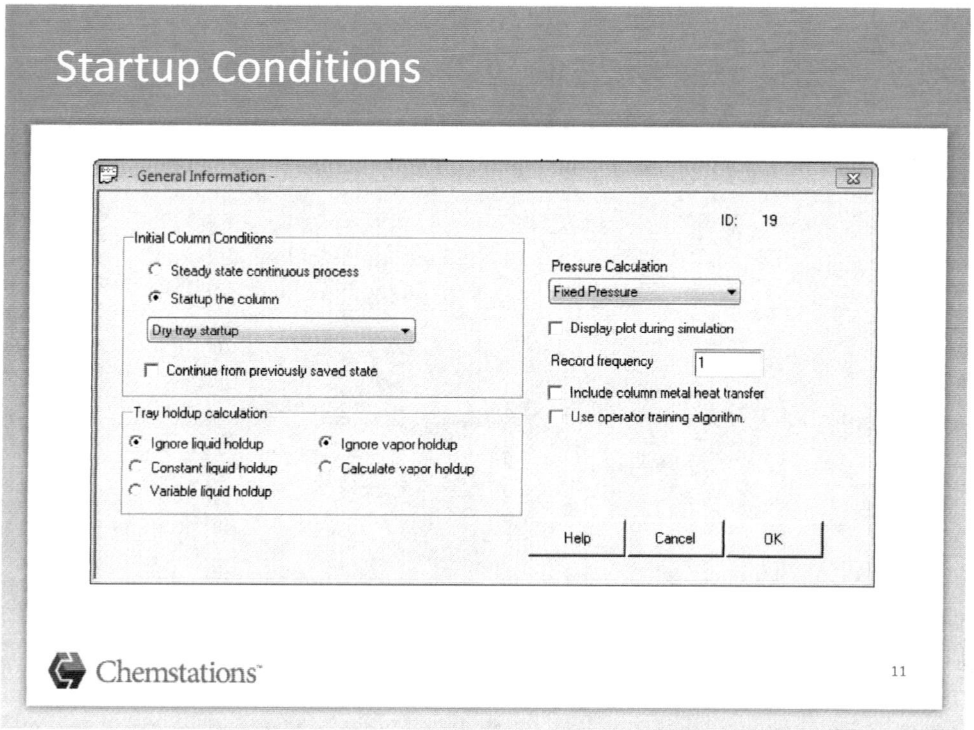
Startup Conditions
- General Information -
ID: 19
Initial Column Conditions
Steady state continuous process
Startup the column
Dry tray startup
Continue from previously saved state
Tray holdup calculation
Ignore liquid holdup
Constant liquid holdup
Variable liquid holdup
Ignore vapor holdup
Calculate vapor holdup
Pressure Calculation
Fixed Pressure
Display plot during simulation
Record frequency
1
Include column metal heat transfer
Use operator training algorithm
Help
Cancel
OK
Chemstations
11

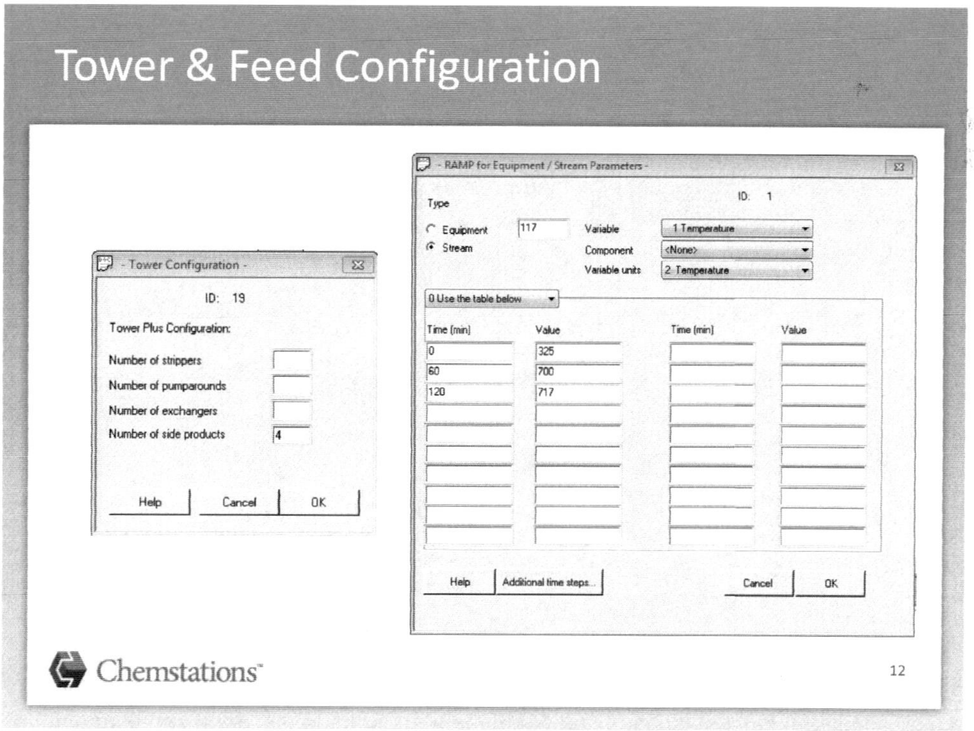
Tower & Feed Configuration
- Tower Configuration -
ID: 19
Tower Plus Configuration:
Number of strippers
Number of pumparounds
Number of exchangers
Number of side products
4
Help
Cancel
OK
- RAMP for Equipment / Stream Parameters -
ID: 1
Type
Equipment
Stream
117
Variable
1 Temperature
Component
<None>
Variable units
2 Temperature
0 Use the table below
Time (min)
Value
Time (min)
Value
0
325
60
700
120
717
Help
Additional time steps...
Cancel
OK
Chemstations
12

Actual Data Determined Ramp Settings

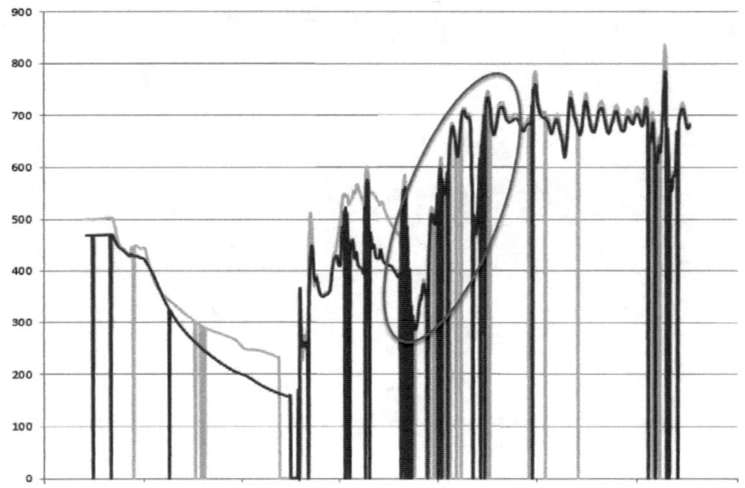

Chemstations

13

Scope of Model

1) Crude charge introduced to tower at increasing temp & flow rate ✓
2) Target level reached in overhead accumulator
3) Reflux system placed in service
4) Stripping steam introduced to tower
5) Product streams reach specification

Chemstations

14

Accumulator Dimensions & Reflux

Set reflux to zero initially

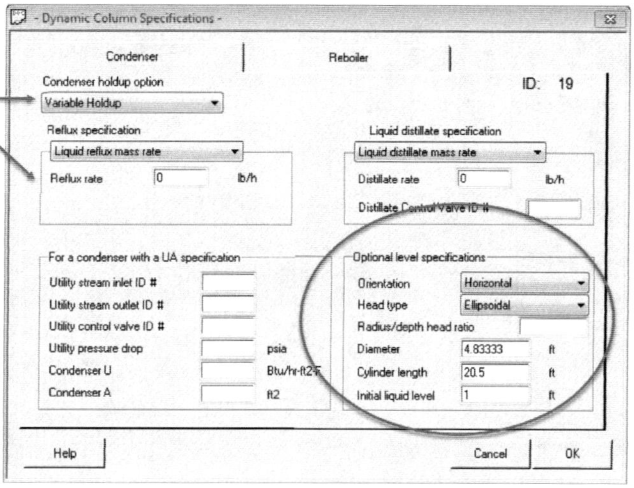

Chemstations™

15

Accumulator Level Target Specification

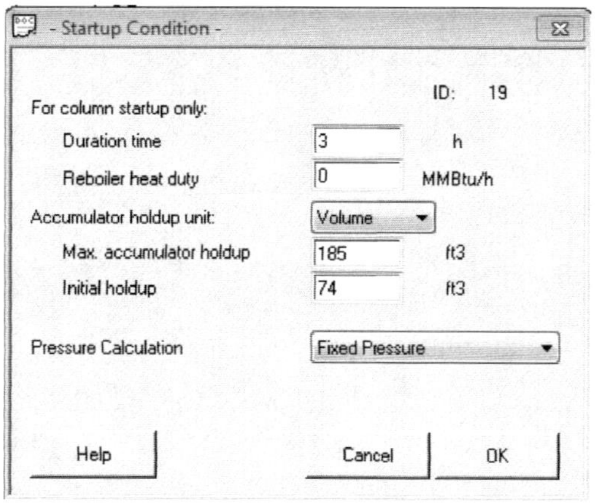

Chemstations™

16

Get Dynamic Model to Converge

- Set initial run time based on actual data if available
- Modify step size:
  - If needed to assist with convergence
  - Depending on target parameter's rate of change
- Run one time step manually & view results
- Other convergence adjustments:
  - # of iterations
  - tolerances

Chemstations

17

Run Time Parameters

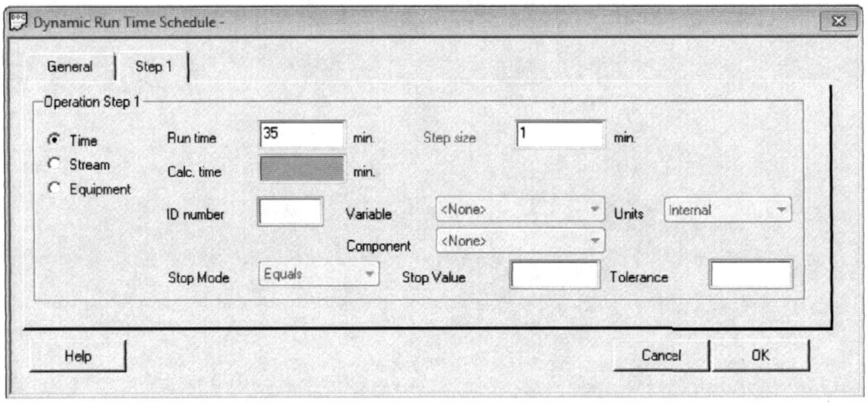

Chemstations

18

Accumulator Level Plot

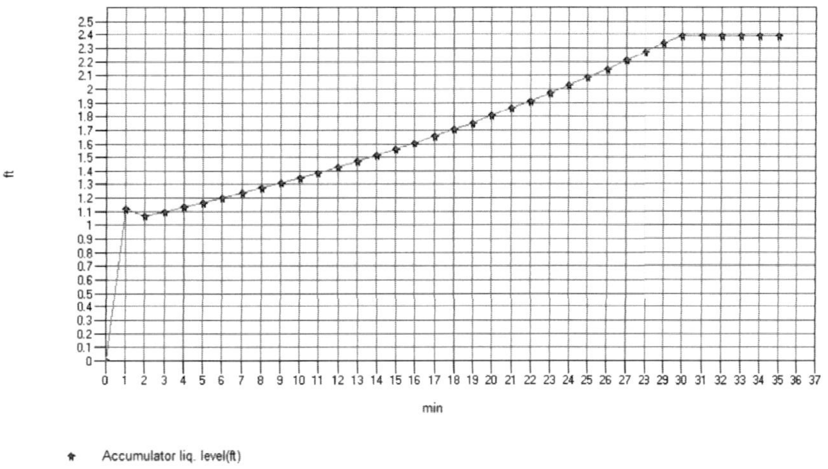

Chemstations™ 19

Accumulator Level Target Reached

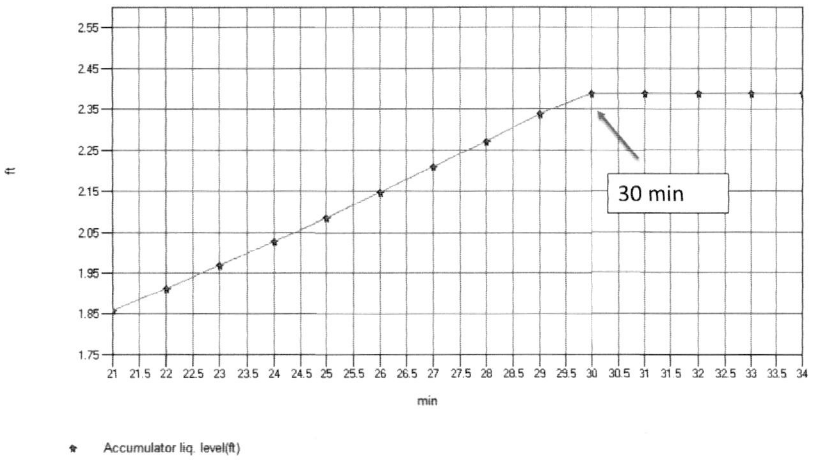

Chemstations™ 20

Scope of Model

1) Crude charge introduced to tower at increasing temp & flow rate ✓
2) Target level reached in overhead accumulator ✓
3) **Reflux system placed in service**
4) Stripping steam introduced to tower
5) Product streams reach specification

Chemstations 21

Place Reflux System In Service

Switch to maintain level & adjust reflux rate

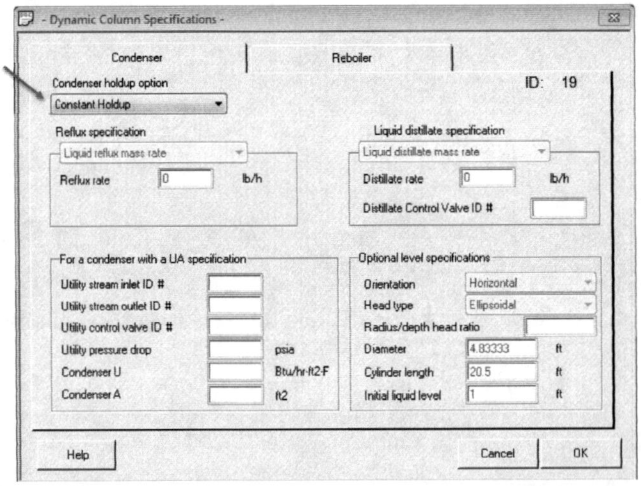

Chemstations 22

Scope of Model

1) Crude charge introduced to tower at increasing temp & flow rate ✓
2) Target level reached in overhead accumulator ✓
3) Reflux system placed in service ✓
4) Stripping steam introduced to tower
5) Product streams reach specification

Chemstations™ 23

Start Stripping Steam to Tower

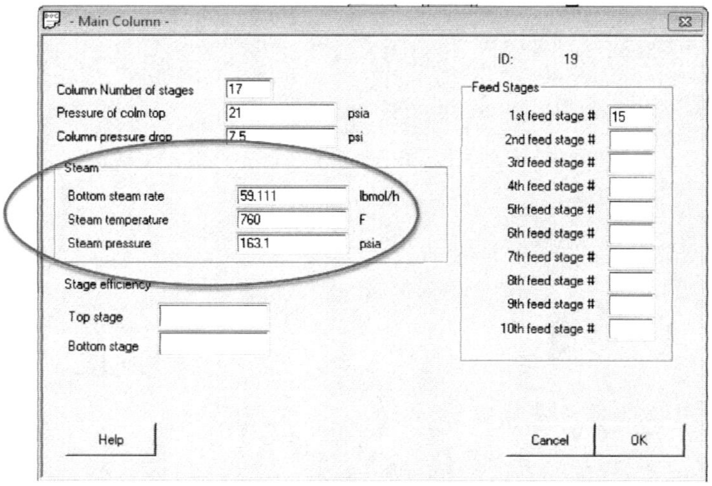

Chemstations™ 24

Scope of Model

1) Crude charge introduced to tower at increasing temp & flow rate ✓
2) Target level reached in overhead accumulator ✓
3) Reflux system placed in service ✓
4) Stripping steam introduced to tower ✓
5) Product streams reach specification

Chemstations™

25

Product Stream Reaching Specification

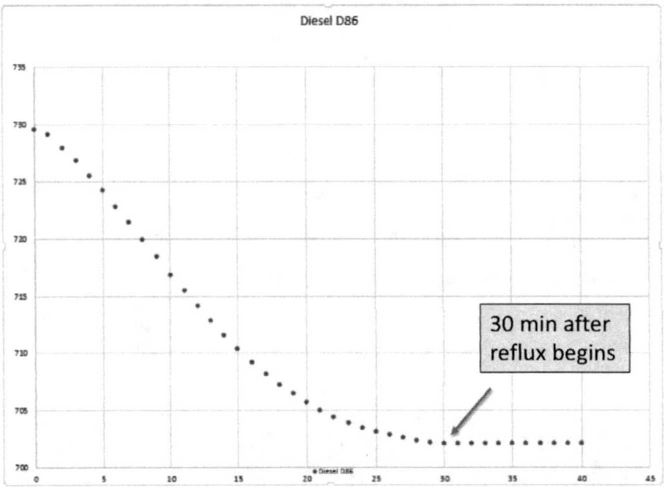

Chemstations™

26

Scope of Model

1) Crude charge introduced to tower at increasing temp & flow rate ✓
2) Target level reached in overhead accumulator ✓
3) Reflux system placed in service ✓
4) Stripping steam introduced to tower ✓
5) Product streams reach specification ✓

Chemstations™ 27

Validating Results

- Time to reach accumulator level:
  - Model → 30 min
  - Actual → 20 min
- Time to side product specs after reflux begins:
  - Model → 30 min
  - Actual → 40 min

Chemstations™ 28

Further Validation of Results

- Compare to additional sets of startup data
  - Is the delta between the model & actual consistent?
- Add complexity as required to get desired results
  - How close is close enough?

Chemstations™ 29

Add Complexity As Required

- Fired heater
- Reflux/distillate control valves
- Pumparounds/side strippers
- Relief valve studies
- Additional upstream / downstream units

Chemstations™ 30

Switch Back to Steady State Model

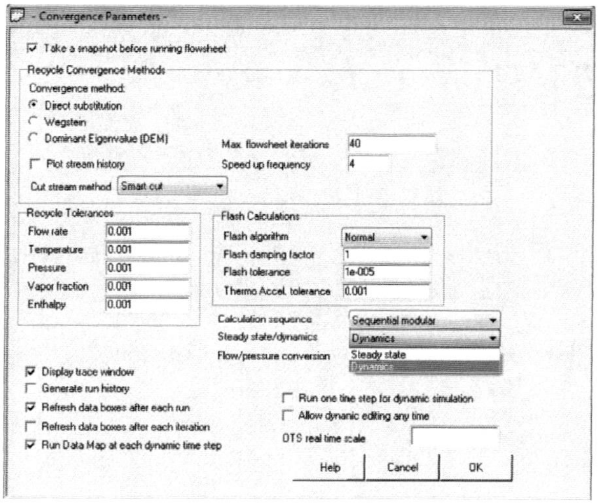

Chemstations™ 31

Review Simulation Guidelines*

- Start with a simple model
  - Fix what you can, i.e. pressures, etc.
  - Ignore what you can, i.e. utilities, etc.
- Get model to converge
- Validate results
- Add complexity as required to reach desired engineering solution
  - more detail may not be better!
  - i.e. don't add utilities controls systems to flowsheet just because they're on the P&ID

* True for steady state & dynamic simulations!

Chemstations™ 32

Dynamic Modeling Specifics

- Set initial run time based on actual data if available
- Modify step size:
  - If needed to assist with convergence
  - Depending on target parameter's rate of change
- Run one time step manually & view results
- Other convergence adjustments:
  - # of iterations
  - tolerances

Chemstations™ 33

Special Thanks

Ron Colwell, PE – Engineering Manager

David Fleming – Senior Process Engineer

Calumet Montana Refining, LLC

1900 10th St NE
Great Falls, MT 59404

Calumet Montana Refining processes Canadian heavy crude oil to produce gasoline, middle distillates and asphalt which they market primarily into local markets in Washington, Montana, Idaho and Alberta, Canada.

Chemstations™ 34

FRI Studies on Kettle Reboiler Hydraulics and Distillation Diagnostics

Tony Cai, Mike Resetarits and Ken McCarley

AIChE Southwest Process Technology Conference

Fractionation Research, Inc.

AIChE Southwest Process Technology Conference

7th AIChE Southwest Process Technology Conference | October 1-2, 2015 | Galveston, TX

Overview

- FRI Test Procedures
- Why investigate the Kettle Reboiler?
- Description of the FRI Unit
- What did we see inside the reboiler?
- Conclusions and Future Work

- improve control and performance of distillation processes?
- Transfer knowledge from heat transfer experts
- Not a study of heat transfer or exchanger design

AIChE
Southwest Process Technology Conference

FRI Test Procedures

- Category 1 – Trays and Packings selected by members
- Category 2 – Proprietary tests released to FRI members
- Category 3 – Proprietary tests
- Research Tests – Distillation projects proposed by FRI Members and Staff
- Parallel Research - Collaborative projects that benefit the members that are not completed using the FRI columns

AIChE
Southwest Process Technology Conference

Experimental Unit & Kettle Reboilers

AIChE Southwest Process Technology Conference

7th AIChE Southwest Process Technology Conference | October 1-2, 2015 | Galveston, TX

Why Investigate the Reboiler?

- An integral part of the distillation process
- Contribute to up to half of distillation issues**
- Hydraulic and entrainment limitations
- Affect control of a distillation process
- Hydraulic limitations were predicted for the FRI kettle reboiler by third-party software

** HZ Kister, Chemical Engineering, Feb 2010

AIChE Southwest Process Technology Conference

7th AIChE Southwest Process Technology Conference | October 1-2, 2015 | Galveston, TX

Kettle Reboiler

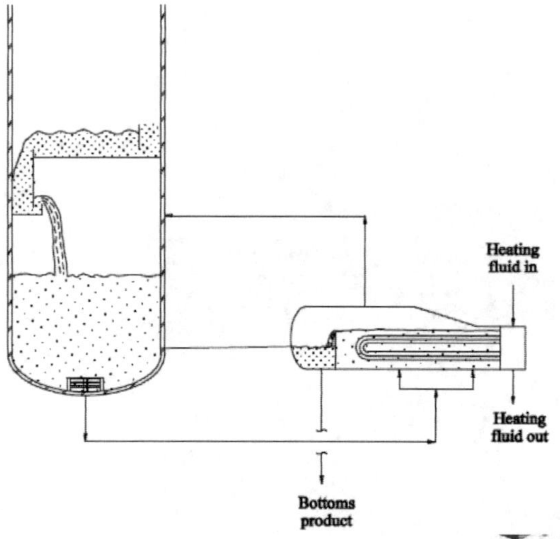

Scope of Reboiler Studies

- Two-phase flow and heat transfer characteristics
- Visual observations and measurements of the phenomena occurring inside the reboiler
  - Pool depth
  - Pressure drop
  - Entrainment
- Develop predictive correlations based on FRI experimental data of the kettle reboiler

FRI Low-Pressure Reboiler

AIChE
Southwest Process Technology Conference

7th AIChE Southwest Process Technology Conference | October 1-2, 2015 | Galveston, TX

Windows Near Steam Inlet

AIChE
Southwest Process Technology Conference

7th AIChE Southwest Process Technology Conference | October 1-2, 2015 | Galveston, TX

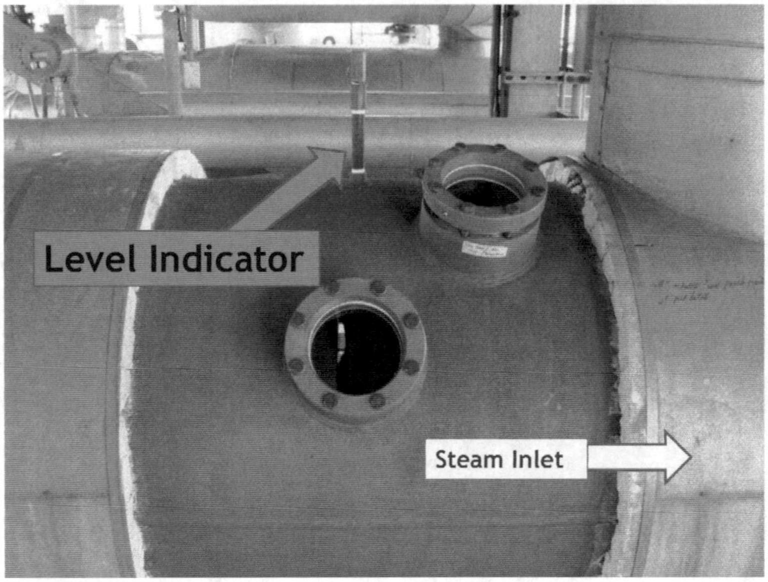

Windows & Level Indicator

AIChE
Southwest Process Technology Conference

7th AIChE Southwest Process Technology Conference | October 1-2, 2015 | Galveston, TX

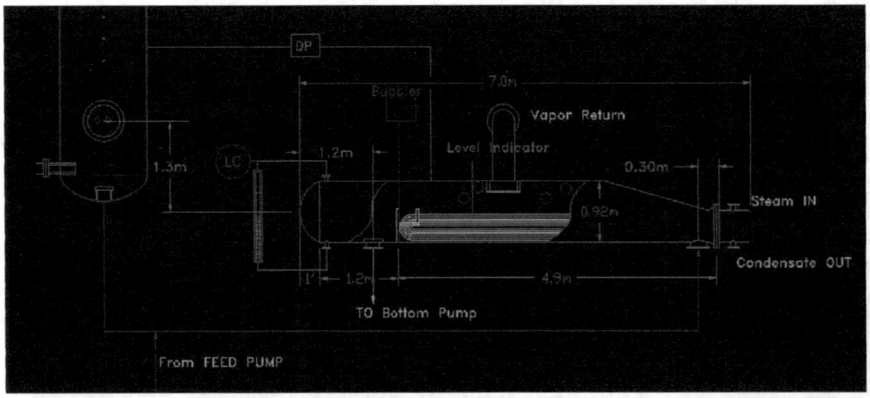

Kettle Reboiler Circuit

AIChE
Southwest Process Technology Conference

7th AIChE Southwest Process Technology Conference | October 1-2, 2015 | Galveston, TX

Kettle Reboiler Dimensions

Shell ID, mm	921
Over-flow baffle/weir, mm	508
Vapor nozzle, mm	457
U-tube bundle length, m	5
Tubes	183-16mm OD, 16GA
Surface Area, m^2	89

AIChE Southwest Process Technology Conference

Level Indicator

AIChE Southwest Process Technology Conference

7th AIChE Southwest Process Technology Conference | October 1-2, 2015 | Galveston, TX

What can We See inside the Kettle?

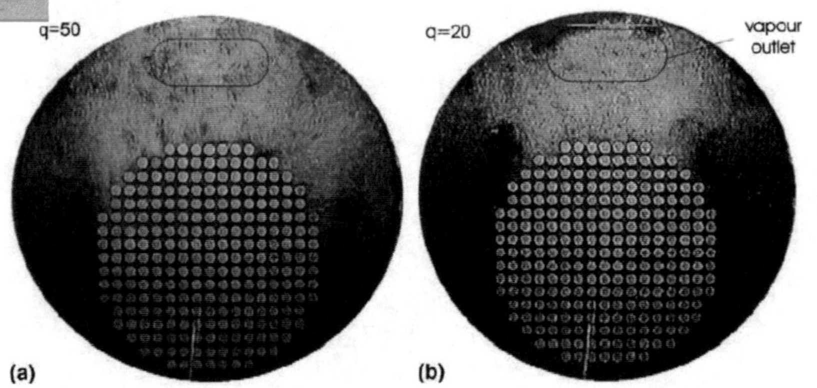

Fig. 1. Boiling pentane at 1 atm [9]. (a) $q = 50\,kW/m^2$; (b) $q = 20\,kW/m^2$.

Courtesy of B.M. Burnside *, K.M. Miller 1, D.A. McNeil, T. Bruce from International Journal of Heat and Mass Transfer 48 (2005) 1000–1016

AIChE Southwest Process Technology Conference

Nucleate Boiling

- At low heat fluxes, nucleate pool boiling was clearly seen first on segmental support baffles and then spacing rods

AIChE Southwest Process Technology Conference

Nucleate Boiling on Baffle and Spacing Rod

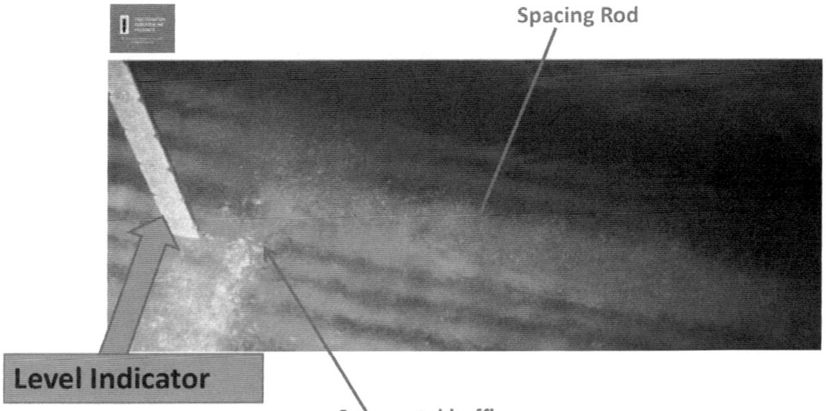

AIChE
Southwest Process Technology Conference

7th AIChE Southwest Process Technology Conference | October 1-2, 2015 | Galveston, TX

C4's @11bara with 5.0MW Heat Duty (54kW/m^2 Heat Flux)

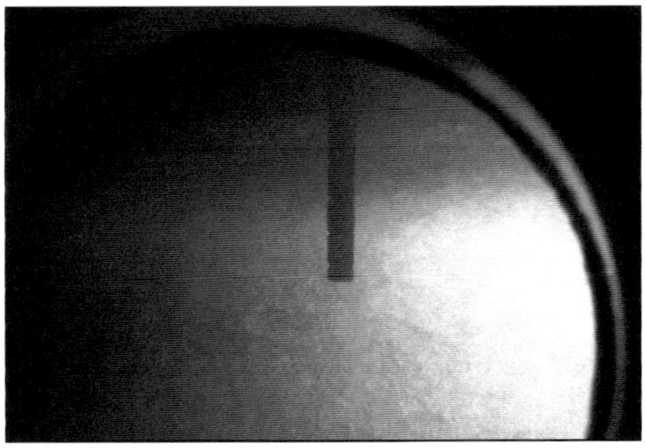

Third-party model predicted over 23 m^3/h liquid entrainment

AIChE
Southwest Process Technology Conference

7th AIChE Southwest Process Technology Conference | October 1-2, 2015 | Galveston, TX

C6/C7 @1.62bara with 24 kW/m² Heat Flux

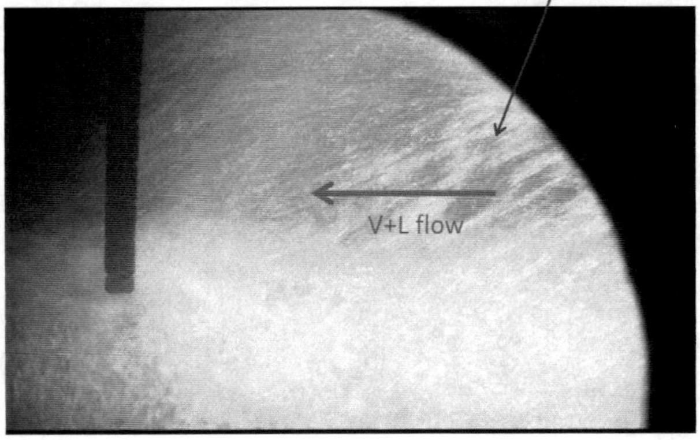

Southwest Process Technology Conference

7th AIChE Southwest Process Technology Conference | October 1-2, 2015 | Galveston, TX

C6/C7 @1.62bara with 36kW/m² Heat Flux

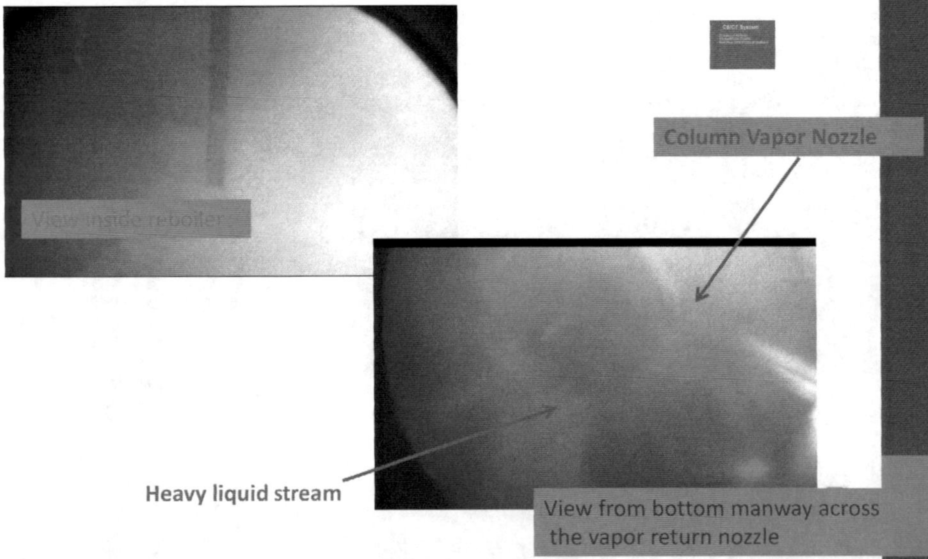

7th AIChE Southwest Process Technology Conference | October 1-2, 2015 | Galveston, TX

Uneven Boiling

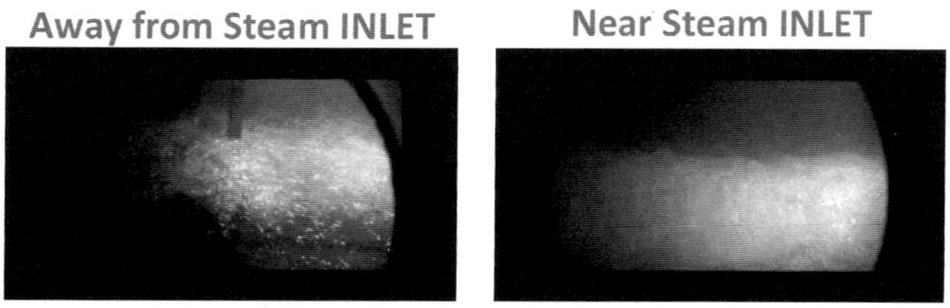

C_6/C_7 @ 23.5 psia (1.62 bara)

Heat Duty: 2.7 MBtu/hr (0.8MW)

Heat Flux: 2.7 x 10^3 Btu/hr-ft^2 (8.5kW/m^2)

Entrainment Measurements

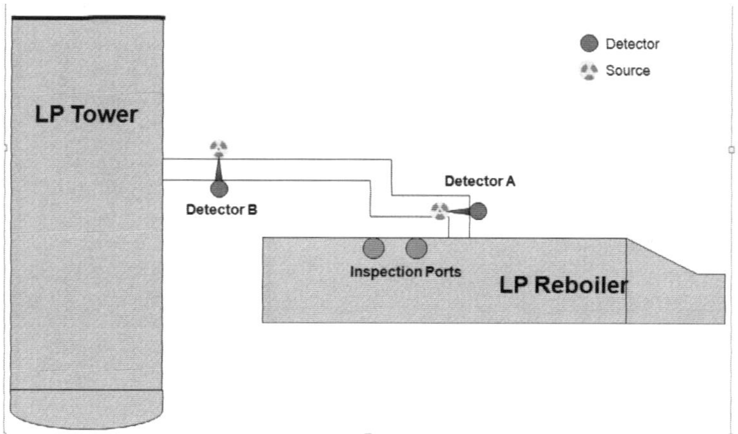

The vapor return line of the kettle reboiler was gamma scanned by Tracerco

Gamma Scan Locations

Location A

Location B

AIChE
Southwest Process Technology Conference

7th AIChE Southwest Process Technology Conference | October 1-2, 2015 | Galveston, TX

Gamma Scan Results

Run No.	Heat Flux	Measured Process Density	Entrainment Fraction
	kW/m^2	kg/m^3	kg-liquid/kg-vapor
1	9.7	5.68	0.00
2	15.5	5.73	0.01
3	23.4	5.95	0.05
4	25.4	6.14	0.09

- **C6/C7 at 1.62 bara (23.5 psia)**
- **The entrainment measured by gamma scanning was consistent with visual observations**

AIChE
Southwest Process Technology Conference

7th AIChE Southwest Process Technology Conference | October 1-2, 2015 | Galveston, TX

Pool Depth Measurements

- The flow of the boiling pool towards the overflow weir was usually weaker near the side of the kettle
- Pool depths
  - Stronger function of loading over the outlet weir
  - Not significantly affected by the reboiler heat fluxes

AIChE Southwest Process Technology Conference

7th AIChE Southwest Process Technology Conference | October 1-2, 2015 | Galveston, TX

Pressure Drops across the Vapor Return Line

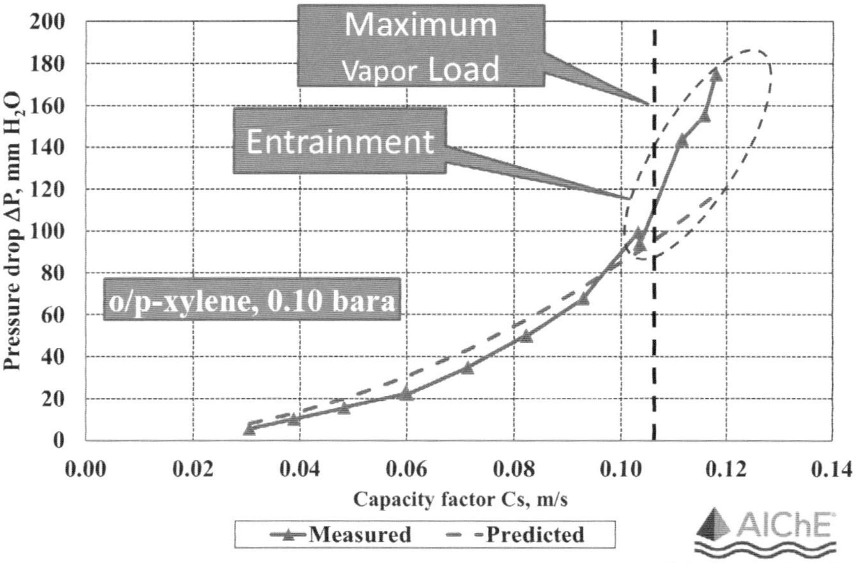

AIChE Southwest Process Technology Conference

7th AIChE Southwest Process Technology Conference | October 1-2, 2015 | Galveston, TX

Kettle Reboiler - Depropanizer

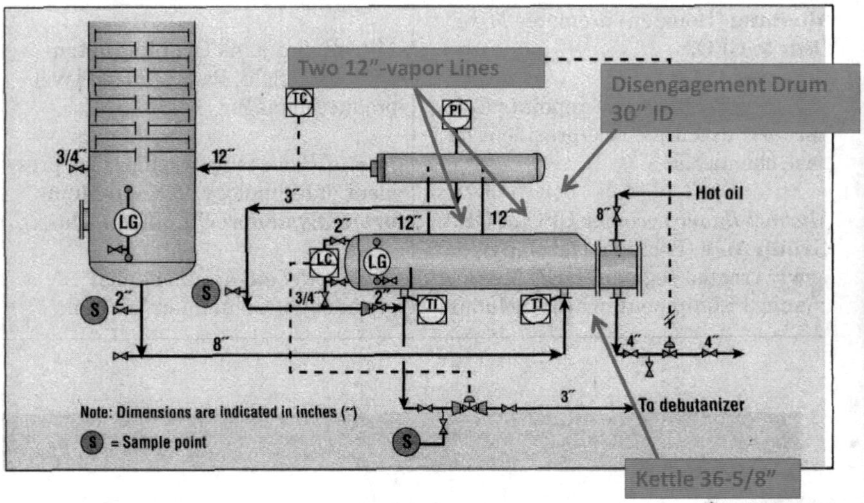

Courtesy of Kister & Chaves

Case Study

- At feed rate of less than 160gal/min, the column operated well
- As the feed rates were raised from 160 to 205 gal/min, column bottom level was observed to first rise slowly, then faster
- Above 205 gal/min, the level shot right up and above the vapor return inlet
- Gamma scans confirmed that flooding initiated at the reboiler circuit

- The case was from a paper published by Henry Kister etc:
  - In Chemical Engineering, February 2010
  - Title : Kettle Troubleshooting

Reboiler Process information

TABLE 3. REBOILER PROCESS DATA

	140 gal/min test	195 gal/min test		140 gal/min test	195 gal/min test
Heat duty, MM Btu/h	7.79 [3]	10.74 [3]	Drum vapor out, °F	257 [1]	272 [1]
C_4+ flowrate, lb/h	21,060 [1]	32,640 [1]	Reboiler process liquid out, °F	270 [1]	295 [1]
Amount vaporized, lb/h	70,050 [3]	96,070 [3]	Heating medium in, °F	489 [1]	484 [1]
Reboiler feed [excluding entrainment], lb/h	91,110 [3]	128,710 [3]	Heating medium out, °F	ND	283 [2]
			Heating medium flow, lb/h	ND	81,000 [5]
Entrainment knocked out in drum, lb/h	ND	13,600 [2]	Reboiler pressure, psia	280 [1]	280 [1]
			LMTD, °F	ND	83 [4]
Entrainment from drum, lb/h	ND	30,800 [6]	Heat transfer coefficient, MM Btu/h-ft²°F	ND	96 [5]
Reboiler process in, °F	245 [1]	256 [1]			

- Enough hardware and process information was available to apply FRI correlations to the kettle reboiler
- Both feed rates were checked

Courtesy of Kister & Chaves

AIChE
Southwest Process Technology Conference

Field Measurement vs. Prediction

Feed Rate, gal/min	140	195
Measured Entrainment, Mlb/hr	NA	13.6
Predicted Vapor Loading, %	70	109

- Predictions based on correlations to FRI data
- Correlations for maximum vapor loading are to be validated

AIChE
Southwest Process Technology Conference

Conclusions

- Visual observations provide valuable information and data
- Gamma scanning enabled measurement of entrainment
- Early indication that new correlations show promise for design guidance and trouble shooting

AIChE
Southwest Process Technology Conference

7th AIChE Southwest Process Technology Conference | October 1-2, 2015 | Galveston, TX

Future Work

- Continue to consult with heat transfer experts
- Investigate
  - Pool boiling curve in the kettle reboiler
  - Circulation ratio
- Study the performance of de-entrainment devices
- Develop correlations to predict entrainnment

AIChE
Southwest Process Technology Conference

7th AIChE Southwest Process Technology Conference | October 1-2, 2015 | Galveston, TX

Acknowledgements

- Authors would like to extend their great appreciations to:
  - FRI memberships for their continuous support to FRI research program

AIChE
Southwest Process Technology Conference

7th AIChE Southwest Process Technology Conference | October 1-2, 2015 | Galveston, TX

Thank you for your attention!

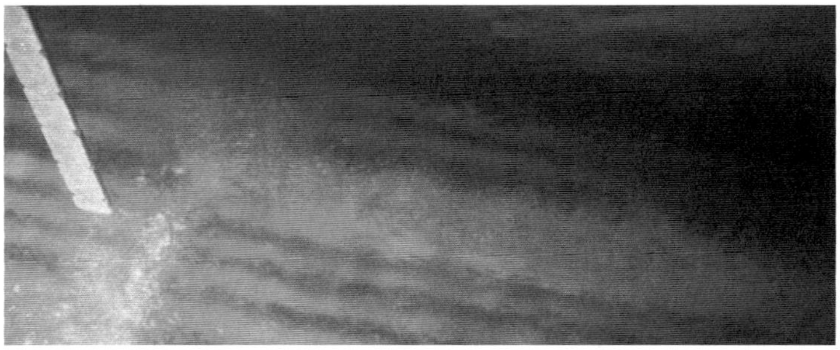

AIChE
Southwest Process Technology Conference

7th AIChE Southwest Process Technology Conference | October 1-2, 2015 | Galveston, TX

Stripper Tower Capacity Problem Resolved Using Gamma Scanning Technology

Pless, L., Tracerco
Davidson, C., Dow Chemical

Originally presented on: 10/1/2015 13:56:00 - 14:22:00

Process Engineers have a wide array of tools to investigate operational problems. These can range from studying process variables such as temperature, pressure, flow rates, and analytical results to more advanced modeling and simulation. On many occasions the scope or priority of the troubleshooting effort is governed by a critical situation and the potential seriousness of a problem requiring quick action and parallel efforts.

Gamma scanning has gained in popularity over the years because it quickly provides process information that is not available from any other source, is real time, and non-intrusive. The technology is flexible and can be used at the onset of a problem to diagnose internal hydraulic issues, used later in an investigation to support other operating parameters pointing to where a problem may lie or validate a hypothesis related to what a particular problem may be.

This presentation offers a troubleshooting case study concerning a waste water stripper tower. A gamma scan was used to help find the root cause of a loss in capacity prior to shut down and entry. The gamma scan determined where problems existed and allowed focus during shutdown resulting in a shorter outage.

This presentation includes system description, explanation of trouble-shooting techniques used, gamma scan results and pictures from inside the tower during the outage inspection corresponding to features detected by the gamma scan.

SimulFlow™ Devices Maximize Ethylene Fractionator Production

Nowak, B., UOP LLC

Originally presented on: 10/1/2015 14:22:00 - 14:48:00

The UOP SimulFlow™ device is a co-current flow distillation internal with an extremely high capacity. Tested at F.R.I. and verified in a commercial C2 Splitter application, the SimulFlow device is a proven and cost-effective option for debottlenecking distillation towers for increased capacities and/or recoveries. This presentation will highlight the technical features of the technology and its successful implementation to produce a high purity ethylene product during an initial revamp, only limited by the auxiliary equipment. Four and a half years later, a second revamp was implemented to take advantage of the full potential of originally installed SimulFlow devices, finally establishing its true capability in a commercial column.

Recent Advances in SRP Packing Hydraulic and Mass Transfer Measurements

Seibert, F., University of Texas at Austin
Wang, C., University of Texas at Austin
Perry, M., University of Texas at Austin

Originally presented on: 10/1/2015 15:12:00 - 15:38:00

High performance packings are often used in absorption, stripping and distillation because of their relatively low cost, high turndown, low pressure drop and improved capacity. The Separations Research Program (SRP) has undertaken a significant study involving the consistent measurements of pressure drop, flooding, mass transfer area and the film mass transfer coefficients. In this presentation, hydraulic, measured mass transfer area and film mass transfer coefficient data on a variety of structured and random packings will be provided. These data and models have been obtained and developed using the air/water or air/caustic chemical test systems for application to carbon dioxide capture. In this presentation, the mass transfer results and models will also be compared with distillation HETP data obtained by the SRP.

Choosing The Right Valve For Cross-Flow Trays

Nieuwoudt, I., Koch-Glitsch

Originally presented on: 10/1/2015 15:38:00 - 16:04:00

Valves for cross-flow trays are typically chosen based on vapor handling capacity, turndown, pressure drop and fouling resistance. A large variety of valves, including valves developed recently, are available. In this presentation the performance of a number of valves will be compared and conclusions will be drawn on the range of applicability of different valves.

UFM Trays Exhibit Capacity Gains for 2 Tower Revamps

Summers, D. R., Sulzer Chemtech USA, Inc.

Originally presented on: 10/1/2015 16:04:00 - 16:30:00

The recently developed UFM™ tray, has received wide industry acceptance over the past 2 years. This paper will document two applications where the superior capacity and performance characteristics of this tray has enabled two towers to achieve higher than expected capacity and performance. Operating data will be presented and tray efficiency will be discussed.

INTEGRATION OF PROCESS SAFETY ENGINEERING AND FIRE PROTECTION ENGINEERING FOR BETTER SAFETY PERFORMANCE

Dr. M. Sam Mannan, PE, CSP, DHC
Mary Kay O'Connor Process Safety Center
Artie McFerrin Department of Chemical Engineering
Texas A& University, College Station, Texas 77843-3122, USA
mannan@tamu.edu, (979) 862-3985

7th AIChE Southwest Process Technology Conference, October 1, 2015

MARY KAY O'CONNOR
PROCESS SAFETY CENTER
TEXAS A&M ENGINEERING EXPERIMENT STATION

Outline

- Introduction
- Origins of the Fields
- Areas of Common Interest and Concern
- Points of Disparity
- Opportunities for Integration and Synergistic Cooperation

2

MARY KAY O'CONNOR
PROCESS SAFETY CENTER
TEXAS A&M ENGINEERING EXPERIMENT STATION

Establishment of MKOPSC

"Converting a personal tragedy into an opportunity to improve process safety"

- **Mary Kay O'Connor**
  - Chemical engineer
  - Operations Superintendent
  - 1989 Chemical plant accident-Phillips 66- Pasadena, TX
- **1995-Center established**
- **Funding**
  - Endowment
  - Contract projects
  - Consortium funding
  - Continuing Education

3

TEES | MARY KAY O'CONNOR PROCESS SAFETY CENTER
TEXAS A&M ENGINEERING EXPERIMENT STATION

Vision

Making Safety Second Nature

To be an international leader in minimizing losses within the process industry - through safer processes, management, equipment, and procedures.

4

TEES | MARY KAY O'CONNOR PROCESS SAFETY CENTER
TEXAS A&M ENGINEERING EXPERIMENT STATION

High Points

- Improve safety in process industries
- Provide good science based counsel
- Develop and disseminate best practices
- Provide benchmarking
- Conduct research

5

TEES | MARY KAY O'CONNOR PROCESS SAFETY CENTER
TEXAS A&M ENGINEERING EXPERIMENT STATION

High Points

- Provide industry
  - Forum at different levels for exchange of ideas
  - Best practices
  - Advance technologies to keep competitive
- Educate a new generation of engineers
  - principles of process safety are *second nature*
- Train personnel
  - OSHA, CSB, DOT, EPA, DHS
  - Industry/ Universities

6

TEES | MARY KAY O'CONNOR PROCESS SAFETY CENTER
TEXAS A&M ENGINEERING EXPERIMENT STATION

Origins of FPE

- Field developed over hundreds of years in response to major catastrophic events:
  - Burning of Rome (64 AD)
  - Great Fire of London (1666)
  - Great Fire of Pittsburg (1845)
  - Great Chicago Fire (1871)
  - Great Boston Fire (1872)
  - Great Fire and Earthquake of San Francisco (1906)

- Some of these events destroyed or significantly damaged entire cities.

7

MARY KAY O'CONNOR PROCESS SAFETY CENTER
TEXAS A&M ENGINEERING EXPERIMENT STATION

Origins of PSE

- Developed in the second half of the 20th century in response to growing numbers of industrial disasters:
  - Texas City (1947)
  - Flixborough (1974)
  - Seveso (1976)
  - Bhopal (1984)
  - Piper Alpha (1988)
  - Phillips 66 (1989)
  - BP Texas City (2005)
  - Deepwater Horizon (2010)

Aftermath of the Phillips 66 disaster

8

MARY KAY O'CONNOR PROCESS SAFETY CENTER
TEXAS A&M ENGINEERING EXPERIMENT STATION

Comparative Origins

- PSE is a comparatively young and new field.
- FPE developed gradually in response to a societal need to control fire hazards
- PSE developed rapidly to address incidents mostly relating to the chemical manufacturing industry
- These origins continue to influence the roles of both fields and their interactions with each other

9

TEES | MARY KAY O'CONNOR PROCESS SAFETY CENTER
TEXAS A&M ENGINEERING EXPERIMENT STATION

Areas of Common Interest and Concern

- Design and Life-Cycle Approach
- Hazard Analysis (PHA and FHA)
- Documentation and Information Management
- Inspection, Testing, Preventative Maintenance and Mechanical Integrity
- Training and Education
- Auditing
- Emergency Planning and Preparedness
- Employee Participation
- Permitting and Work Authorization
- Management of Change
- Incident Investigation

10

TEES | MARY KAY O'CONNOR PROCESS SAFETY CENTER
TEXAS A&M ENGINEERING EXPERIMENT STATION

Hazard Analysis (PHA and FHA)

- Both fields use various forms of hazards analysis to systematically identify and address safety concerns

- Both fields use lists of representative scenarios to assess possible consequences

11

MARY KAY O'CONNOR PROCESS SAFETY CENTER
TEXAS A&M ENGINEERING EXPERIMENT STATION

Documentation and Information Management

- Meticulous and careful record keeping is essential to the success of a PSE or FPE program.
  - Audit, inspection, maintenance records
  - PHA/FHA records and action items
  - Incident investigation records
  - Designs and plans
  - Training records
- Share the information!

12

MARY KAY O'CONNOR PROCESS SAFETY CENTER
TEXAS A&M ENGINEERING EXPERIMENT STATION

Inspection, Testing, Preventative Maintenance and Mechanical Integrity

- Both fields require acceptance and commissioning tests
- Both fields use periodic inspections
- Because PSE deals with a greater variety of systems, there is greater flexibility and variability in inspection and maintenance programs than FPE, which normally requires inspections or tests every 6-12 months

13

TEES | MARY KAY O'CONNOR PROCESS SAFETY CENTER
TEXAS A&M ENGINEERING EXPERIMENT STATION

Training and Education

Training must be provided to <u>all personnel</u> responsible for operating the facility and equipment in both fields

- Initial training
- Refresher training
- Contractor training
- Communication of changes
- Verification of Understanding
  - Onsite demonstration
  - Verbal demonstration
  - Simulation

14

TEES | MARY KAY O'CONNOR PROCESS SAFETY CENTER
TEXAS A&M ENGINEERING EXPERIMENT STATION

Emergency Planning and Preparedness

- Both fields require significant planning and drilling to prepare for incidents
- The systems and personnel need to be ready to perform at the rare times they are needed
- Dialogue with first responders

15

MARY KAY O'CONNOR PROCESS SAFETY CENTER
TEXAS A&M ENGINEERING EXPERIMENT STATION

Employee Participation

- Intimate knowledge of the worksite
- Foster increased and potentially lasting awareness, responsibility and support.
- Employees must know about safety and health program and understand how it works for them
- Open communication systems
- Management should take prompt action for employee safety and health concerns

16

MARY KAY O'CONNOR PROCESS SAFETY CENTER
TEXAS A&M ENGINEERING EXPERIMENT STATION

Permitting and Work Authorization

- Both fields share concerns with:
  - Planning and authorizing work
  - Assessing the hazards involved in the work
  - Tracking progress
  - Tracking deactivated and disabled equipment
  - Authorizing equipment return to service

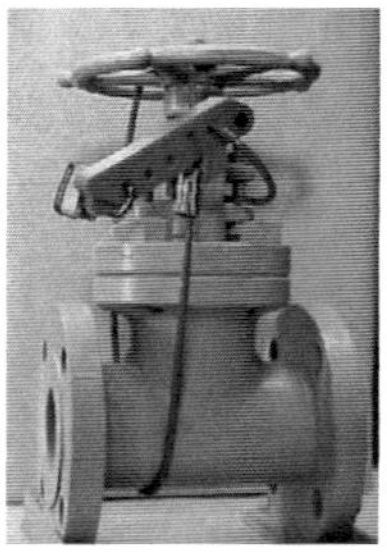

17

MARY KAY O'CONNOR PROCESS SAFETY CENTER
TEXAS A&M ENGINEERING EXPERIMENT STATION

Management of Change

Both fields require practitioners to be careful when making modifications

PSE:
- Materials of Construction
- Relief System Capacity
- Toxic Inventory
- Decommissioning

FPE
- Materials of Construction
- Waste Water Containment
- Flammable Inventory
- Land grade and congestion issues

... and that changes be communicated to stakeholders

18

MARY KAY O'CONNOR PROCESS SAFETY CENTER
TEXAS A&M ENGINEERING EXPERIMENT STATION

Incident Investigation

- Both fields have their origins in investigating and attempting to learn from past incidents
- Root cause analysis principally used
- Incident Investigation Report:
  - **Date of Incident**
  - **Date** *(and time)* **Investigation Began**
  - **Description of Incident**
  - **Contributing Factors**
  - **Cause(s)**
  - **Recommendations**
  - **Retain for 5 years**

19

TEES | MARY KAY O'CONNOR PROCESS SAFETY CENTER
TEXAS A&M ENGINEERING EXPERIMENT STATION

Points of Disparity

- Hazard Analysis Scope
- Application
- Consequence Only vs Risk-Based Approaches
- Prescriptive vs Performance-Based Regulations and Standards
- Many of these differences arise at least in part from the fact that the fields developed in response to different needs and at different times when the available technology and knowledge was starkly different

20

TEES | MARY KAY O'CONNOR PROCESS SAFETY CENTER
TEXAS A&M ENGINEERING EXPERIMENT STATION

Hazard Analysis Scope

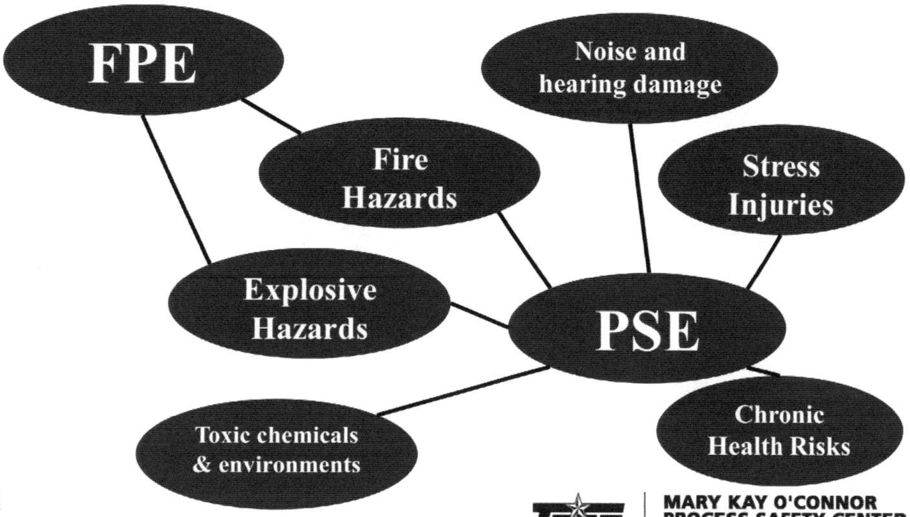

21

TEES | MARY KAY O'CONNOR PROCESS SAFETY CENTER
TEXAS A&M ENGINEERING EXPERIMENT STATION

Typical Application of Field

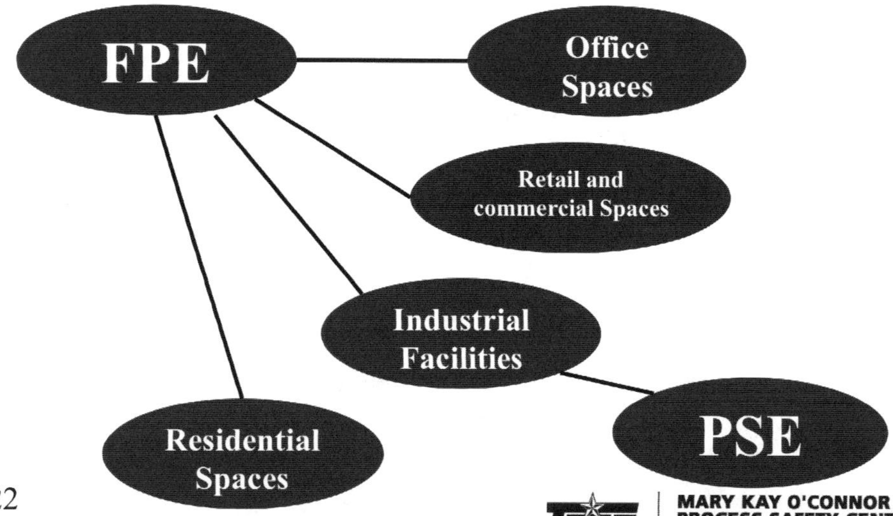

22

TEES | MARY KAY O'CONNOR PROCESS SAFETY CENTER
TEXAS A&M ENGINEERING EXPERIMENT STATION

Consequence-Only vs Risk-Based Approaches

- FHA's tend to place far more emphasis on the consequences of a fire scenario than the probability of it or the overall risk – with emphasis on model

- PHA's tend to be risk-based

23

MARY KAY O'CONNOR PROCESS SAFETY CENTER
TEXAS A&M ENGINEERING EXPERIMENT STATION

Prescriptive vs Performance-Based Regulations and Standards

- Fire protection standards and regulations tend to be highly prescriptive
  - Codes tend to be very explicit in stating requirements and leave little flexibility
  - Maintenance and inspection schedules are largely specified by codes
- Process Safety uses both prescriptive and performance-based standards, but seems to be trending towards more performance-based standards and regulations
- FPE's reliance on prescriptive standards may be attributable to their broad application to residential and commercial spaces, where there is usually less variability in the hazards present.

24

MARY KAY O'CONNOR PROCESS SAFETY CENTER
TEXAS A&M ENGINEERING EXPERIMENT STATION

Opportunities for Integration

- Integrate FPE and PSE into professional education programs
- Integrate FPE personnel into PHA teams
- Insure the consequence analysis in FHAs are incorporated into the PHA process
- Apply the diverse set of hazard identification and risk assessment techniques used in PHAs more fully in FHAs
- Apply PHA and PSE techniques more broadly to aid in Fire prevention

25

MARY KAY O'CONNOR
PROCESS SAFETY CENTER
TEXAS A&M ENGINEERING EXPERIMENT STATION

Likely Reasons Integration Has Not Yet Occurred

- Separate histories and origins of the fields
  - They developed separately, at different times, and have continued to work independently
- Lack of personnel/resources
  - Many facilities and companies have only a very limited number of FPE personnel
  - These personnel can not just spend all day in HAZOPs

26

MARY KAY O'CONNOR
PROCESS SAFETY CENTER
TEXAS A&M ENGINEERING EXPERIMENT STATION

MKO and Safety Engineering at Texas A&M

- **Process Safety Engineering is part of the core Chemical Engineering curriculum**
- **Safety Engineering is a Major under the Chemical Engineering Department**
- **Degree Programs**
  - Master of Science in Safety Engineering
    - Distance Education Option
  - PhD programs
- **Safety Engineering Certificate- Undergraduate**
- **Safety Engineering Certificate- Graduate**
- **Process Safety Practice Certificate- Industry practitioners**

27

TEES | MARY KAY O'CONNOR PROCESS SAFETY CENTER
TEXAS A&M ENGINEERING EXPERIMENT STATION

Thank You!

QUESTIONS

28

TEES | MARY KAY O'CONNOR PROCESS SAFETY CENTER
TEXAS A&M ENGINEERING EXPERIMENT STATION

Improving Operational Risk Management: Connecting the Impact of Process Safety Management Performance to Daily Operations

Neill, M., Petrotechnics USA Inc.

Originally presented on: 10/1/2015 14:00:00 - 14:30:00

Operators need to understand the cumulative impact operating activities, plant conditions, and working processes have on fundamental process safety barriers in order to make better, safer operating decisions.

Validating the Risk Reduction Claimed for IPLs in LOPA
DuPont, D., aeSolutions
Originally presented on: 10/1/2015 14:30:00 - 15:00:00

Layer of Protection Analysis (LOPA) is a widely used semi-quantitative tool that can be used to determine the safety integrity level in designing the protections of a safety instrumented system (SIS). It is often used to meet the objectives of:

1. Allocating safety functions to protection layers;
2. Determining the required safety instrumented functions;
3. Determine, for each safety instrumented function, the associated safety integrity level (SIL).

Inherent to this process is having the procedures that lead to a consistent consequence analysis and determination of failure frequencies. This allows one to determine if the frequency of occurrence of a hazardous event meets the numerical risk tolerance criteria established by the organization.

This talk will focus on the importance of having a specific set of rules that specify the default data for values assigned to independent protective layers (IPLs), as well as the process of assigning values to IPLs that do not exist in your library; in addition to the maintenance, testing and inspection programs required to achieve the values claimed in LOPA.

Next Generation Root Cause Investigation (RCI) and Analysis

7th AIChE Southwest Process Technology Conference
October 1-2, 2015
Galveston, TX

The Dow Chemical Company

Bill Efaw

AIChE
Southwest Process Technology Conference

Biography – Bill Efaw

- Associate Process Safety Director, The Dow Chemical Company – Gulf Coast & Latin America responsibility
- 34 Years Industry Experience (Polyolefins R&D, Dowlex™ LLDPE Mfg, Tubular LDPE, Saran, Quality, Site Logistics/Supply Chain Leader, Global EH&S leader, Dow Fiber Solutions start-up business, Process Safety Technology Leader
- 7 years in Process Safety
- BS Chemical Engineering – University of Cincinnati

AIChE
Southwest Process Technology Conference

7th AIChE Southwest Process Technology Conference | October 1-2, 2015 | Galveston, TX

Contents

- Introduction & Background
- RCI Standard and work process
- RCI effectiveness reviews and repetitive incident analysis
- Leveraging corrective actions
- Learning Experience Reports
- Conclusions

AIChE Southwest Process Technology Conference

7th AIChE Southwest Process Technology Conference | October 1-2, 2015 | Galveston, TX

Introduction & Background

Next Generation Root Cause Investigation (RCI) and Analysis

Improvements that engage leadership and change the culture to prevent repetitive incidents

- ✓ Corporate RCI Standard and work process
- ✓ RCI effectiveness reviews and repetitive incident analysis
- ✓ Leveraging corrective actions
- ✓ Learning Experience Reports (LERs) that reinforce the proper execution of management systems

AIChE Southwest Process Technology Conference

7th AIChE Southwest Process Technology Conference | October 1-2, 2015 | Galveston, TX

Introduction & Background

Generational goals to reduce PS and LOPC Incidents

- 1996 → 2005: 90% Reduction
- 2005 → 2015: 75% Reduction
- 2015 → 2025: Eliminate

Process Safety Incidents

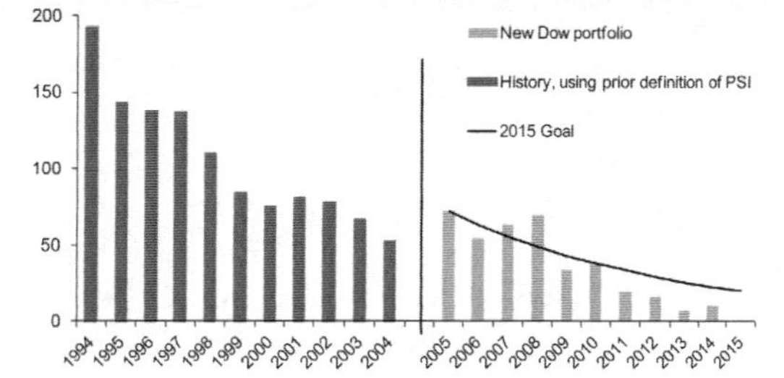

2005 → 2008 Plateaued performance

- Repetition
- Same Failure Modes / Management Systems
- Action Plan

AIChE Southwest Process Technology Conference

RCI Standard & Work Process

RCI: Identification of Management System Opportunities

- Legacy RCI Standard
  - ✓ Required finding corrective and preventive actions for unplanned and potential unplanned events
- Legacy RCI Work Process:
  - ✓ Find effective solution, not having the solution was the root cause
  - ✓ Did not formally recognize required protection layers and their failure

AIChE Southwest Process Technology Conference

RCI Standard & Work Process

RCI: Identification of Management System Opportunities

- New investigation process
  - ✓ Identify the cause for every required protection layer and management system failure, including both:
    - ➢ Dow standard required protection layers and
    - ➢ Risk assessment validated protection layers
  - ✓ Involve the function responsible for the local implementation of each management system failure
  - ✓ Establish corrective actions for each protection layer and management system failure
  - ✓ Include the appropriate level of functional ownership for management system corrective actions

AIChE Southwest Process Technology Conference

7th AIChE Southwest Process Technology Conference | October 1-2, 2015 | Galveston, TX

RCI Standard & Work Process

RCI: Identification of Management System Opportunities

- Management system opportunities
  - ✓ A management system may cover hundreds if … layers
  - ✓ If the man… ent system is not fixed, …
  - ✓ Even if an … fixed there is no prote…
- Involving th… f the manageme…
  - ✓ Identificati…
  - ✓ Proper ow…
  - ✓ The appropriate … a management system and drives the improvement

A management system may need to be corrected at a single facility, in other cases a management system needs to be corrected across a site, a business or for the entire company

AIChE Southwest Process Technology Conference

7th AIChE Southwest Process Technology Conference | October 1-2, 2015 | Galveston, TX

RCI Standard & Work Process

Repetitive Incident Analysis:

- RCI team reviews past incidents in a facility, site or business.
- Key questions:
  - ✓ Was this a repetitive incident within the plant by equipment type and protection layer failure type?
  - ✓ Was this a repetitive incident within the plant by protection layer failure type or management system failure on a different equipment type?
  - ✓ Was this incident type a historical, significant event that is reviewed as part of the plant process hazard assessment or a scenario in LOPA? Or is this new to the plant?
  - ✓ The same questions are assessed for a site and business.
- This process is greatly aided by a searchable database of incidents.

RCI Standard & Work Process

RCI Effectiveness Review:

LEADERSHIP

- Manufacturing and Process Safety Leadership Reviews
  - ✓ All protection layers and management system failures were identified and corrected.
  - ✓ Appropriate leveraging of management system opportunities
    - ➢ Are appropriate actions being taken across a facility, site or business, consistent with the level of management system failure
    - ➢ Are corporate corrective actions warranted
    - ➢ Should the incident should be communicated for learning value
- Leveraging to Tier 2 Process Safety Events and High Potential Process Safety Near Misses

Leveraging Corrective Actions

What Happened: Case Study – Corporate Wide Actions
Ethylene piping failure due to Corrosion Under Insulation (CUI) – Picture 1
Note that CUI is particularly aggressive where operating temperatures cause frequent or continuous condensation and re-evaporation of atmospheric moisture.

Management System Failure
Corporate Mechanical Integrity (MI) Standard and Work Process was not specific on performing CUI inspections

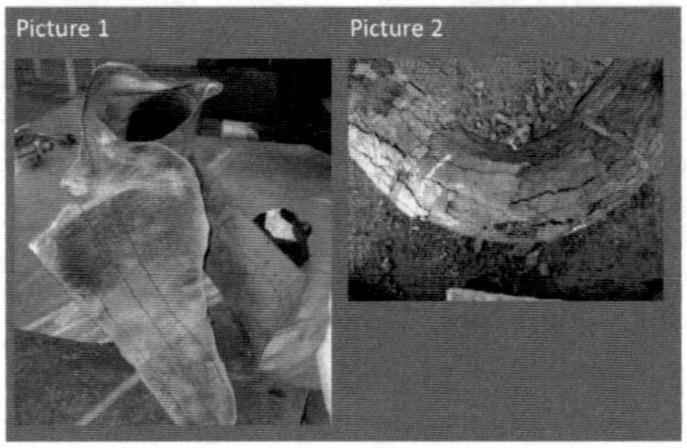

Corporate Corrective Actions
- Update MI Standard and Work Process to clearly articulate CUI Requirements
- Train Maintenance and Production Leadership
- Technology Centers define and prioritize susceptible areas for CUI
- Facilities to carry out CUI Inspections within a defined time frame and report back findings

Results
Additional finding of severe CUI, prevention of repeat incidents – Picture 2

AIChE
Southwest Process Technology Conference

Leveraging Corrective Actions

What Happened: Case Study – Business Actions
The investigation of a product quality issue uncovered that additional raw material had been added inadvertently (and automatically) five hours after the termination of the feed step. Further investigation uncovered issues with the code for the reactors' double-block-and-buffers that led to the event. This case was classified as a HP PSNM due to the high learning value and the potential to have a PSI under slightly different circumstances.

Management System Failure
Inadequate checkout of the process control code conditions for allowing raw material feed upstream valve opening outside of raw material feed step.

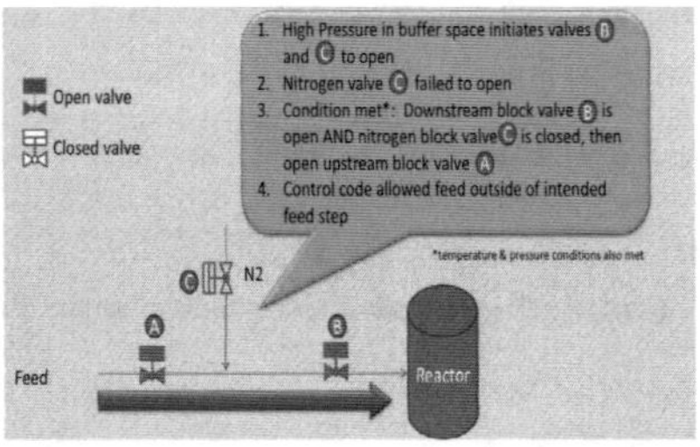

Business Corrective Actions
Review all plants utilizing standard software Code. Upgrade software code validation protocols

Results
Found same programming at several other plants, immediate action was taken, potentially averting a repeat incident. A LER was generated for review across the company.

AIChE
Southwest Process Technology Conference

7th AIChE Southwest Process Technology Conference | October 1-2, 2015 | Galveston, TX

Learning Experience Reports: Reinforcing Strong Management Systems

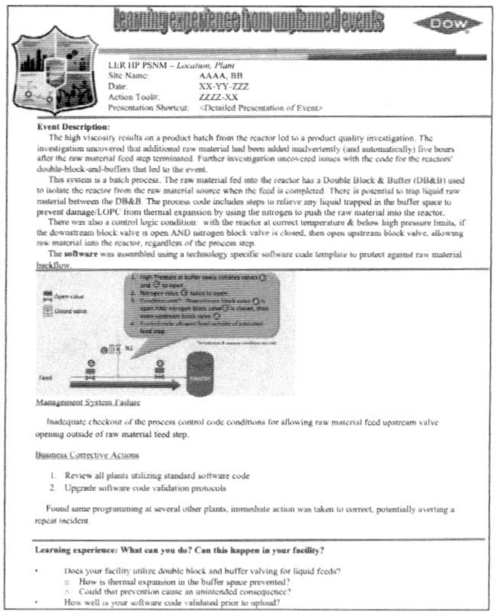
learning experience from unplanned events

DOW

LER HP PSNM – *Location, Plant*

Site Name: AAAA, BB

Date: XX-YY-ZZZ

Action Tool#: ZZZZ-XX

Presentation Shortcut: <Detailed Presentation of Event>

Event Description:

The high viscosity results on a product batch from the reactor led to a product quality investigation. The investigation uncovered that additional raw material had been added inadvertently (and automatically) five hours after the raw material feed step terminated. Further investigation uncovered issues with the code for the reactors' double-block-and-buffers that led to the event.

This system is a batch process. The raw material fed into the reactor has a Double Block & Buffer (DB&B) used to isolate the reactor from the raw material source when the feed is completed. There is potential to trap liquid raw material between the DB&B. The process code includes steps to relieve any liquid trapped in the buffer space to prevent damage/LOPC from thermal expansion by using the nitrogen to push the raw material into the reactor.

There was also a control logic condition: with the reactor at correct temperature & below high pressure limits, if the downstream block valve is open AND nitrogen block valve is closed, then open upstream block valve, allowing raw material into the reactor, regardless of the process step.

The **software** was assembled using a technology specific software code template to protect against raw material backflow.

Management System Failure

Inadequate checkout of the process control code conditions for allowing raw material feed upstream valve opening outside of raw material feed step.

Business Corrective Actions

1. Review all plants utilizing standard software code
2. Upgrade software code validation protocols

Found same programming at several other plants, immediate action was taken to correct, potentially averting a repeat incident.

Learning experience: What can you do? Can this happen in your facility?

- Does your facility utilize double block and buffer valving for liquid feeds?
  - How is thermal expansion in the buffer space prevented?
  - Could that prevention cause an unintended consequence?
- How well is your software code validated prior to upload?

"Learning Event Reports" (LERs), are published on all process safety incidents & HP-PSNM, leveraging to Tier 2 PSE.

One-page summaries that are broadly communicated across the company to reinforce the importance of:

- the proper definition, design, and implementation of protection layers
- operating within the constraints and maintaining all protection layers

The goal is to prevent repetitive failures by providing learning in a simplified manner that reinforces the appropriate behaviors in all functions of the company.

Leaders review the learnings and determine if they are applicable to any of their facilities, sites, or businesses.

AIChE Southwest Process Technology Conference

Learning Experience Reports: Key Elements

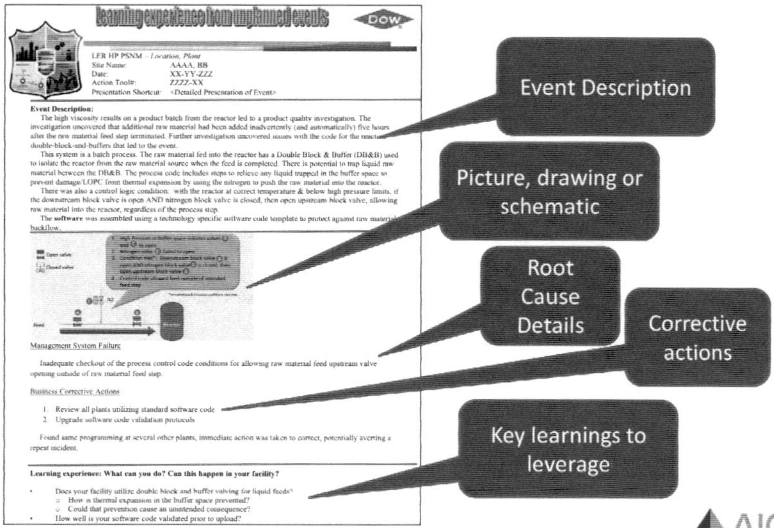

learning experience from unplanned events

DOW

LER HP PSNM – *Location, Plant*

Site Name: AAAA, BB

Date: XX-YY-ZZZ

Action Tool#: ZZZZ-XX

Presentation Shortcut: <Detailed Presentation of Event>

Event Description:

The high viscosity results on a product batch from the reactor led to a product quality investigation. The investigation uncovered that additional raw material had been added inadvertently (and automatically) five hours after the raw material feed step terminated. Further investigation uncovered issues with the code for the reactors' double-block-and-buffers that led to the event.

This system is a batch process. The raw material fed into the reactor has a Double Block & Buffer (DB&B) used to isolate the reactor from the raw material source when the feed is completed. There is potential to trap liquid raw material between the DB&B. The process code includes steps to relieve any liquid trapped in the buffer space to prevent damage/LOPC from thermal expansion by using the nitrogen to push the raw material into the reactor.

There was also a control logic condition: with the reactor at correct temperature & below high pressure limits, if the downstream block valve is open AND nitrogen block valve is closed, then open upstream block valve, allowing raw material into the reactor, regardless of the process step.

The **software** was assembled using a technology specific software code template to protect against raw material backflow.

Management System Failure

Inadequate checkout of the process control code conditions for allowing raw material feed upstream valve opening outside of raw material feed step.

Business Corrective Actions

1. Review all plants utilizing standard software code
2. Upgrade software code validation protocols

Found same programming at several other plants, immediate action was taken to correct, potentially averting a repeat incident.

Learning experience: What can you do? Can this happen in your facility?

- Does your facility utilize double block and buffer valving for liquid feeds?
  - How is thermal expansion in the buffer space prevented?
  - Could that prevention cause an unintended consequence?
- How well is your software code validated prior to upload?

AIChE Southwest Process Technology Conference

7th AIChE Southwest Process Technology Conference | October 1-2, 2015 | Galveston, TX

Conclusion

Effective RCIs:

- Correcting all protection layer failures and their associated management systems
- Appropriately leveraging the corrective actions and learnings
- Engaging leadership are foundational to building a strong culture.

AIChE Southwest Process Technology Conference

7th AIChE Southwest Process Technology Conference | October 1-2, 2015 | Galveston, TX

HAZOP & LOPA execution and efficiency

- Kathy Pearson, BP Upstream Engineering Centre
- HAZOP & LOPA Advisor

AIChE
Southwest Process Technology Conference

7th AIChE Southwest Process Technology Conference | October 1-2, 2015 | Galveston, TX

Kathy Pearson

bp

- Upstream Process Safety with BP – 6 years
- Prior companies – Rohm and Haas/Dow – 25 years
- HAZOP/LOPA, incident investigation, human factors
- Projects, technical support
- Active in AIChE

Overview

- Focus on Upstream HAZOP & LOPA execution and efficiency
  - occurring over 5+ years
- Assess HAZOP & LOPA processes
- Optimize HAZOP LOPA execution
- Implement common facilitator process
- Training classes targeted to specific audiences
- Written guidance with examples for common issues

3

Assess HAZOP & LOPA processes

- Important factors for success
  - Committed leadership & personnel
  - Belief in intent
  - Good procedures
- Opportunities
  - Primary opportunities change over time
  - Wide variation in facilitator practices
  - LOPA technical aspects - very difficult to apply to complex systems

4

Challenges

- Global workforce
  - Alaska, Houston, Trinidad, UK, Norway, Azerbaijan, Egypt, Algeria, Angola, Oman, Jakarta
- Personnel availability – time zones, shift rotations, travel
- Languages and cultures
- Varying levels of expertise

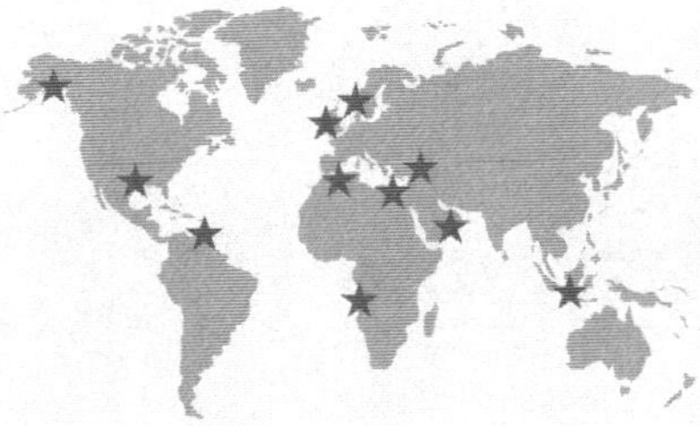

Continuous improvement process

- Gather frequently asked questions
- Network to:
  - Review answers to questions
  - Share best practices between regions
- Write informal guidance as common issues surface
  - Share guidance through internal network and web
  - Share with facilitators via email
- Continue to assess status each year

Example Issues

- How long should it take to study an FPSO?
- What is the best way to HAZOP pig launchers and receivers?
- How do you HAZOP a ballast system? Should you?
- What is the best way to revalidate a study?
- What failure rates and PFD's apply to subsea kit?

HAZOP & LOPA Training

- Classes
  - Interactive classes – about half exercises
  - Develop local trainers
- Classes targeted by audience
  - Awareness level – wide audience, computer based
  - Basics – new engineers and support people, local delivery
  - Advanced – key support engineers for studies or ready to develop as facilitators

Assess Facilitators

- Common interview process
  - Reduce rework and multiple interviews
  - Two interviewers – usually from different work groups
  - Develop trust between interviewers
- Common induction process and package
- Central database of facilitators
- Formally assess first studies
  - Track coaching done

Efficiency

- Small to medium teams – 7-8 optimal
- Good preparation – nodes defined, descriptions written
- PSI up-to-date, team can quickly find information
- Right size nodes – not too small
- Reasonable schedule
- Little team member turnover
- Appropriate application of HAZOP and other tools

Study flexibility

- Revalidation
  - Techniques to speed update and renewal
  - Move towards evergreen
- Initial vs final design in projects
- Does every system need a HAZOP?
  - Flexibility to use appropriate techniques

11

Written guidance

- Guidance on Layer of Protection Analysis (LOPA)
- Fault Tree Analysis Supporting LOPA
- HAZOP and LOPA Action Prioritization Guide
- Project HAZOP and LOPA Implementation Strategy
- HAZOP Update and Revalidation
- What-if as an Alternative to HAZOP
- Effective Facilitation Techniques
- Brainstorming tables & checklists
- HAZOP and LOPA procedures – simple and clear

12

LOPA guidance

- Examples relevant to business
- Examples increase in complexity
- Each example includes:
  - Scenario description and facts
  - Sketch
  - Explanation of factors used
  - Resulting LOPA table

LOPA guidance
simple examples to start

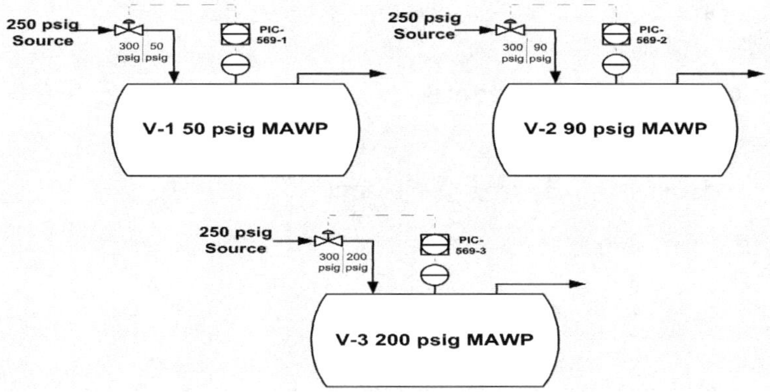

LOPA guidance
examples increase in complexity

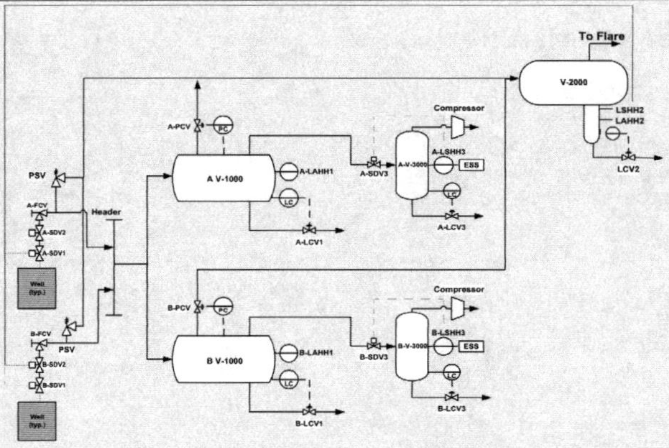

Conclusions & path forward

- Most HAZOP issues relate to management of process and appropriate application - not the methodology
- Develop systems to share experience
- Common understanding of what "Good" looks like
- LOPA is difficult on many systems, often iterative
- Path forward
  - HAZOP and LOPA templates for common equipment
  - Leverage database technology

Questions

Adrift: Does America Have a Future in Space?

Berger, E.,

Originally presented on: 10/1/2015 18:00:00 - 21:00:00

Over the course of 2014 science writer Eric Berger looked at NASA's building of a big new rocket, the collapse of the Constellation program, Congressional infighting for funds, shifting priorities of successive White House administrations, the promise of private space companies and, ultimately, the fate of Houston as Space City. This is what he found.

Professional Ethics for Chemical Engineers

Alan Rossiter
President, Rossiter & Associates
alan@rossiters.org

Southwest Process Technology Conference

A. Introduction to Engineering Ethics

Basis/Definition:

Ethics: 1. a discipline dealing with good and evil and with moral duty 2. moral principles or practice

The Merriam-Webster Dictionary, 1974, G. & C. Merriam Co.

Southwest Process Technology Conference

Professional Engineers' Continuing Education Program (CEP)

"Ethics Requirement" varies between jurisdictions.

Example (Texas):
1 PDH "in the area of professional ethics, roles and responsibilities of professional engineering, or review of the Texas Engineering Practice Act and Board Rules."

The Need for a Code of Engineering Ethics (1)

There is no good and evil, only power, and those too weak to seek it…

J. K. Rowling, *Harry Potter and the Sorcerer's Stone*

The Need for a Code of Engineering Ethics (1)

There is no good and evil, only power, and those too weak to seek it…

J. K. Rowling, *Harry Potter and the Sorcerer's Stone*

With great power comes great responsibility

Stan Lee, Peter Parker *(Spider-Man)*

AIChE Southwest Process Technology Conference

5

7th AIChE Southwest Process Technology Conference | October 1-2, 2015 | Galveston, TX

The Need for a Code of Engineering Ethics (2)

In a culturally diverse society, a code of ethics provides:

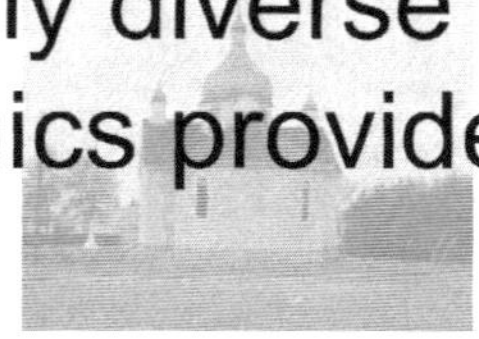

- a standard of action for those who subscribe to it
- a common basis for evaluating the actions of others

AIChE Southwest Process Technology Conference

6

7th AIChE Southwest Process Technology Conference | October 1-2, 2015 | Galveston, TX

AIChE Code of Ethics

Members of the American Institute of Chemical Engineers shall uphold and advance the integrity, honor and dignity of the engineering profession by:

- Being honest and impartial and serving with fidelity their employers, their clients, and the public;
- Striving to increase the competence and prestige of the engineering profession;
- Using their knowledge and skill for the enhancement of human welfare.

AIChE
Southwest Process Technology Conference

7

AIChE Code of Ethics - Continued

To Achieve these Goals, Members shall:

1. Hold paramount the safety, health and welfare of the public and protect the environment in performance of their professional duties.
2. Formally advise their employers or clients (and consider further disclosure, if warranted) if they perceive that a consequence of their duties will adversely affect the present or future health or safety of their colleagues or the public.
3. Accept responsibility for their actions, seek and heed critical review of their work and offer objective criticism of the work of others.

AIChE
Southwest Process Technology Conference

8

7th AIChE Southwest Process Technology Conference | October 1-2, 2015 | Galveston, TX

AIChE Code of Ethics - Continued

4. Issue statements or present information only in an objective and truthful manner.
5. Act in professional matters for each employer or client as faithful agents or trustees, avoiding conflicts of interest and never breaching confidentiality.
6. Treat fairly and respectfully all colleagues and co-workers, recognizing their unique contributions and capabilities.
7. Perform professional services only in areas of their competence.

AIChE Code of Ethics - Continued

8. Build their professional reputations on the merits of their services.
9. Continue their professional development throughout their careers, and provide opportunities for the professional development of those under their supervision.
10. Never tolerate harassment.
11. Conduct themselves in a fair, honorable and respectful manner.

Engineering and Life…

Engineering projects are about building and improving things.

Life is about interacting with people.

Professional Excellence and Ethics in Engineers Blends the Two.

Key People Principle…

…respect for the other person.

Applications vary, but the principle doesn't change

Case Study 1

- Consultant at Client's Site with Major Unreported Pollution Issue
- Repeatedly Advised Management to Report and Correct Problems
- Management Refused
- Which Section(s) of the AIChE Code Apply?

AIChE
Southwest Process Technology Conference

Case Study 1 - Continued

Dilemma: Conflict within AIChE Code of Ethics

5. Act in professional matters for each employer or client as faithful agents or trustees, avoiding conflicts of interest and never breaching confidentiality.

1. Hold paramount the safety, health and welfare of the public and protect the environment in performance of their professional duties.

AIChE
Southwest Process Technology Conference

Case Study 1 - Continued

Resolution

Safety, health, public welfare and environment are "paramount."

Consultant terminates contract with client and reports pollution to relevant authority.

AIChE
Southwest Process Technology Conference

15

Case Study 1 - Continued

Be Aware:

- Not always a clear or simple decision.
- Check the facts.
- Don't overreact. Seek impartial advice.
- Possible negative consequences: loss of business, legal issues, etc.
- Whistleblower laws may protect you.

AIChE
Southwest Process Technology Conference

16

Case Study 2

- Small Engineering Firm A in Project Discussions with Client Company X.
- 3 Employees, 2 of Them Familiar with the Discussions, Leave Firm A to Start Firm B.
- Firm B Bids on Project with Company X, and Wins Contract.

AIChE Southwest Process Technology Conference

17

7th AIChE Southwest Process Technology Conference | October 1-2, 2015 | Galveston, TX

Case Study 2 - Continued

- Complex legal issues
  - Contracts
  - Statutes – vary by jurisdiction
- Which Section(s) of the AIChE Code Apply?

AIChE Southwest Process Technology Conference

18

7th AIChE Southwest Process Technology Conference | October 1-2, 2015 | Galveston, TX

Case Study 2 - Continued

Ethical Guideline

5. Act in professional matters for each employer or client as faithful agents or trustees, avoiding conflicts of interest and never breaching confidentiality.

Also: NSPE code of ethics; case studies

Southwest Process Technology Conference

19

Case Study 2 - Continued

Consensus from literature cases:

- Engineers who serve a given client while employed by A may serve same client after leaving A.
- Engineers may not use specialized knowledge obtained while in employment to secure and execute projects after leaving their employer.

Southwest Process Technology Conference

20

Case Study 2 - Continued

Conclusion

- Based on facts presented, Firm B engineers acted unethically because they used specialized knowledge obtained while in employment to secure their contract with Company X.

AIChE Southwest Process Technology Conference

21

7th AIChE Southwest Process Technology Conference | October 1-2, 2015 | Galveston, TX

Case Study 2 - Continued

- Difficult, stressful situations.
- Balance right to seek new opportunities vs. proprietary rights of parent company.
- Where possible, seek amicable agreements.

AIChE Southwest Process Technology Conference

22

7th AIChE Southwest Process Technology Conference | October 1-2, 2015 | Galveston, TX

Case Study 3: Conflict in the Work Place

Conflict: 1. war
2. a clash between hostile or opposing elements or ideas

The Merriam-Webster Dictionary, 1974, G. & C. Merriam Co.

AIChE Southwest Process Technology Conference

1. The Set Up

- I Just Received an Email from Colleague "A"
- Accuses me of Unprofessional Conduct
- Copied to Several Other People

- How Should I Respond??

AIChE Southwest Process Technology Conference

7th AIChE Southwest Process Technology Conference | October 1-2, 2015 | Galveston, TX

2. The Response…

- Hit "Reply to All"
- Justify My Actions
- Point Out Colleague A's OWN Lack of Professionalism

- Hit Delete. Sit Back. Start Again…

AIChE
Southwest Process Technology Conference

25

7th AIChE Southwest Process Technology Conference | October 1-2, 2015 | Galveston, TX

A Word to the Wise…

- Delete is the most important key on your computer.

AIChE
Southwest Process Technology Conference

26

7th AIChE Southwest Process Technology Conference | October 1-2, 2015 | Galveston, TX

Key People Principle…

…respect for the other person.

Applications vary, but the principle doesn't change

27

Principles to Consider (1)

- Respect the Other Person
  - Where is he coming from?
  - How can I build him up (affirmation)?

Seek first to understand, then to be understood.
Habit 5 in Stephen Covey's "Seven Habits of Highly Successful People"

- Self-Control
  - Manage your anger
  - Be honest with yourself
  - Don't respond out of insecurity

28

Principles to Consider (2)

- Relationships Before Issues
  - Keep the immediate issue in perspective
  - Think long-term
- Pursue Honest Dialog
  - Honest disagreements can be healthy
  - Conflict, handled honesty, can open up new opportunities (Functional Conflict)
  - Conflict, handled dishonestly, destroys relationships and organizations (Dysfunctional Conflict)

AIChE Southwest Process Technology Conference

29

7th AIChE Southwest Process Technology Conference | October 1-2, 2015 | Galveston, TX

3. Start Over (1)

- Hit Reply, NOT Reply to All
  - Don't defend yourself or humiliate him. Remember – relationships before issues.
- Start With Affirmation
  - e.g. "I appreciate your contribution to the organization, and especially XXX."

I will speak ill of no man, and speak all the good I know of everybody.
– Benjamin Franklin

AIChE Southwest Process Technology Conference

30

7th AIChE Southwest Process Technology Conference | October 1-2, 2015 | Galveston, TX

3. Start Over (2)

- Thank Him for Drawing Your Attention to the Problem
- Express My Level of Understanding
  - e.g., "I think I can see your perspective"
- Explain My View of the Cause of Conflict
  - e.g., "I had to cut that program because spending had exceeded the budget."

Southwest Process Technology Conference

31

7th AIChE Southwest Process Technology Conference | October 1-2, 2015 | Galveston, TX

3. Start Over (3)

- Be Willing to Acknowledge Your Oversights
  - e.g., "I apologize for not informing you of my decision sooner."
- Look to the Future
  - e.g. "I look forward to working with you as we continue to expand the reach of our organization"
  - If possible, move the dialog on to a specific different activity that you are doing or will be doing together.

Southwest Process Technology Conference

32

7th AIChE Southwest Process Technology Conference | October 1-2, 2015 | Galveston, TX

4. The Outcome

- Colleague "B", who was also on the distribution list for the original email, "replied to all" to defend my professionalism.
- Colleague "A" replied to my private email.
  - acknowledged he had overreacted in his original email
  - raised additional concerns, some of which I consider valid
- We have resolved some of Colleague A's issues, and continue to work on some others.
- We continue to work together.

AIChE Southwest Process Technology Conference

33

A Final Word on Ethics and Integrity...

...the supreme quality for a leader is unquestionably integrity. Without it, no real success is possible, no matter whether it is on a section gang, a football field, in an army, or in an office. If a man's associates find him guilty of being a phony, if they find that he lacks forthright integrity, he will fail. His teachings and his actions must square with each other. The first great need, therefore, is integrity and high purpose. - Dwight Eisenhower

AIChE Southwest Process Technology Conference

34

PROCESS SIMULATION IN SUPPORT OF MATERIALS TESTING & SELECTION

Ray French - Shell International Exploration & Production, Inc.

Brian Chambers - Shell Global Solutions (US) Inc.

7th AIChE Southwest Process Technology Conference/ October 1-2, 2015/ Galveston, TX

1

DEFINITIONS & CAUTIONARY NOTE

Reserves: Our use of the term "reserves" in this presentation means SEC proved oil and gas reserves.

Resources: Our use of the term "resources" in this presentation includes quantities of oil and gas not yet classified as SEC proved oil and gas reserves. Resources are consistent with the Society of Petroleum Engineers 2P and 2C definitions.

Organic: Our use of the term Organic includes SEC proved oil and gas reserves excluding changes resulting from acquisitions, divestments and year-average pricing impact.

Resources plays: Our use of the term 'resources plays' refers to tight, shale and coal bed methane oil and gas acreage.

The companies in which Royal Dutch Shell plc directly and indirectly owns investments are separate entities. In this presentation "Shell", "Shell group" and "Royal Dutch Shell" are sometimes used for convenience where references are made to Royal Dutch Shell plc and its subsidiaries in general. Likewise, the words "we", "us" and "our" are also used to refer to subsidiaries in general or to those who work for them. These expressions are also used where no useful purpose is served by identifying the particular company or companies. ''Subsidiaries'', "Shell subsidiaries" and "Shell companies" as used in this presentation refer to companies in which Royal Dutch Shell either directly or indirectly has control. Companies over which Shell has joint control are generally referred to as "joint ventures" and companies over which Shell has significant influence but neither control nor joint control are referred to as "associates". The term "Shell interest" is used for convenience to indicate the direct and/or indirect ownership interest held by Shell in a venture, partnership or company, after exclusion of all third-party interest.

This presentation contains forward-looking statements concerning the financial condition, results of operations and businesses of Royal Dutch Shell. All statements other than statements of historical fact are, or may be deemed to be, forward-looking statements. Forward-looking statements are statements of future expectations that are based on management's current expectations and assumptions and involve known and unknown risks and uncertainties that could cause actual results, performance or events to differ materially from those expressed or implied in these statements. Forward-looking statements include, among other things, statements concerning the potential exposure of Royal Dutch Shell to market risks and statements expressing management's expectations, beliefs, estimates, forecasts, projections and assumptions. These forward-looking statements are identified by their use of terms and phrases such as ''anticipate'', ''believe'', ''could'', ''estimate'', ''expect'', ''intend'', ''may'', ''plan'', ''objectives'', ''outlook'', ''probably'', ''project'', ''will'', ''seek'', ''target'', ''risks'', ''goals'', ''should'' and similar terms and phrases. There are a number of factors that could affect the future operations of Royal Dutch Shell and could cause those results to differ materially from those expressed in the forward-looking statements included in this presentation, including (without limitation): (a) price fluctuations in crude oil and natural gas; (b) changes in demand for Shell's products; (c) currency fluctuations; (d) drilling and production results; (e) reserves estimates; (f) loss of market share and industry competition; (g) environmental and physical risks; (h) risks associated with the identification of suitable potential acquisition properties and targets, and successful negotiation and completion of such transactions; (i) the risk of doing business in developing countries and countries subject to international sanctions; (j) legislative, fiscal and regulatory developments including potential litigation and regulatory measures as a result of climate changes; (k) economic and financial market conditions in various countries and regions; (l) political risks, including the risks of expropriation and renegotiation of the terms of contracts with governmental entities, delays or advancements in the approval of projects and delays in the reimbursement for shared costs; and (m) changes in trading conditions. All forward-looking statements contained in this presentation are expressly qualified in their entirety by the cautionary statements contained or referred to in this section. Readers should not place undue reliance on forward-looking statements. Additional factors that may affect future results are contained in Royal Dutch Shell's 20-F for the year ended 31 December, 2014 (available at www.shell.com/investor and www.sec.gov). These factors also should be considered by the reader. Each forward-looking statement speaks only as of the date of this presentation, 1 October, 2015. Neither Royal Dutch Shell nor any of its subsidiaries undertake any obligation to publicly update or revise any forward-looking statement as a result of new information, future events or other information. In light of these risks, results could differ materially from those stated, implied or inferred from the forward-looking statements contained in this presentation. There can be no assurance that dividend payments will match or exceed those set out in this presentation in the future, or that they will be made at all.

We use certain terms in this presentation, such as discovery potential, that the United States Securities and Exchange Commission (SEC) guidelines strictly prohibit us from including in filings with the SEC. U.S. Investors are urged to consider closely the disclosure in our Form 20-F, File No 1-32575, available on the SEC website www.sec.gov. You can also obtain this form from the SEC by calling 1-800-SEC-0330.

OUTLINE

- ➢Materials challenges for Deep Water oil & gas exploration & production
- ➢Modeling Field environments
- ➢Modeling Lab environments
- ➢Establishing Target Environments in the Lab
- ➢Wrap-up

For fuller discussion of these issues, please see CORROSION 2014 papers:
4285: "Laboratory Simulation of Oil & Gas Field Conditions: Important Phase Behavior Considerations and Approaches", B. Chambers, S. Huizinga, M. Yunovich, W.Grimes,M. Wilms and R. French
3870: "The Physical Chemistry Nature of Hydrogen Sulfide Gas as it Affects Sulfide Stress Crack Propagation in Steel" W. Grimes, B. Miglin, R. French, M. Gonzalez and B. Chambers

MATERIALS CHALLENGES FOR DEEP WATER OIL & GAS EXPLORATION & PRODUCTION

- Accelerated failure of materials in oil and gas environments
  - Environmentally-assisted cracking (EAC)
  - Corrosion
- Aggressive species
  - H_2S, CO_2
  - Chloride
  - Organic acids
- Degradation mechanisms include:
  - **Sulfide Stress Cracking**
  - **Stress Corrosion Cracking**
  - Acid gas corrosion
  - Localized corrosion

Presentation focuses on H_2S-related cracking evaluations

MOTIVATION

- EAC failure modes can result in rapid, catastrophic failure
- Material selection and risk assessment improved by laboratory tests
- Laboratory evaluation/qualification of material resistance to EAC
- Need to simulate reasonable worst-case environment for testing
  - What field parameters are recreated in laboratory?

FIELD PARAMETERS → LABORATORY

- Laboratory tests for EAC typically recreate worst-case environment
  - Environment under most scrutiny is worst-case production
    - Water-full system (no reliance on oil wetting, corrosion inhibitors)
    - Maximum chloride content anticipated in production
    - H_2S/CO_2 representative of worst-case in production
    - Lowest pH anticipated in service (typically dictated by CO_2)
  - Other environments (e.g. acidizing) often investigated separately
    - Shorter-duration, transient exposures

FIELD PARAMETERS → LABORATORY

- H_2S-related parameters for EAC exposures
  - For laboratory test at lower pressure than field, choose parameter
  - Publications, research, and experience concerning various parameters
    - **Acid gas partial pressure**
      - Historical parameter of choice and prescribed by ISO 15156
      - Most material qualification work for sour service completed per partial pressure targets
    - **Acid gas fugacity**
      - CORROSION 2014 Paper 3870 (Grimes et al.)
    - **Dissolved acid gas concentration**
      - CORROSION 2014 Paper 4243 (Kumar et al.)

7

DEEPWATER OIL & GAS PRODUCTION – FIELD ENVIRONMENT

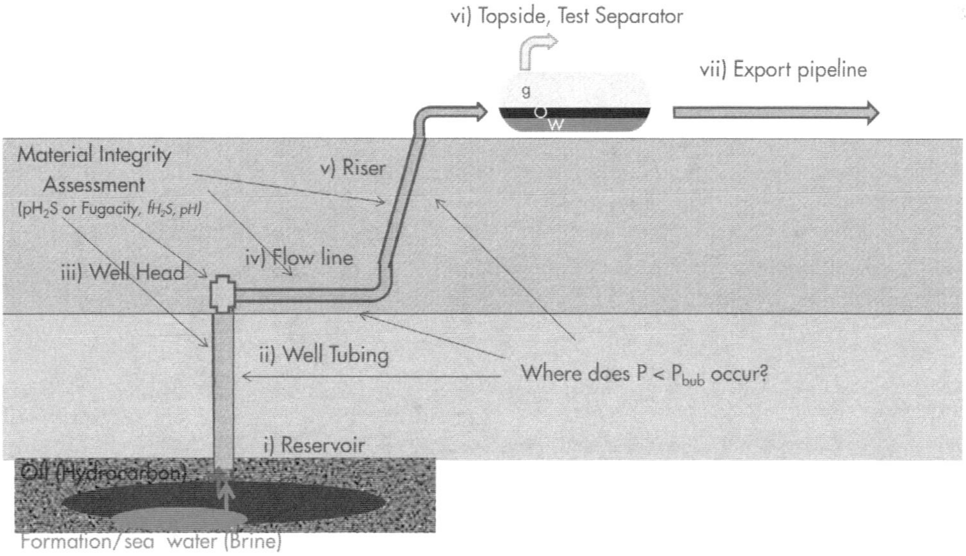

8

H_2S – Why It Matters for Deep Water Oil & Gas Production

- HSE NTL No. 2009-G31*: if H_2S > 20ppm at atmospheric conditions (1atm, ~77°F) ⇒ facility classified as "H_2S present"

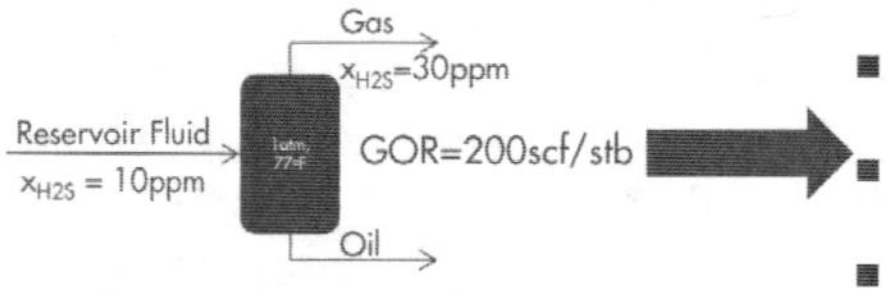

- H_2S in atmospheric gas will depend on Gas/Oil Ratio
- significant concentration effect for low GOR fluids!
- H_2S in atmospheric gas increases with increasing water cut

- Materials (wells, subsea system, facilities) ISO 15156/NACE MR0175:
  - If pH_2S < 0.05 psia (0.3 kPa), no sour service material required
  - If P < Pbub, pH_2S = mole fraction$_{H2S\text{-}Gas}$ x Pressure$_{(total)}$
  - If P > Pbub, pH_2S = mole fraction$_{H2S\text{-}Gas}$ x Pressure$_{(bubble\ point)}$ ← HP undersaturated oils No gas in wellbore

pH_2S=0.05 psia → 5 ppm H_2S at 10,000 psia
pH_2S=0.05 psia → 2.5 ppm H_2S at 20,000 psia

*Notice To Lessees for Gulf Of Mexico issued by US Bureau of Safety & Environmental Enforcement

MODELING FIELD ENVIRONMENTS - FLUIDS

- (Hydrocarbon) Fluid Characterization
  - PVT report → combined reservoir composition of hydrocarbon phase(s)
  - Real + Standard Carbon Number + Hypothetical hydrocarbons
  - For oil, match P_{bub} (as per ISO standard) and Gas/Oil Ratio
  - For gas condensate, match P_{dew} (for condensate dropout) and Condensate/Gas Ratio
- Water
  - Formation water (need to specify water cut) → high Cl^-, higher pH (buffering from HCO_3^-)
    - Analyses of surface samples – needs QC/reconciliation step
    - Equilibrate formation water with minerals (e.g. calcite) at reservoir conditions (T, P, CO_2 concentration in oil/gas) → adjust HCO_3^- value → needed for pH predictions
  - Condensed water (need to saturate reservoir oil/gas with H_2O in equilibrium with brine phase) → low Cl^-, lower pH
  - Seawater – from waterflooding (need to mix with formation water)
  - Well/formation treatments (e.g. acid stimulation) introduces another water source

MODELING FIELD ENVIRONMENTS – H2S & CO2

- H_2S Content in DeepWater Gulf Of Mexico
  - Typical H_2S may be < 50 ppmmole (in oil/gas)
  - But as low as 2 ppmmole (in oil/gas) can be a concern
  - Simulate to convert H_2S in surface sample gas to H_2S in reservoir fluid
  - Large uncertainties in measuring these low ppm levels
  - Native H_2S increases with increasing reservoir T (can sometimes be estimated)
  - Need to assess possible souring due to waterflood (via Sulfate Reducing Bacteria)
- CO_2 Content in DeepWater Gulf Of Mexico
  - CO_2 typically 0.1 to 10 %mole
  - Usually reliable values from PVT report
  - Key to predicting pH of aqueous phases

11

MODELING FIELD ENVIRONMENTS - THERMODYNAMICS

- Thermodynamic models/tools
  - Cubic EOS, e.g. Peng Robinson (USD, PVTSIM, etc.)
  - Mixed Solvent Electrolyte or AQ models in OLI StreamAnalyzer
  - ScaleSoftPitzer (Brine Chemistry Consortium, Rice University)
  - (Ensemble) Henry's law
  - Need parameters validated for field T, P, composition, fluid
- Ionic Effects (Speciation of H_2S & CO_2 , Salting effect)
  - Include in single model?
  - Apply hybrid approach (mix & match models)?
  - Ignore?

-Approximate distribution of H_2S, HS^-, and $S^=$ as a function of pH[11]

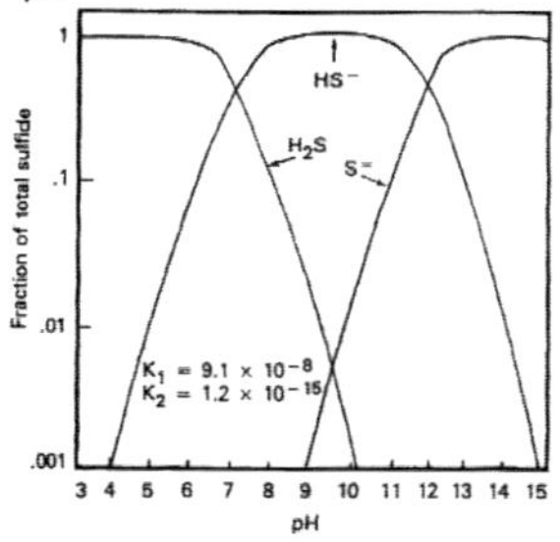

12

MODELING FIELD ENVIRONMENTS - APPLICATION

- Key aspects to represent
  - Partitioning of H_2S & CO_2 among gas/oil/brine
  - Aqueous phase dew point → condensed water
  - H_2S Fugacity, partial pressure (for $P < P_{bub}$), concentration in aq phase, pH of aq phase

- Uncertainty increases with increasing T, P, salt, oil MW & aromaticity

- Scenarios
  - Normal (steady-state) production
  - Shut-in conditions (uncertainty due to transient & nonequilibrium effects)
  - Worst-case (uncontrolled) discharge (uncertainty due to transient & nonequilibrium effects)

13

LAB ENVIRONMENT - TEST SETUP

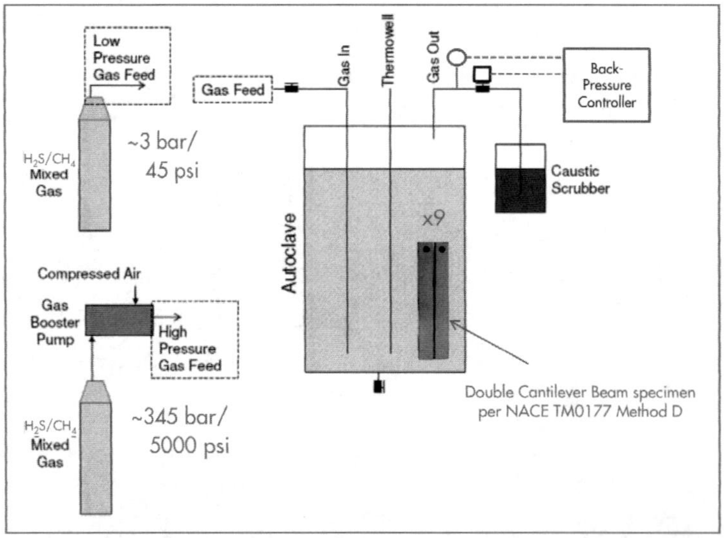

14

MODELING LAB ENVIRONMENTS – KEY DIFFERENCES FROM FIELD

- Two-phase system (Gas + Brine) vs Three-phase (Gas + Oil/Condensate + Brine)
- No Oil phase → no fluid characterization. (Sometimes a model hydrocarbon, e.g. decane, is used to, for example, account for inhibitor partitioning between oil & brine)
- Gas phase simpler than field: may be N_2 or CH_4 + CO_2, H_2S
- Brine simpler than field: typically only NaCl
- pH controlled at target value using a buffer
- Solutions defined by NACE may be used
- Usually run at low pressures
- SSC tests often run at ambient T (cracking risk increases with decreasing T)
- Corrosion & inhibitor testing often run at elevated T
- Tests run in Autoclave → Fixed total volume → V/L volume ratio varies with T changes
- Closed system → possible depletion of H_2S & CO_2 during test

15

MODELING LAB ENVIRONMENTS - THERMODYNAMICS

- OLI StreamAnalyzer includes all key features needed
  - H_2S & CO_2 solubility in brine
  - Speciation of H_2S & CO_2 & modeling of acid-base buffer
  - Partial pressure of water (above brine)
  - Solubility of N_2 & CH_4 in brine
  - Density of brine → thermal expansion → V/L ↓ as T ↑ → impact on total P
  - Nonideal behavior of gas phase as P↑
- Other models/tools can be used with appropriate adjustment of results
  - Cubic EOS, e.g. PR (USD, PVTSIM, etc.)
  - ScaleSoftPitzer (Brine Chemistry Consortium, Rice University)
- Uncertainty increases with increasing T, P, salt

16

Setting targets in Fit For Purpose testing

Partial Pressure Approach - Match $P(H_2S)$ but not P(tot)

Fugacity Approach – Match $f(H_2S)$*

Solubility Approach – Match ppmm(H_2S) in Brine ***

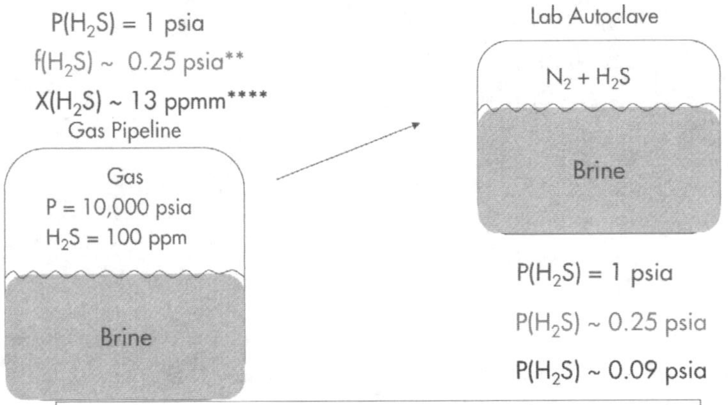

**Calculated for 100 ppmmol H_2S in C1 at 75F using REFPROP v9.1 from NIST

*Corrosion 2014 (Shell) "Physical Chemistry Tests of Hydrogen Sulfide Gas and Sulfide Stress Cracking Results at Elevated Pressure", Grimes et al.

****Calculated for 100 ppmmol H_2S in C1 at 75F in equilibrium with pure water using Extended Henry's law model

***SPE 97576 Nelson & Reddy, Exxon (2005) 'Selecting Representative Laboratory Test Conditions for Fit-for-Purpose OCTG Materials Evaluations'

CASE STUDY: OIL WELL – WELLHEAD CASE

FIELD	Example Case:	High pressure Oil Well @ Wellhead		
	Field Fluid Molar Composition[a]:	0.1% H_2S, bal. 50:50 CH_4:Decane		
	Field Pressure, Temperature:	1000 bar_a, 25°C		
	Laboratory Capacity:	1 bar_a @ 25°C		
	Phase parameter, units :	pH_2S, kPa_a	fH_2S, kPa_a	xH_2S,mg/l
	Field Value for above parameter:	5.9[c]@ P_{BP}	12.1	104
LAB	Laboratory gas H_2S mole % to duplicate field & parameter above[b]:	5.9%	12.1%	3.2%
	Lab pH_2S value; Factor	5.9, 1.0	12.1; 2.0	3.2; 1/1.8
	Lab fH_2S value; Factor	5.9, 1/2.1	12.1, 1.0	3.3; 1/3.8
	Lab xH_2S_{aq} value; Factor	191, 1.8	392; 3.8	104, 1.0

- Choosing different parameters:
  - Fugacity is the most conservative parameter
  - Focusing on either partial pressure or dissolved H_2S concentration result in significantly less conservative tests

18

ESTABLISHING TARGET ENVIRONMENTS IN LABORATORY

- Once the target environment is selected, it must be established in lab
- Standardized tests exist for stressing and exposing specimens for SSC/SCC evaluation
- Standards are written with regards to environments under ambient conditions
- Lack of consistent procedures across industry to establish custom H_2S/CO_2 environments in the laboratory

ESTABLISHING TARGET ENVIRONMENTS IN LABORATORY

- Large number of procedural approaches are used to attempt to establish the target environment
  - Some general categories include:

Method of gas load	Details	Gas load temperature
Purge	Initial purge and lock-in.	At ambient or initial other temperature
		At test temperature
	Continuous purge.	At test temperature
Batch	Apply components and lock-in. Repeat.	At ambient or initial other
		At test temperature
Fixed inventory loading	Load mixed gas and lock-in.	At ambient or initial other
		At test temperature
	Sequentially load different components and lock-in.	At ambient or initial other
		At test temperature

ESTABLISHING TARGET ENVIRONMENTS IN LABORATORY

Critical common errors include:

- Not accounting for water vapor pressure
  - $P_{Total,\ gauge,\ absolute} = P_{H2O} + P_{H2S} + P_{CO2} + P_{other....}$
  - Note that steam tables or other simplistic water vapor calculations can result in large errors in HTHP environments
- Failure to address volume change created by liquid thermal expansion
- Not accounting for selective adsorption if charging mixed gas (e.g. H_2S/N_2)
- During gas load via a batch or fixed inventory process, insufficient time for H_2S & CO_2 adsorption on equipment and saturation of the brine
- Not documenting, verifying & maintaining a consistent de-oxygenation procedure
- Not addressing consumption of corrodents
  - Particularly for low applied pH_2S in SSC/SCC testing

ESTABLISHING TARGET ENVIRONMENTS IN LABORATORY

- Case studies were examined for errors associated with varying laboratory procedures and common mistakes
  - H_2S partial pressure accuracy was examined assuming this parameter was the target
  - OLI Stream Analyzer utilized to predict environments
  - Other assumptions and bases for the modeling detailed in CORROSION 2014 paper 4285 (Chambers et al.)

ESTABLISHING TARGET ENVIRONMENTS IN LABORATORY

Case	Laboratory Setup			Experimental Approach		
	Temp.	Pressure	Target H_2S PP	Continuous Purge at Test Pressure and Temperature	Purge-and-Heat	Sequential Loading of Gases at Test Temperature
High Pressure Gas Pipeline	25 °C	1 bara	10 kPa	9.7 kPa	N/A	9.7 kPa
High Pressure Oil well at Wellhead	25 °C	1 bara	5.8 kPa	5.6 kPa	N/A	5.6 kPa
NACE Level VII	205 °C	88 bara	3447 kPa [a]	3166.3 kPa	2394.5 kPa	3025 kPa

- Simple atmospheric pressure, room temperature experiments exhibit little to no error with varying approach when done properly
- Each type of approach results in some level of error when used to attempt HTHP laboratory environments

ESTABLISHING TARGET ENVIRONMENTS IN LABORATORY

- Use of proper thermodynamic models can increase accuracy of environmental loading in laboratory tests
  - Becoming more common among labs and requested from clients
  - Approach permits loading quantities as determined in model
  - Can benchmark laboratory observations on pressures at given temperatures from those determined in the model

SUMMARY AND FINAL COMMENTS

- Presentation focused on modeling field and lab environments related to H_2S-related cracking evaluations
  - Approaches are applicable for other testing (e.g. corrosion)
- H_2S-related parameters for EAC exposures
  - Acid gas partial pressure
  - Dissolved acid gas concentration
  - Acid gas fugacity
- Consider controlling parameter (ongoing research) vs. conservatism
- Laboratory practices with regard to establishing test environments need to be improved
  - Carefully document the approach/procedures when reporting
  - Use thermodynamic models as possible to assist in laboratory

Q & A

Back-Up

28

DEEPWATER OIL & GAS PRODUCTION – FIELD ENVIRONMENT

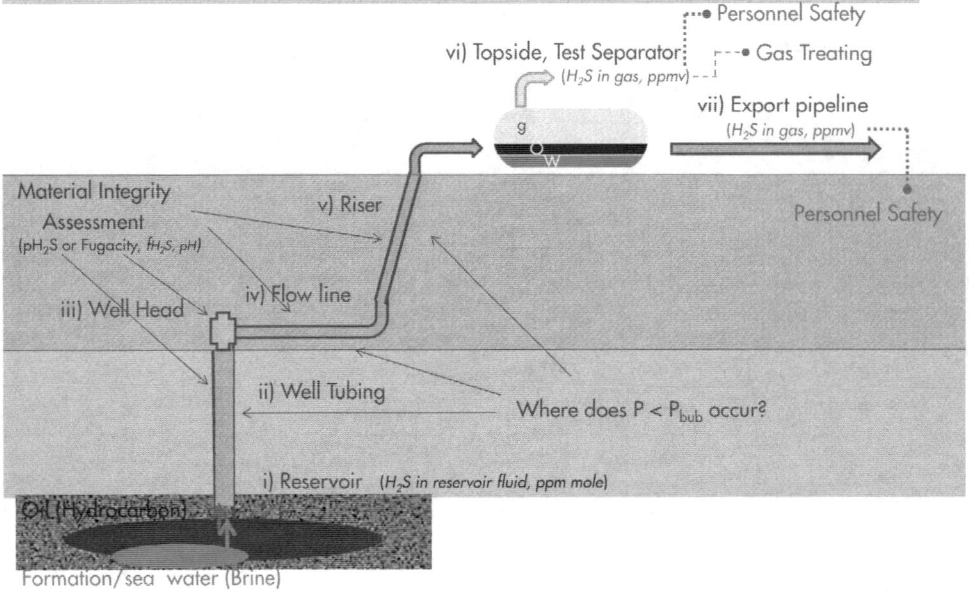

29

Setting targets in Fit For Purpose testing

Partial Pressure Approach - Match $P(H_2S)$ but not P(tot)

Fugacity Approach – Match $f(H_2S)$*

Solubility Approach – Match ppmm(H_2S) in Brine ***

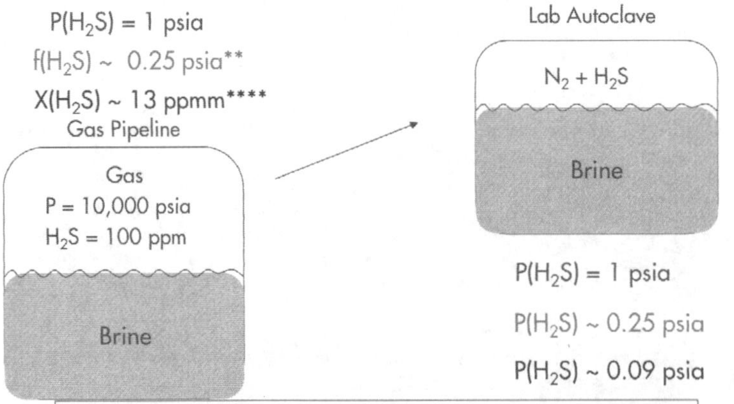

**Calculated for 100 ppmmol H_2S in C1 at 75F using REFPROP v9.1 from NIST

*Corrosion 2014 (Shell) "Physical Chemistry Tests of Hydrogen Sulfide Gas and Sulfide Stress Cracking Results at Elevated Pressure", Grimes et al.

****Calculated for 100 ppmmol H_2S in C1 at 75F in equilibrium with pure water using Extended Henry's law model

***SPE 97576 Nelson & Reddy, Exxon (2005) 'Selecting Representative Laboratory Test Conditions for Fit-for-Purpose OCTG Materials Evaluations'

FIELD PARAMETERS → LABORATORY

- Partial pressure
  - Historically, the commonly used parameter for testing
  - $P_{H2S} = y_{H2S}\,P_t;\quad P_{CO2} = y_{CO2}\,P_t$
- Fugacity
  - Effective pressure and same chemical potential as real gas
  - Fugacity, at equilibrium, is the same value in all phases
    - Relevant for calculating effects in liquid full systems (above bubble point)
  - $f_{H2S} = \phi_{H2S}\,y_{H2S}\,P_t;\quad f_{CO2} = \phi_{CO2}\,y_{CO2}\,P_t$
- Dissolved gas concentration
  - Aqueous concentration of H_2S or CO_2
  - x_{H2S} or xH_2S in mg/L_{aq} or ppm_{aq} (mg/kg)

31

EXAMPLE – PARAMETER DIVERGENCE

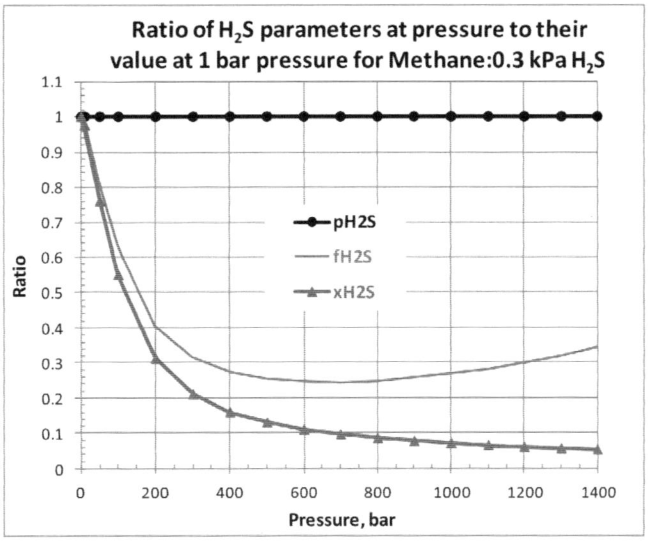

32

OIL & GAS INDUSTRY DEFINITION OF HIGH PRESSURE & HIGH TEMPERATURE (HPHT)

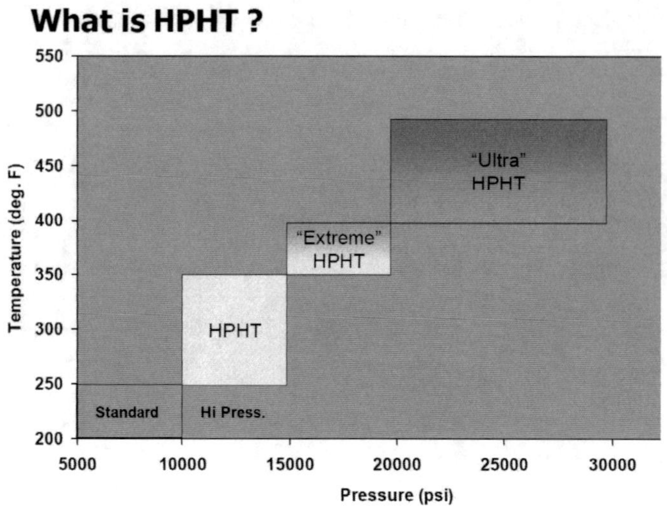

33

OVERVIEW

- Ongoing challenge to characterize EAC and corrosion using laboratory testing to represent oil and gas field environments
  - Equipment limitations restrict total pressures
    - **Laboratory pressure < Field pressure**
  - Representative testing
    - **Select a field condition** to target in laboratory
      - Partial pressure
      - Dissolved concentration
      - Fugacity
  - **Achieve target environment** in laboratory
    - Multiple methods to apply environments (H_2S, CO_2 loading)
    - Approach may result in environments significantly off target

34

Optimizing Simulation of Refinery and Petrochemical Reactors with Molecular Structure Fluid Lumping and Object Oriented Unit Design

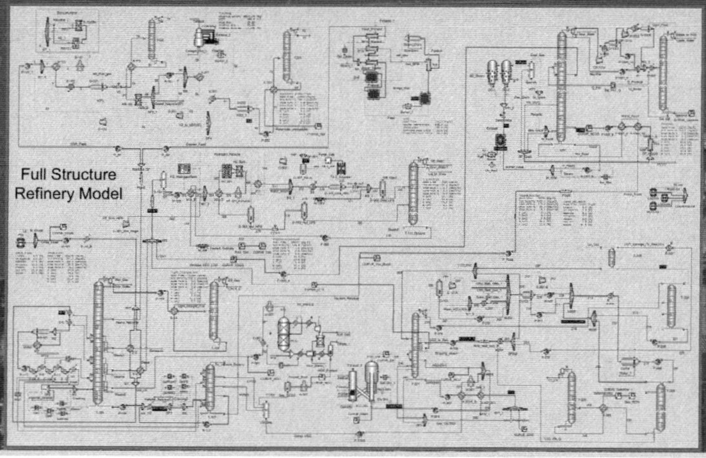

Glen Hay, Gerald Jacobs, Andrew Nathan

October 2nd , 2015

VMG EMPOWERING PROCESS SIMULATION

Expanding Scope of Simulators

Many areas of process engineering:

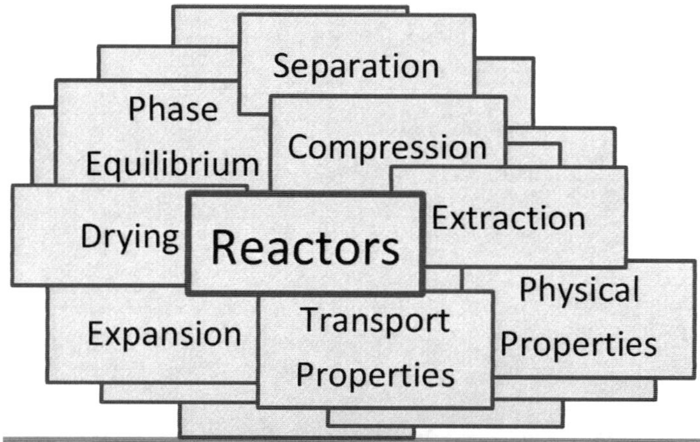

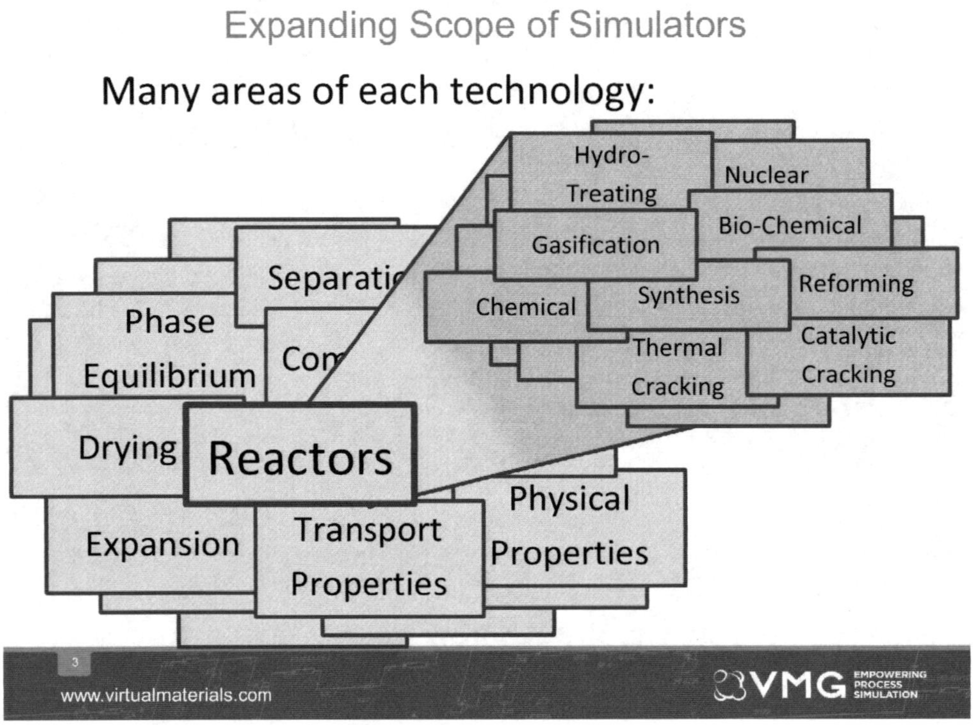
Expanding Scope of Simulators
Many areas of each technology:
Hydro-
Treating
Nuclear
Bio-Chemical
Gasification
Separati
Reforming
Synthesis
Chemical
Phase
Equilibrium
Com
Catalytic
Cracking
Thermal
Cracking
Drying
Reactors
Physical
Properties
Transport
Properties
Expansion
3
www.virtualmaterials.com
VMG
EMPOWERING
PROCESS
SIMULATION

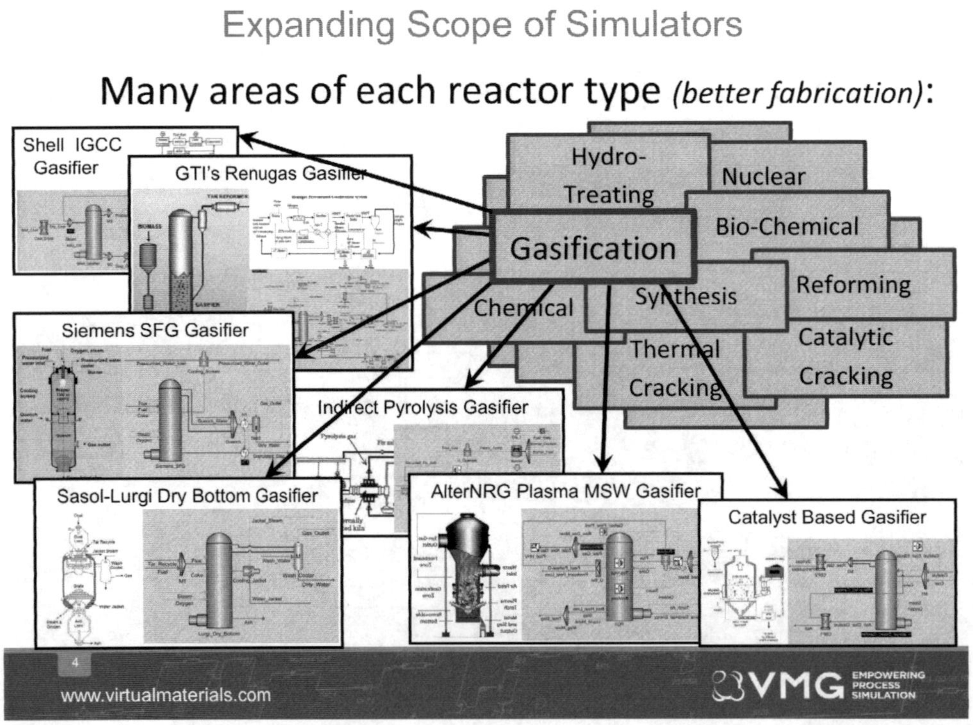
Expanding Scope of Simulators
Many areas of each reactor type (better fabrication):
Shell IGCC
Gasifier
GTI's Renugas Gasifier
Hydro-
Treating
Nuclear
Bio-Chemical
Gasification
Chemical
Synthesis
Reforming
Catalytic
Cracking
Thermal
Cracking
Siemens SFG Gasifier
Indirect Pyrolysis Gasifier
Sasol-Lurgi Dry Bottom Gasifier
AlterNRG Plasma MSW Gasifier
Catalyst Based Gasifier
4
www.virtualmaterials.com
VMG
EMPOWERING
PROCESS
SIMULATION

Start, Good Foundation

Reactor fundamentals are essential:

- Equilibrium or kinetic based reactions
- Mass transfer effects on reactions
- Pressure drop and holdup considerations
- Detailed heat loss in vapor/liquid/solid phase

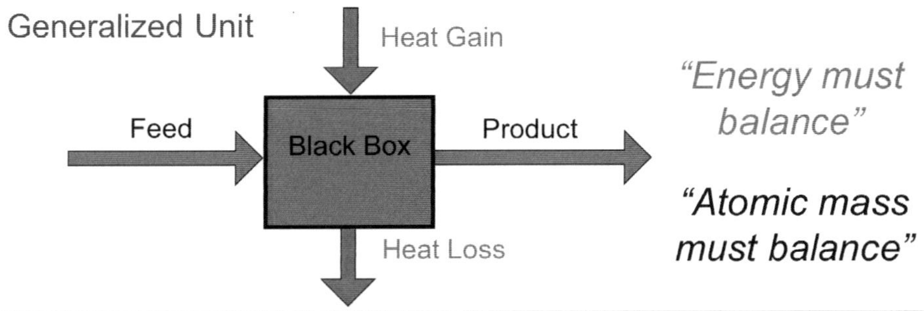

5
www.virtualmaterials.com
VMG EMPOWERING PROCESS SIMULATION

Start, Good Foundation

Reactor fundamentals are essential:

- *Equilibrium or kinetic based reactions*
- Mass transfer effects on reactions
- Pressure drop and holdup considerations
- Detailed heat loss in vapor/liquid/solid phase

Equilibrium Constant

$$\ln\left(K_{eq}\right) = -\frac{\sum_{i}^{N} \upsilon_i g_i^*}{RT}$$

K_{eq} based
or
Gibbs minimization

Reaction Rate

$$k = A\exp\left(\frac{-E}{RT}\right)$$

Basic kinetics
or
Detailed kinetics

6
www.virtualmaterials.com
VMG EMPOWERING PROCESS SIMULATION

Start, Good Foundation

Reactor fundamentals are essential:

- *Equilibrium or kinetic based reactions*
- Mass transfer effects on reactions
- Pressure drop and holdup considerations
- Detailed heat loss in vapor/liquid/solid phase

Specific Kinetics

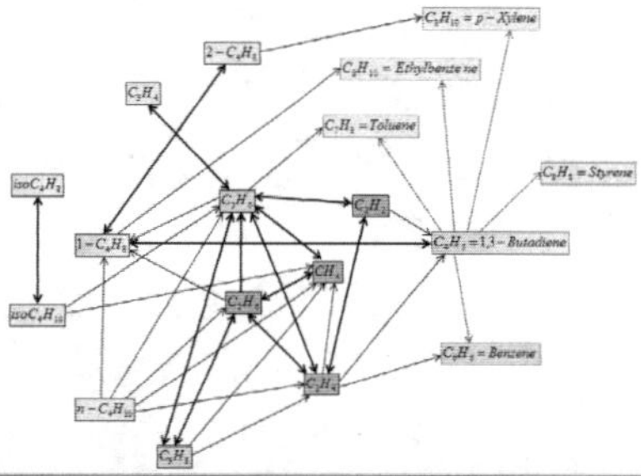

Generalized Kinetics (PIONA)

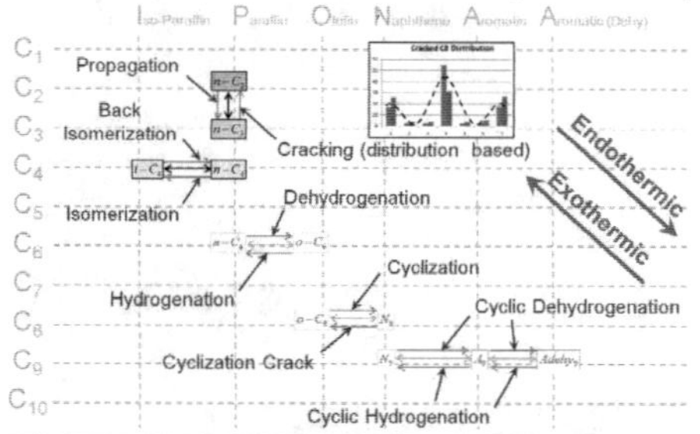

7

www.virtualmaterials.com

Start, Good Foundation

Reactor fundamentals are essential:

- Equilibrium or kinetic based reactions
- *Mass transfer effects on reactions*
- Pressure drop and holdup considerations
- Detailed heat loss in vapor/liquid/solid phase

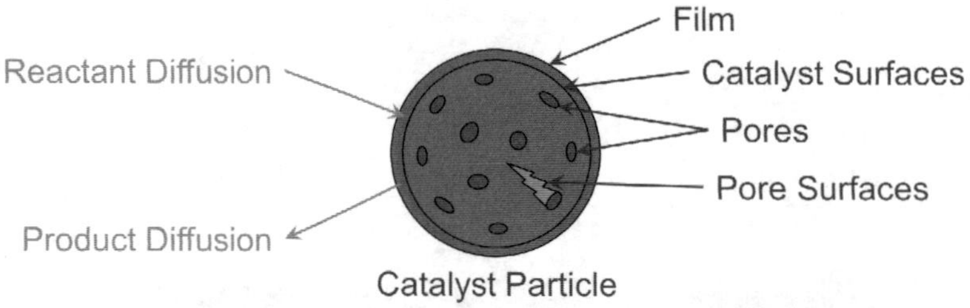

Need an effectiveness factor to account for overall diffusion, or component based diffusion, effect on catalyst activity…

8

www.virtualmaterials.com

Start, Good Foundation

Reactor fundamentals are essential:

- Equilibrium or kinetic based reactions
- Mass transfer effects on reactions
- *Pressure drop and holdup considerations*
- Detailed heat loss in vapor/liquid/solid phase

Catalyst pressure drop…

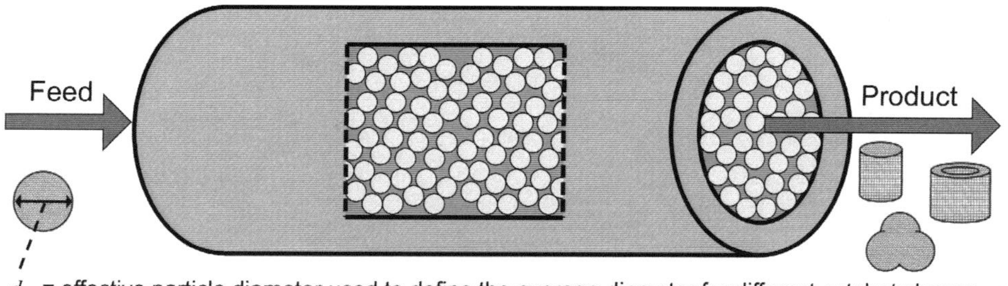

d_{pe} = effective particle diameter used to define the average diameter for different catalyst shapes

Start, Good Foundation

Reactor fundamentals are essential:

- Equilibrium or kinetic based reactions
- Mass transfer effects on reactions
- *Pressure drop and holdup considerations*
- Detailed heat loss in vapor/liquid/solid phase

Catalyst pressure drop… standards are changing...

Correlations are getting better!!!

Understand them! Use them!

~~*Ergun (1952):*~~ ~~$f = \frac{150}{\mathrm{Re}} + 1.75$~~ → *Tallmadge (1970):* $f = \frac{150}{\mathrm{Re}} + \frac{4.2}{\mathrm{Re}^{\frac{1}{6}}}$

Better across higher Reynold's range and catalyst porosity

$$\mathrm{Re} = \frac{velBulk \cdot \rho_{bulk} \cdot d_{pe}}{visc_{bulk}(1 - bedVoid)}$$

$$dGw = \left(\frac{150}{\mathrm{Re}} + \frac{4.2}{\mathrm{Re}^{\frac{1}{6}}}\right) \cdot \frac{(1 - eGw)}{eGw^3 (d_{pe})} \cdot \rho_{vap} \cdot velG^2$$

$$\frac{dP}{L} = \frac{-dGw}{\left(1 - \frac{eL}{bedVoid}\right)^3}$$

Start, Good Foundation

Reactor fundamentals are essential:

- Equilibrium or kinetic based reactions
- Mass transfer effects on reactions
- Pressure drop and holdup considerations
- *Detailed heat loss in vapor/liquid/solid phase*

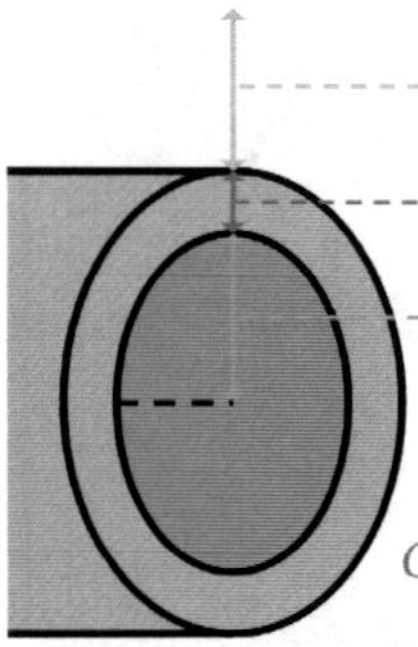

$h_s + h_{rad_s}$

$\frac{x_{tube}}{\lambda_{tube}}$

$h_t + h_{rad_t}$

Heat transfer resistances... Conduction, Convection and Radiant

$$Q = UA(T_o - T)$$

$$U = \frac{1}{\frac{1}{(h_t + h_{rad_t})}\frac{A_O}{A_I} + \frac{x_{tube}}{\lambda_{tube}} + \frac{1}{(h_s + h_{rad_s})}}$$

Start, Good Foundation

Reactor fundamentals are essential:

- Equilibrium or kinetic based reactions
- Mass transfer effects on reactions
- Pressure drop and holdup considerations
- *Detailed heat loss in vapor/liquid/solid phase*

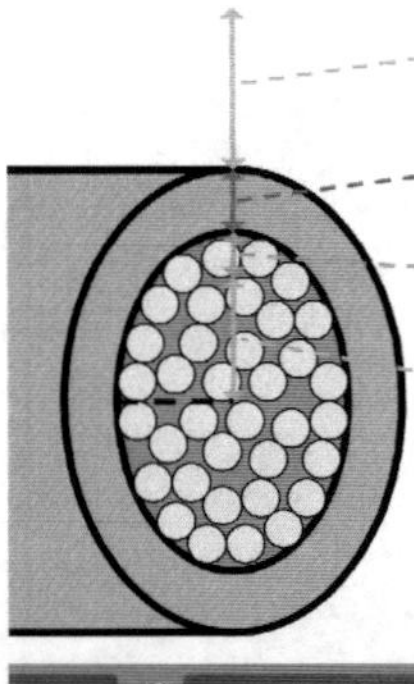

$h_s + h_{rad_s}$

$\frac{x_{tube}}{\lambda_{tube}}$

$h_t + h_{rad_t}$

$\frac{ID_{tube}}{8\lambda_{ER}}$

$$Q = UA(T_o - T)$$

$$U = \frac{1}{\left[\frac{1}{(h_t + h_{rad_t})} + \frac{ID_{tube}}{8\lambda_{ER}}\right]\frac{A_O}{A_I} + \frac{x_{tube}}{\lambda_{tube}} + \frac{1}{(h_s + h_{rad_s})}}$$

Catalyst option terms

$$h_t = \frac{\lambda_G}{d_{eff}}\left(\frac{10.21 \cdot \lambda_{Cat}}{ID_{tube}^{4/3}}\frac{d_{eff}}{\lambda_G} + 0.033\,\mathrm{Re}_G\,\mathrm{Pr}_G\right)$$

Numerical Solvers

Different driving mechanisms are all simultaneous

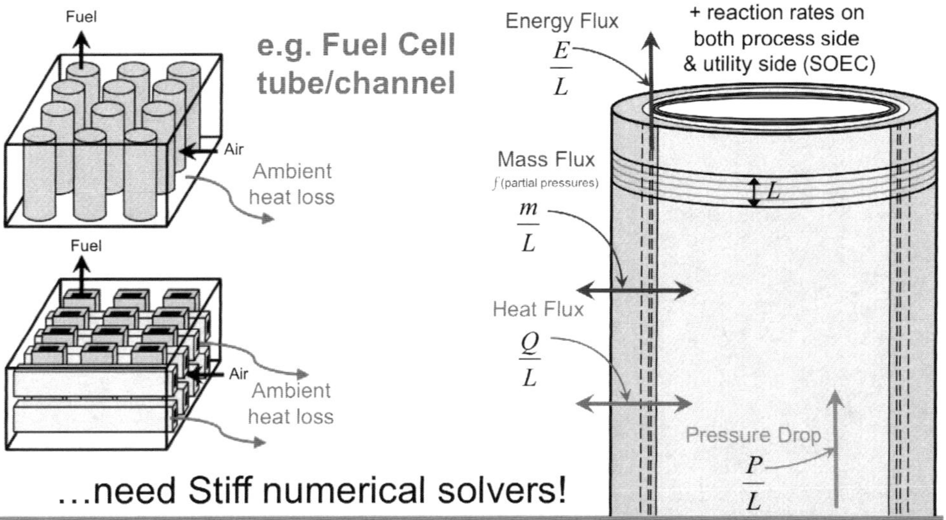

...need Stiff numerical solvers!

13
www.virtualmaterials.com
VMG EMPOWERING PROCESS SIMULATION

PIONA Structure Characterization

Models consist of carbon number & molecular structure component grid... where properties are trended to each class

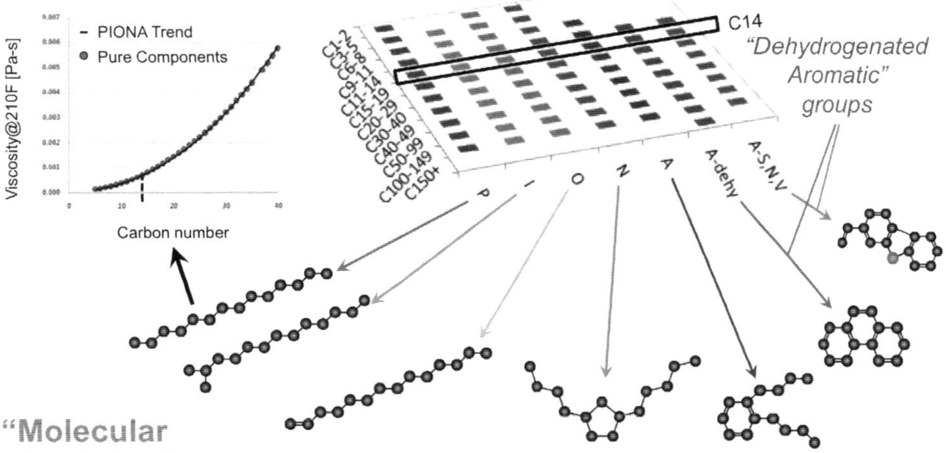

"Molecular Modeling" *(R. J. Quann "Mobile Research", 1980's)*

14
www.virtualmaterials.com
VMG EMPOWERING PROCESS SIMULATION

PIONA Defined Feed Stocks

Regressed component slate to minimize error on provided laboratory information *(distillation curves, bulk properties, etc.)*

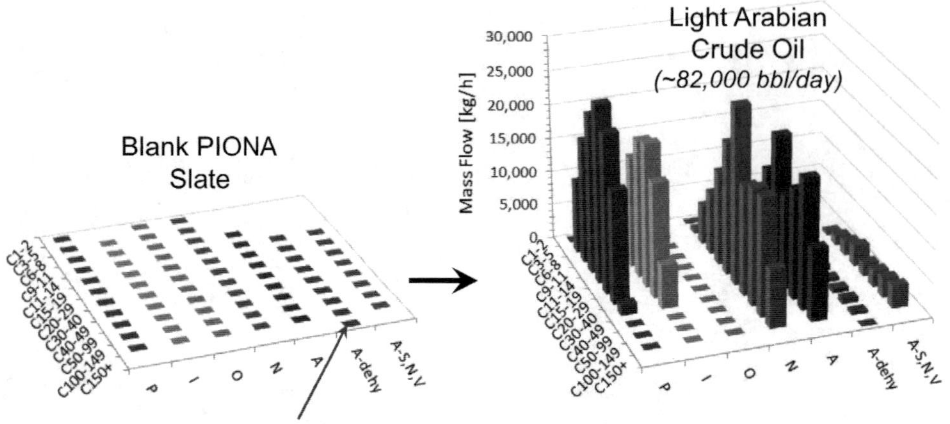

High molecular weight dehydrogenated aromatic components to represent asphaltenes

Molecular Structure Modeling Separation

Consistent PIONA component slate matrix with property predictions available anywhere

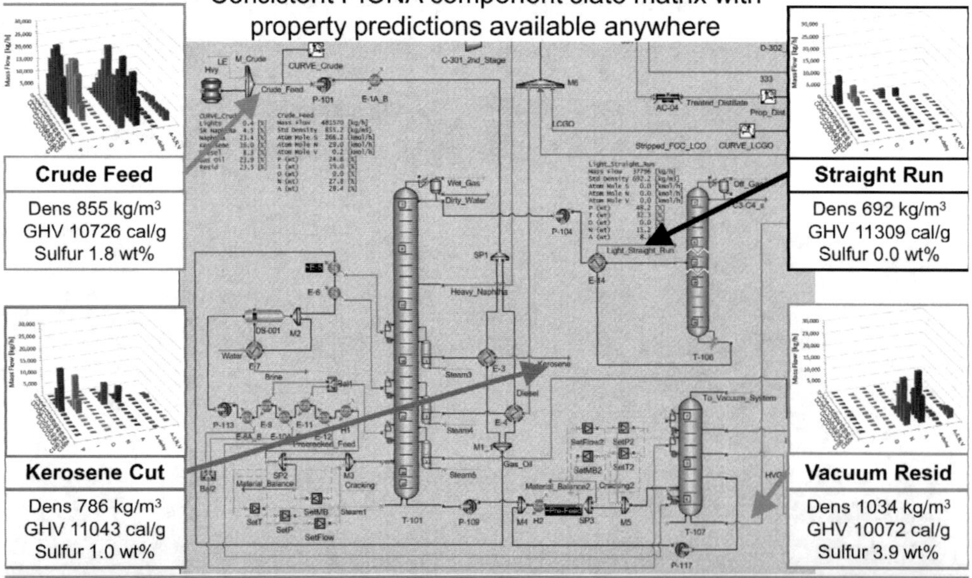

Mapping Catalyst Affects on Reaction Kinetics

PIONA models track shifts in carbon number & molecular structure component grid throughout the reactors (e.g. CCR unit)

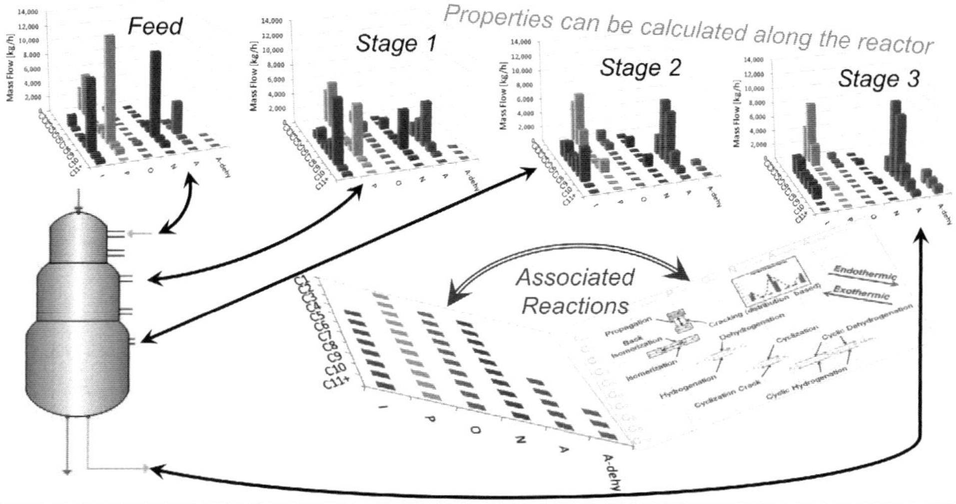

17

www.virtualmaterials.com

VMG EMPOWERING PROCESS SIMULATION

Mapping Catalyst Affects on Reaction Kinetics

Any hydrocarbon reaction pathways can be described (PIONA)

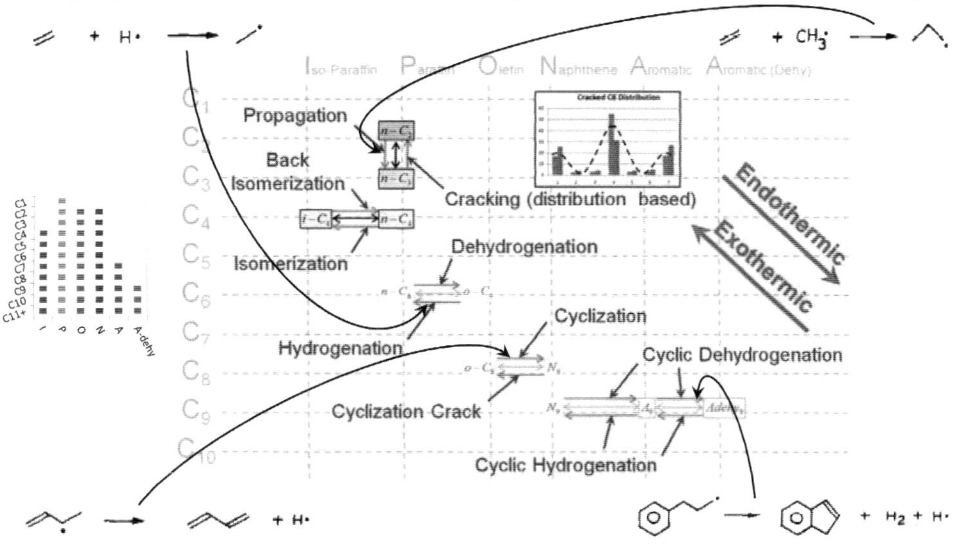

18

www.virtualmaterials.com

VMG EMPOWERING PROCESS SIMULATION

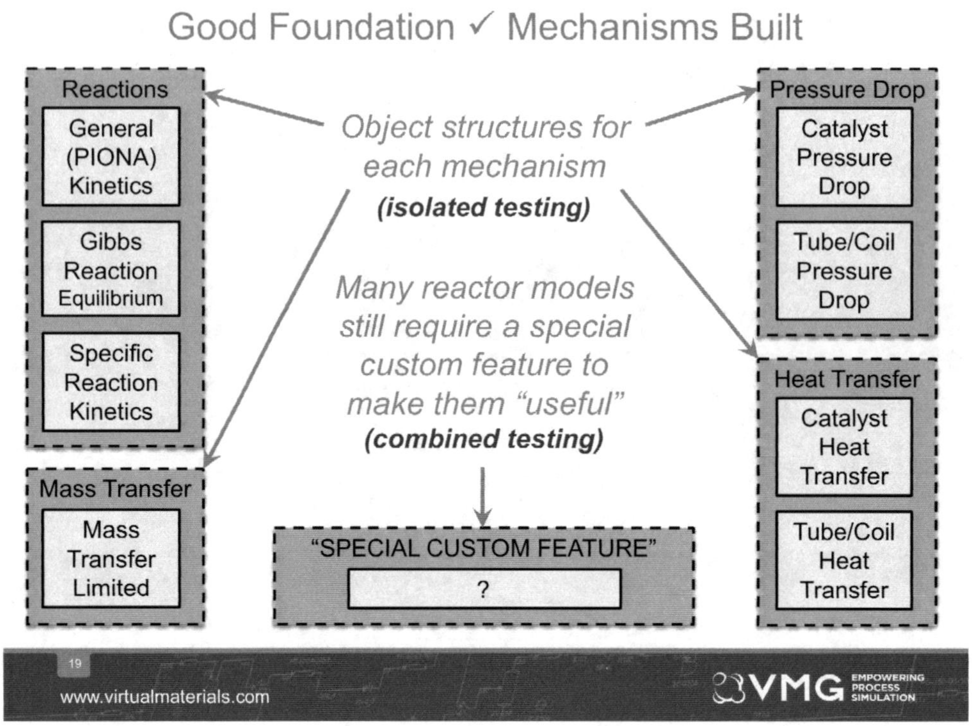

Good Foundation ✓ Mechanisms Built
Reactions
General (PIONA) Kinetics
Gibbs Reaction Equilibrium
Specific Reaction Kinetics
Mass Transfer
Mass Transfer Limited
Object structures for each mechanism
(isolated testing)
Many reactor models still require a special custom feature to make them "useful"
(combined testing)
"SPECIAL CUSTOM FEATURE"
?
Pressure Drop
Catalyst Pressure Drop
Tube/Coil Pressure Drop
Heat Transfer
Catalyst Heat Transfer
Tube/Coil Heat Transfer
19
www.virtualmaterials.com
VMG EMPOWERING PROCESS SIMULATION

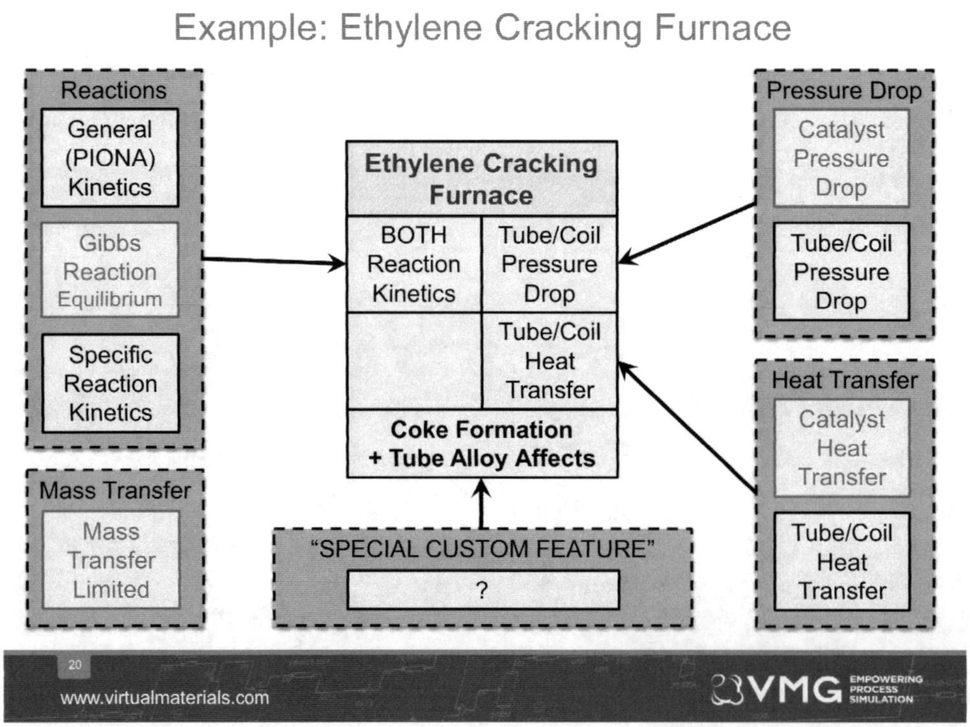

Example: Ethylene Cracking Furnace
Reactions
General (PIONA) Kinetics
Gibbs Reaction Equilibrium
Specific Reaction Kinetics
Mass Transfer
Mass Transfer Limited
Ethylene Cracking Furnace
BOTH Reaction Kinetics
Tube/Coil Pressure Drop
Tube/Coil Heat Transfer
Coke Formation + Tube Alloy Affects
"SPECIAL CUSTOM FEATURE"
?
Pressure Drop
Catalyst Pressure Drop
Tube/Coil Pressure Drop
Heat Transfer
Catalyst Heat Transfer
Tube/Coil Heat Transfer
20
www.virtualmaterials.com
VMG EMPOWERING PROCESS SIMULATION

Example: Ethylene Cracking Furnace

Process simulation with no "Black Box" reactor models!!!

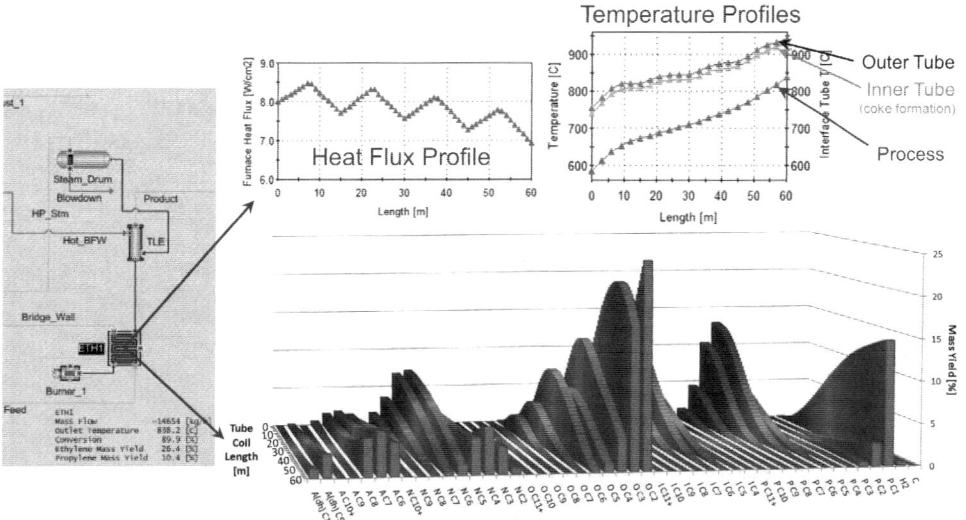

Property profiles for reactors known everywhere rigorously

21
www.virtualmaterials.com
VMG EMPOWERING PROCESS SIMULATION

Example: Ethylene Cracking Furnace

"Special Custom Feature": Rigorous mechanisms still there, but going into rigorous coke <u>affects</u> can be very detailed

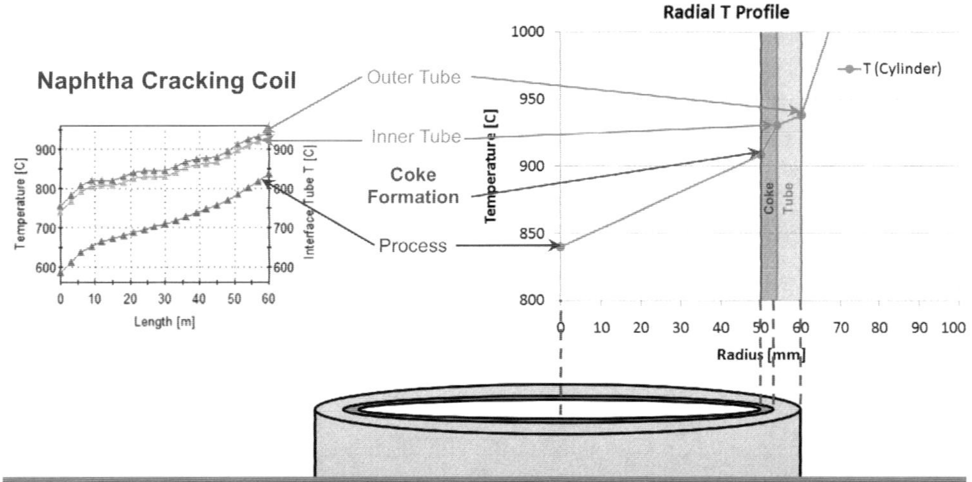

22
www.virtualmaterials.com
VMG EMPOWERING PROCESS SIMULATION

Example: Ethylene Cracking Furnace

e.g. Tube alloys catalyze coke growth at earlier service times when coke thickness is minimal

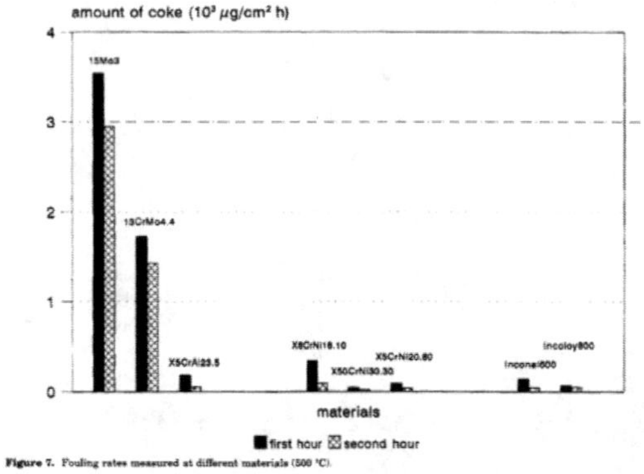

REF: G. Bach, et al. (1995)

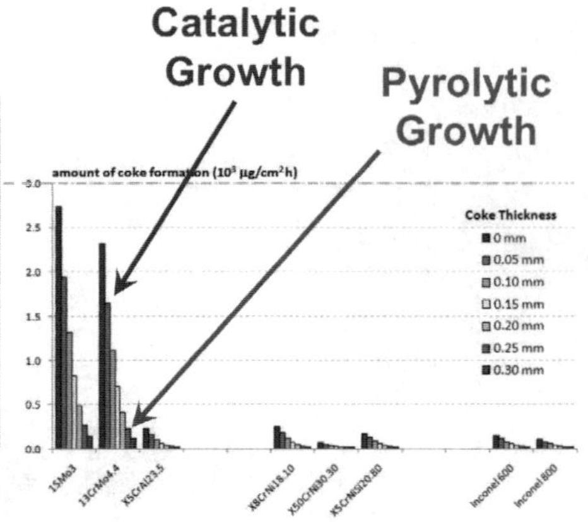

REF: VMGSim v8.0

Example: Urea Stripper Unit

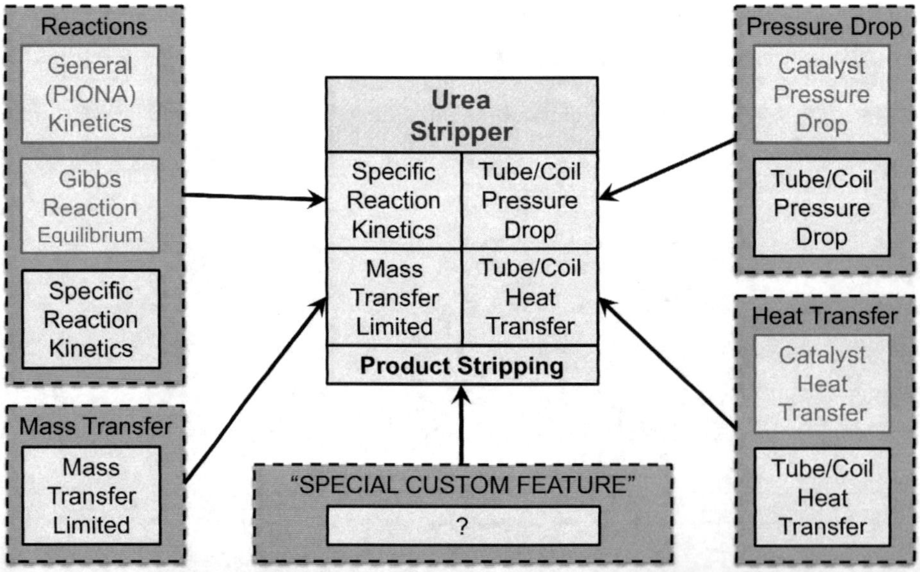

Example: Urea Stripper Unit

Counter flow situation contained within the process side phase flow *(needs simultaneous solution)*

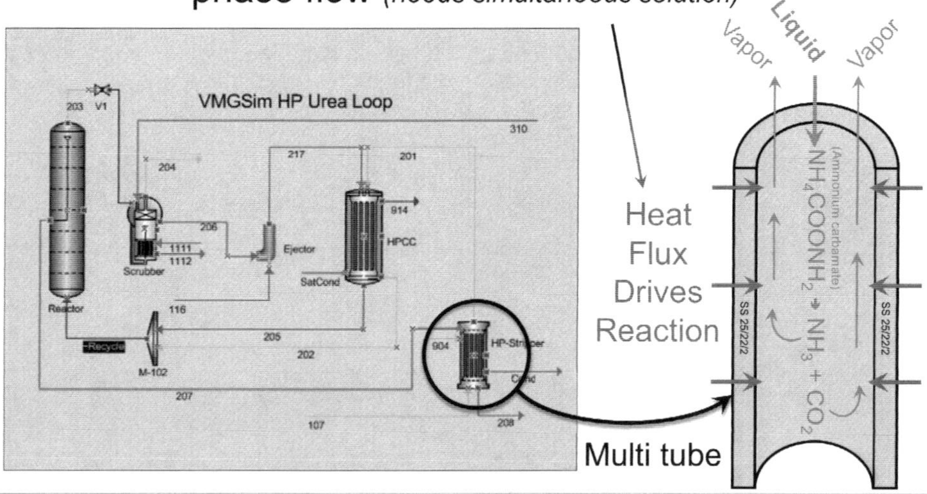

Example: Urea Stripper Unit

Reaction Extent Studies (fouling & tube length):

Unconverted ammonium carbamate flow

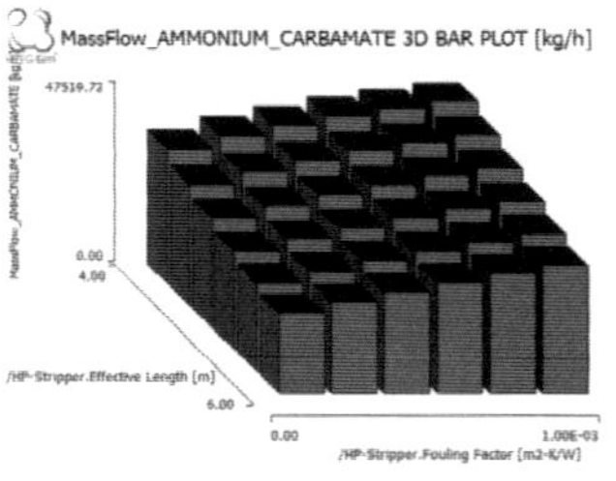

Utility duty exchanged

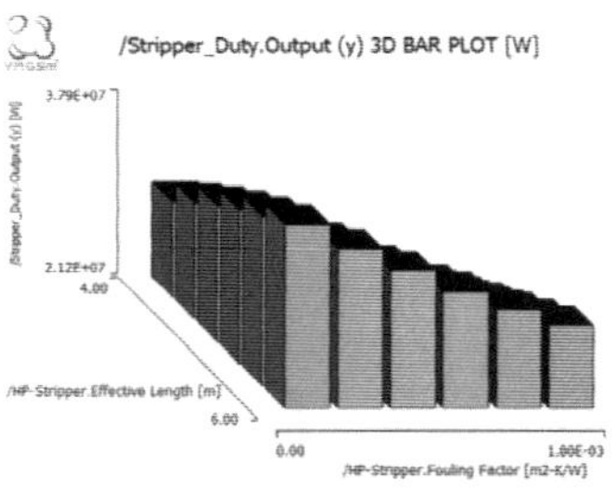

Tube Length

Fouling

"Heat flux drives reaction" - Yes

- *Increased fouling increases heat flux resistance*
- *Reduced tube length reduces surface area*

Example: Fischer-Tropsch Reactor

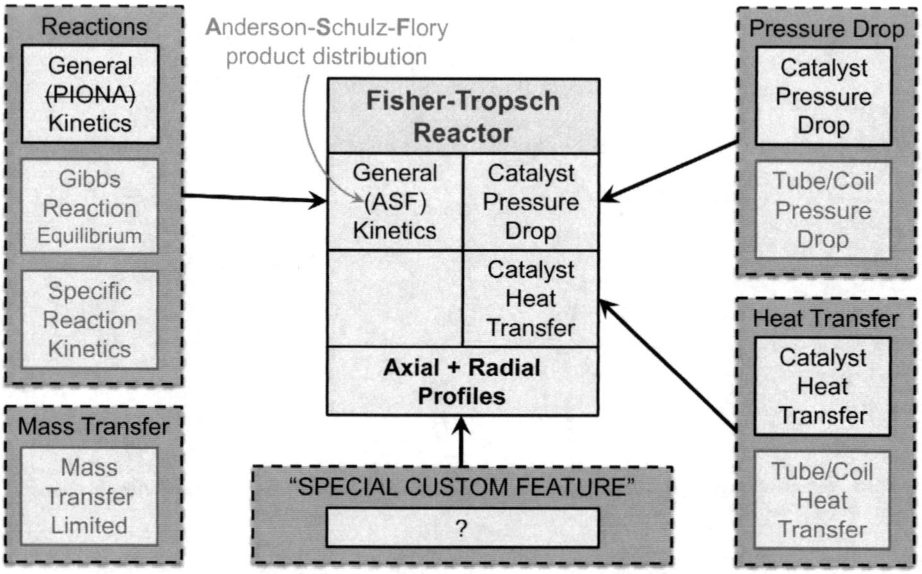

27
www.virtualmaterials.com
VMG EMPOWERING PROCESS SIMULATION

Example: Fischer-Tropsch Reactor

Built-in general kinetic reactions:

Confirm good "**Alpha**" distribution slate…

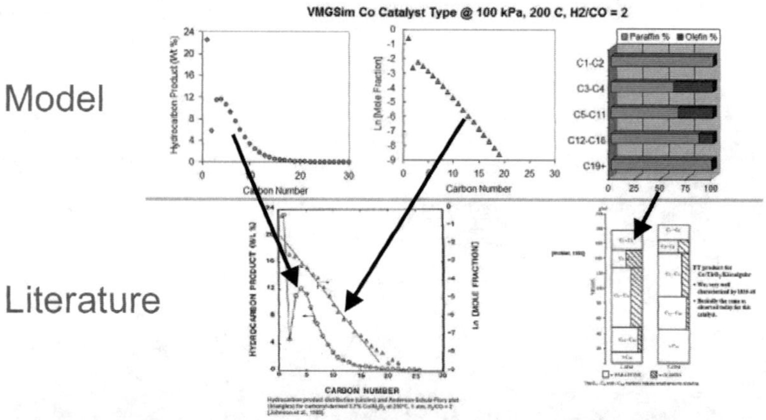

28
www.virtualmaterials.com
VMG EMPOWERING PROCESS SIMULATION

Example: Fischer-Tropsch Reactor

Shell uses shell & tube style reactors with thousands of tubes/reactor !

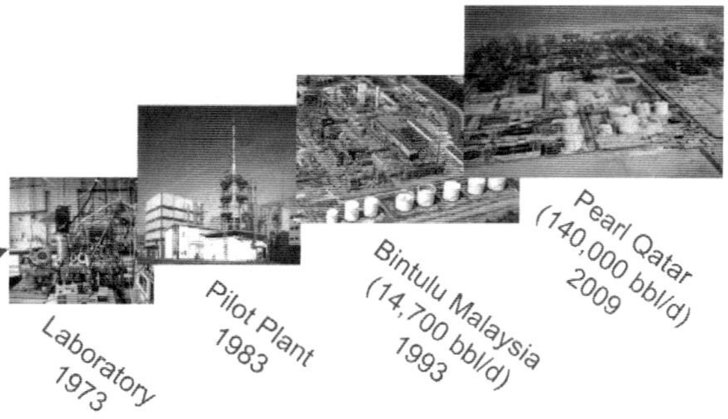

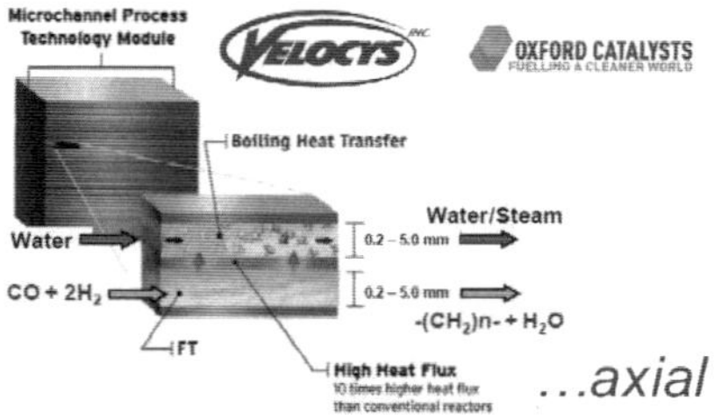

Velocys/Oxford catalyst activity so high shell & tube style reactors do not make sense due to extreme exothermic heats

...axial + radial T model profiles essential

Example: Fischer-Tropsch Reactor

Study reduced flow rate effects in tubes:

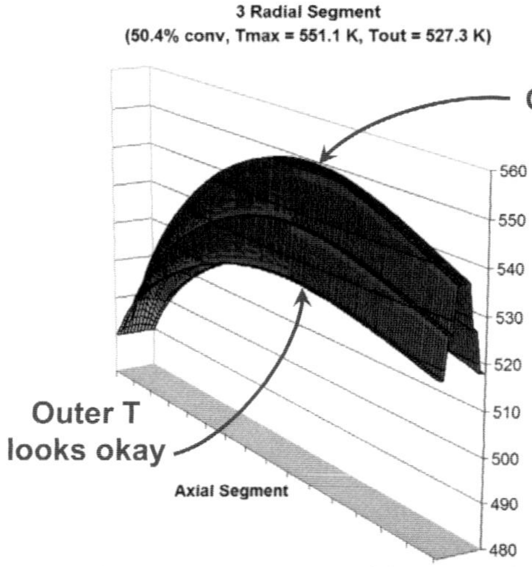

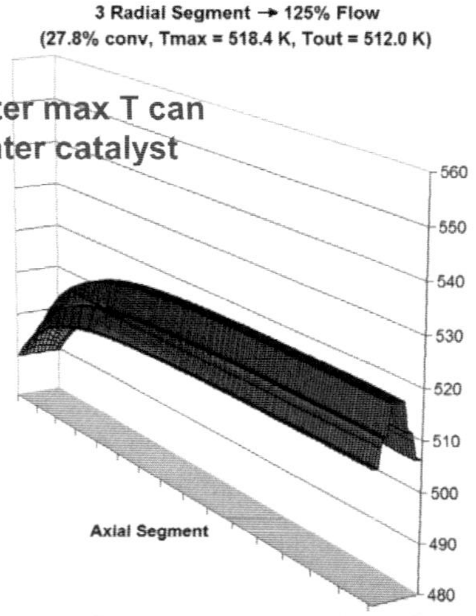

Example: Fischer-Tropsch Reactor

Study increased tube diameter effects in tubes:

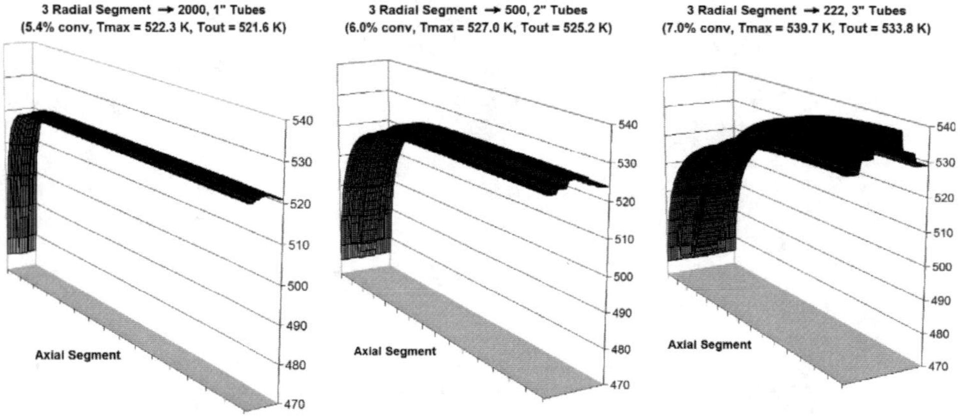

Catalyst axial heat loss decides maximum tube diameter (i.e. number of tubes)

31
www.virtualmaterials.com
VMG EMPOWERING PROCESS SIMULATION

Constant Basis = Full Refinery Model in 1 Flowsheet

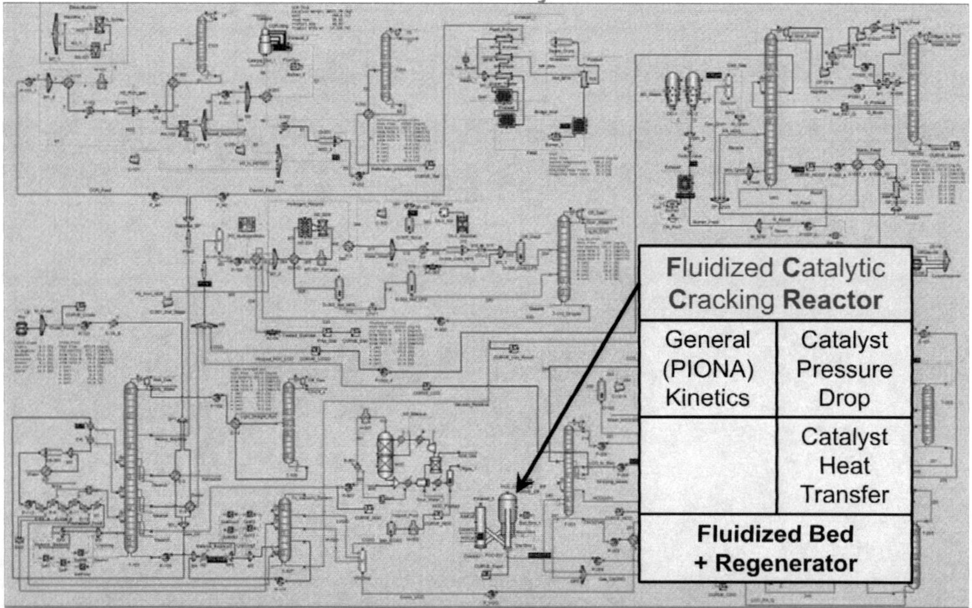

32
www.virtualmaterials.com
VMG EMPOWERING PROCESS SIMULATION

Constant Basis = Full Refinery Model in 1 Flowsheet

Rigorous FCC unit with heat of reaction calculated from component shifts

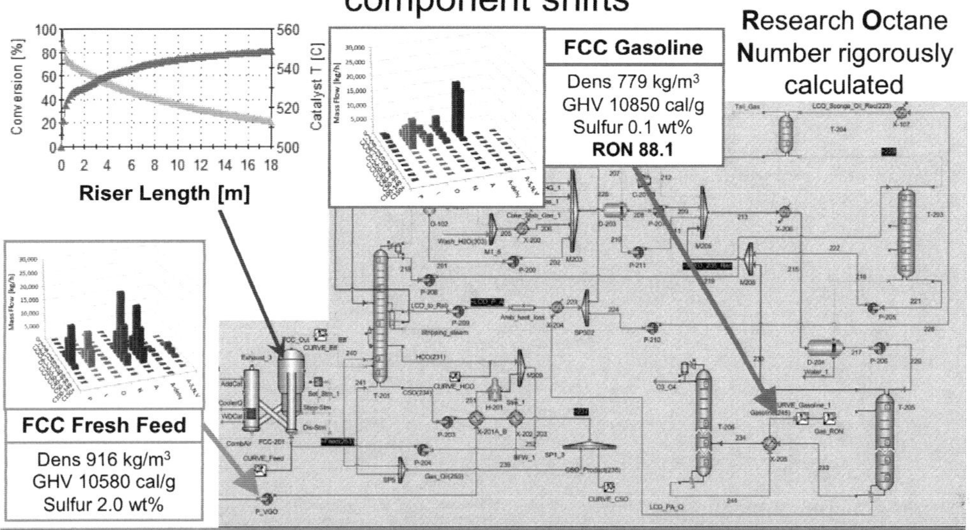

Optimizing Simulation of Refinery and Petrochemical Reactors with Molecular Structure Fluid Lumping and Object Oriented Unit Design

Conclusions:

- Generic object classed mechanisms *(heat transfer, pressure drop, etc.)* has enabled rapid release of advanced reactor models in VMGSim
- Structure based lumped Characterization is the lynchpin of the cracking models and has enabled refinery-wide simulation and more…
- Simulation with a single component slate is enabled from reservoir to gas processing to refining to petrochemicals. Major breakthrough.
- Systematic model has provided path to rigorous reactor models for Urea, Fuel Cell and Propane dehydrogenation

Thank You

Any questions?

Any comments?

Any suggestions?

35

www.virtualmaterials.com

VMG EMPOWERING PROCESS SIMULATION

SS 1

Example: Fuel Cell Unit

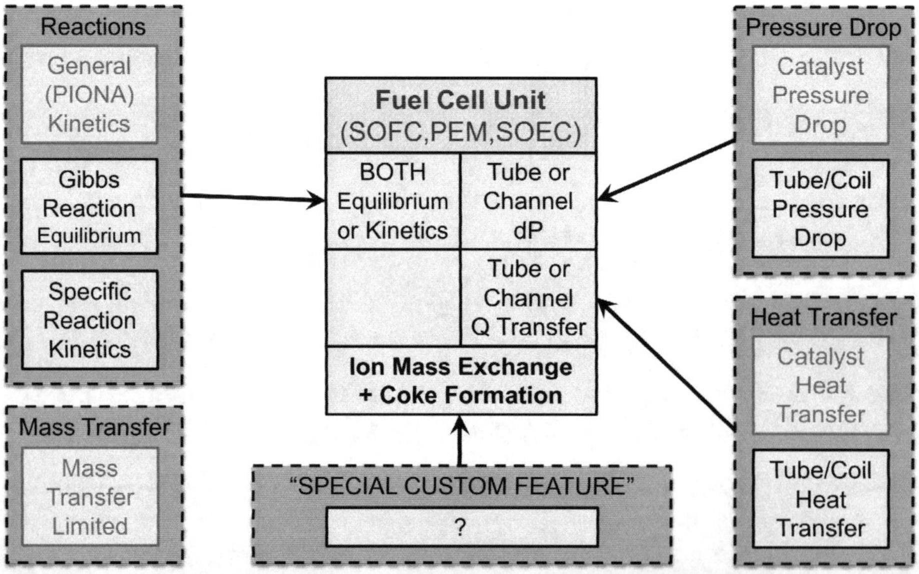

SS 36

www.virtualmaterials.com

VMG EMPOWERING PROCESS SIMULATION

SS 2

Example: Fuel Cell Unit

a = Anode
e = Electrolyte
c = Cathode
s = Support

Rigorous underlying mechanisms still there

$$U_{SOFC} = \frac{1}{\frac{A_O}{h_t A_I} + \left[\frac{x_a}{k_a} + \frac{x_e}{k_e} + \frac{x_c}{k_c} + \frac{x_s}{k_s}\right] + f_t + \frac{1}{h_s}}$$

Convective Heat

$$\frac{Q_{conv}}{L} = U \cdot \pi \cdot OD_{tube} \cdot \Delta T$$

Radiant Heat Flux

$$\frac{Q_{rad}}{L} = \frac{\sigma \cdot F \cdot \pi \cdot OD_{tube} \cdot \left(T^4_{Furnace} - T^4_{Process}\right)}{\frac{1}{\varepsilon_1} + \frac{1}{\varepsilon_2} - 1}$$

$$\frac{Q}{L} = \frac{Q_{conv}}{L} + \frac{Q_{rad}}{L}$$

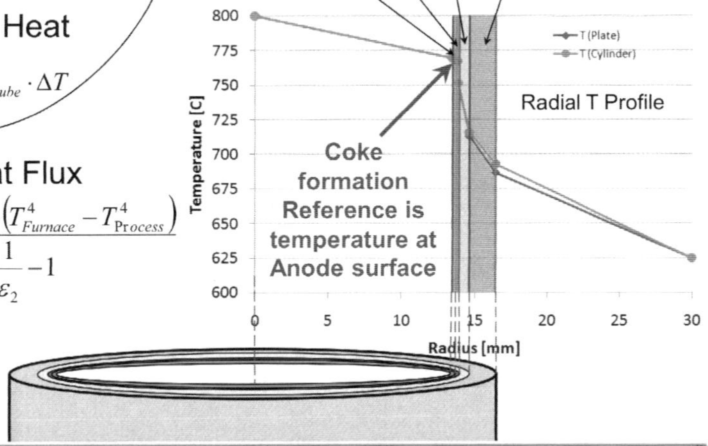

SS 37
www.virtualmaterials.com
VMG EMPOWERING PROCESS SIMULATION

SS 3

Example: Fuel Cell Unit

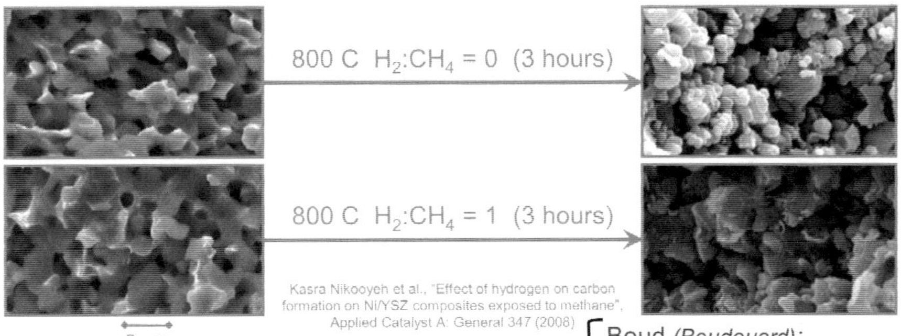

Kasra Nikooyeh et al., "Effect of hydrogen on carbon formation on Ni/YSZ composites exposed to methane", Applied Catalyst A: General 347 (2008)

Carbon Types:

$C_nH_m \rightarrow (CH_2)_n \rightarrow Gum$ → • Gum

• Whisker

$C_nH_m \rightarrow Olefins \rightarrow Coke$ → • Coke

Jens. R. Rostrup-Nielsen *et al.*, "Hydrogen and Syngas by Steam Reforming", Acedemic Press (2002)

Boud *(Boudouard)*:
$$2CO \leftrightarrow C + CO_2$$

Sooting (Methane Dehydrogenation):
$$CH_4 \rightarrow 2H_2 + C$$

Steam (Steam Reforming):
$$C + H_2O \rightarrow CO + H_2$$

Oxidize (Reverse Electro-Chemistry):
$$C + O^{2-} \rightarrow CO + 2e^-$$

SS 38
www.virtualmaterials.com
VMG EMPOWERING PROCESS SIMULATION

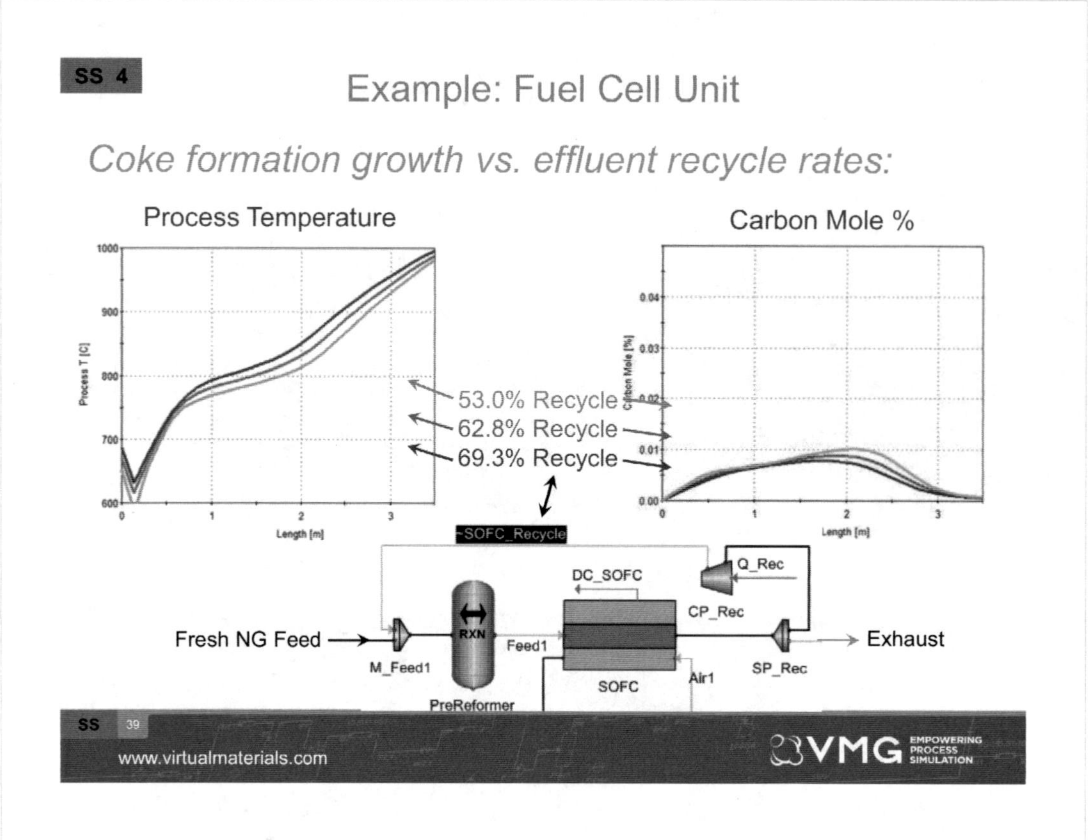
SS 4
Example: Fuel Cell Unit
Coke formation growth vs. effluent recycle rates:
Process Temperature
Carbon Mole %
Process T [C]
Length [m]
Carbon Mole [%]
53.0% Recycle
62.8% Recycle
69.3% Recycle
SOFC_Recycle
DC_SOFC
Q_Rec
CP_Rec
Fresh NG Feed
RXN
Feed1
Exhaust
M_Feed1
SOFC
Air1
SP_Rec
PreReformer
SS 39
www.virtualmaterials.com
VMG
EMPOWERING PROCESS SIMULATION

Optimization Approach for Planning Application of Ethylene and its Downstream Businesses

Bhieng Tjoa
Optience

OPTIENCE

Strategic Solutions through Optimization Science

Background on Ethylene Business

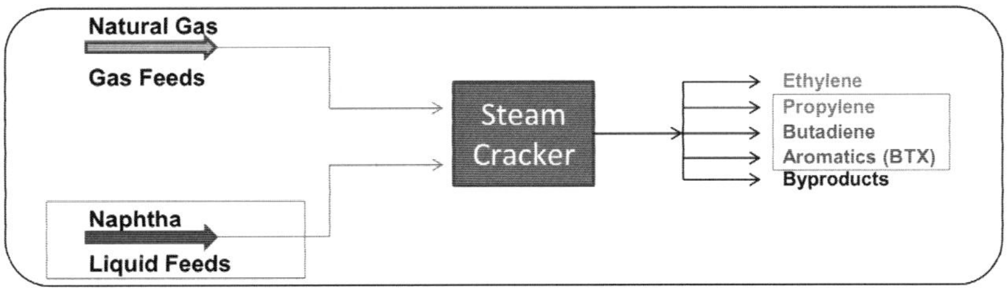

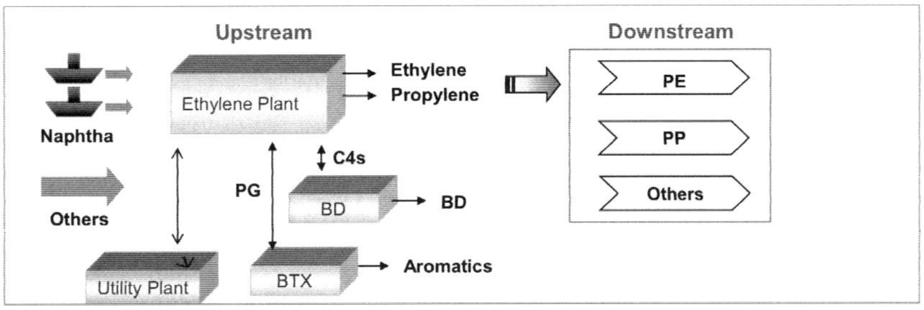

OPTIENCE

Strategic Solutions through Optimization Science

Planning Challenges: Monomers + Polymers

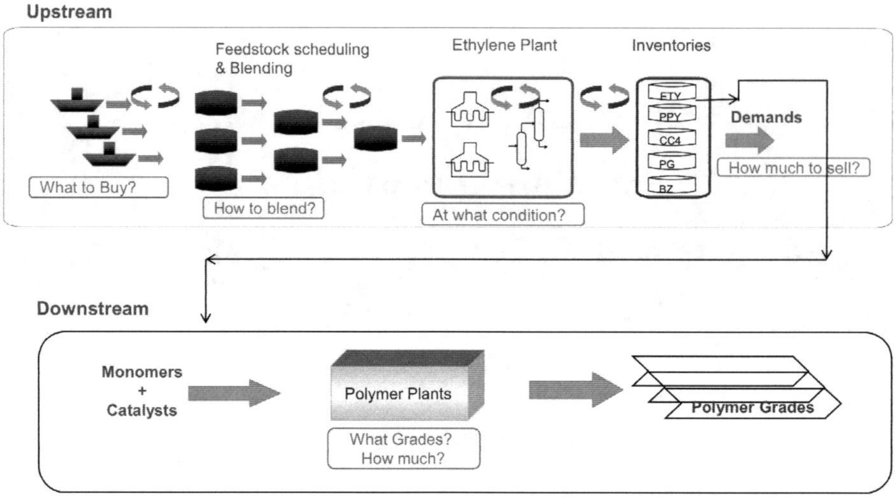

OPTIENCE

Strategic Solutions through Optimization Science

Planning Decision Challenges

Upstream

- Feedstock selection
  - Naphtha → properties?
    - PIONA
  - LPG → properties?
    - PPA, n/i C4?
- Naphtha blending?
- Feedstock to which furnaces?
  - COT, SHC, COP?
- Which streams to recycle?
- Other constraints?
  - Compressors, towers?
- Plant load?
  - Demands, prices

Downstream

- Product allocation
  - Which grades?
    - Demands & prices
    - Inventories
    - Production efficiency?
      - Rate, Sequences, off grades, downtime, etc.
  - Impact of feedstock price?
    - Optimal production?

OPTIENCE

Strategic Solutions through Optimization Science

Planning Decisions

How do we make planning decisions?

→ Optimal profit & Manage risks?

Excel based simulation/ Optimization based planning model?

OPTIENCE

Strategic Solutions through Optimization Science

Challenges in Building Planning Model

Upstream

- Cracker modeling
  - How to track components?
    - COT, SHC, COP?
  - Naphtha
    - Properties → Components?
- Stream compositions
  - How to track components?
    - Tanks, distillations, etc.
- How to model nonlinear processes?
  - Accurate representation → NLP
  - Robust solution method

Downstream

- Reactor modeling → product allocations
  - Production rate
    - Min, Max
  - Production sequences
    - Downtime, off-grades
  - Product recipes
- How to model discrete decisions?
  - Accurate representation → MILP
  - Robust solution method

OPTIENCE

Strategic Solutions through Optimization Science

Example: Integrated Planning Model

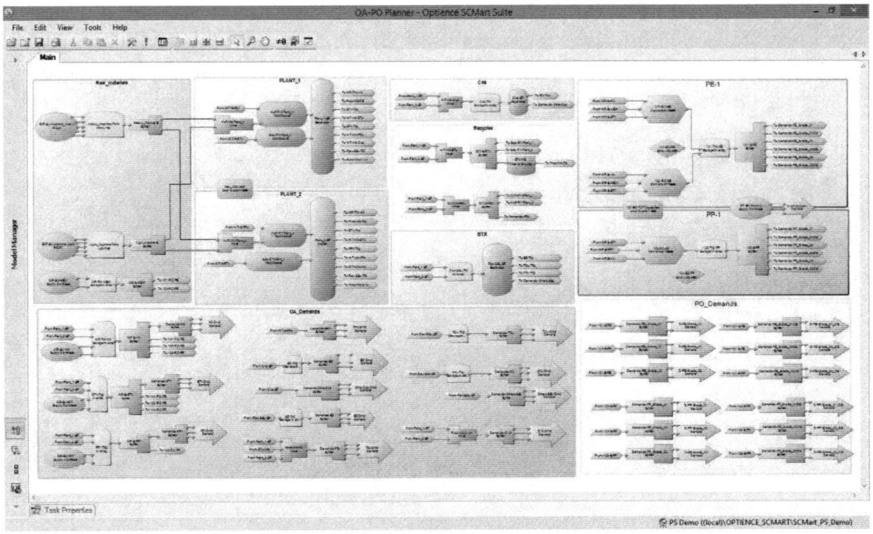

OPTIENCE

Strategic Solutions through Optimization Science

PO Planning – Modeling Product Off-Grades

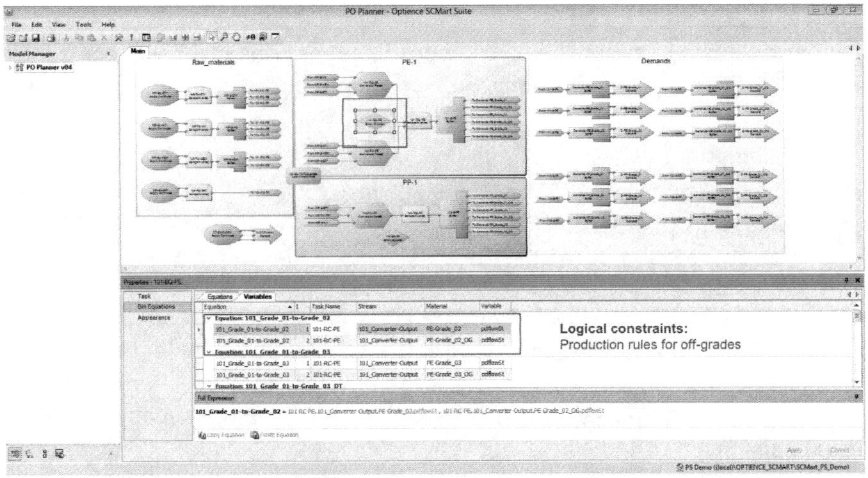

OPTIENCE

Strategic Solutions through Optimization Science

PO Planning – Multi-Product Capacity Constraints

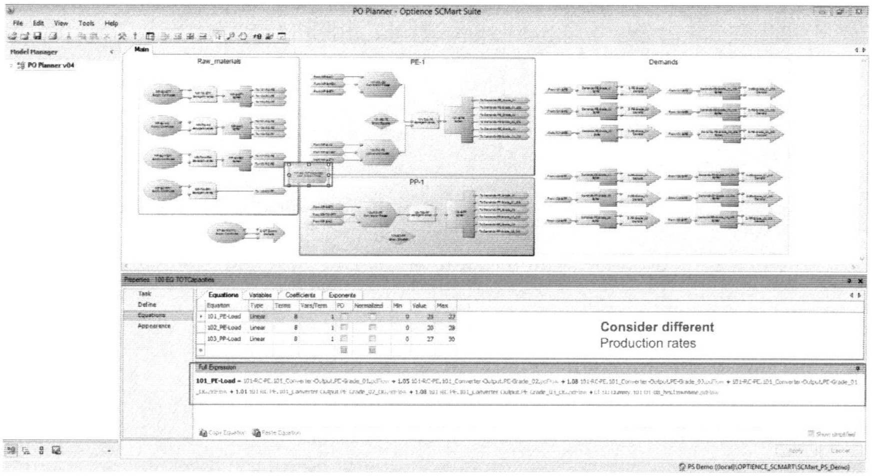

Strategic Solutions through Optimization Science

Planning Application

What other requirements to make it practical?

Strategic Solutions through Optimization Science

Other Useful Features

- Model transparency and easy maintainability
  - Unit operation approach → automatic mathematical model generation
    - » Mass balance, operation constraints
  - Model inconsistency check
- Robust optimization solution method
  - Support single and multi-periods planning with sub-cases
    - » Large scale optimization
- User friendly solution analysis & reporting
  - User friendly platform
    - » Support Charts & powerful grid to display results

OPTIENCE

Strategic Solutions through Optimization Science

Flexible User Interface

- Easy to customize views
  - Flexible grids and graph
    - » Focus on the user needs
      - Incorporate the workflow, Import/export data, Generate reports

Customizable View

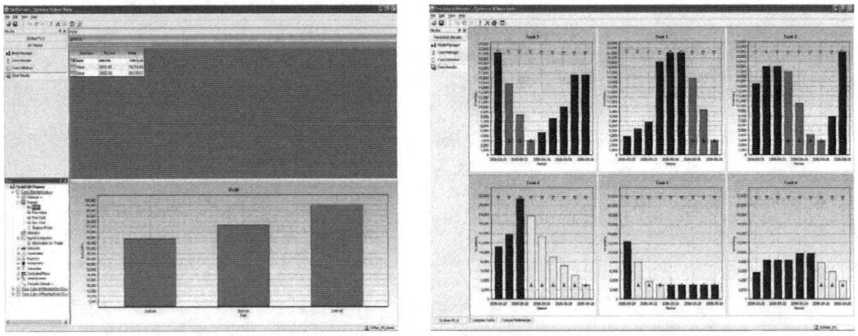

OPTIENCE

Strategic Solutions through Optimization Science

Optimization Results – Capacities

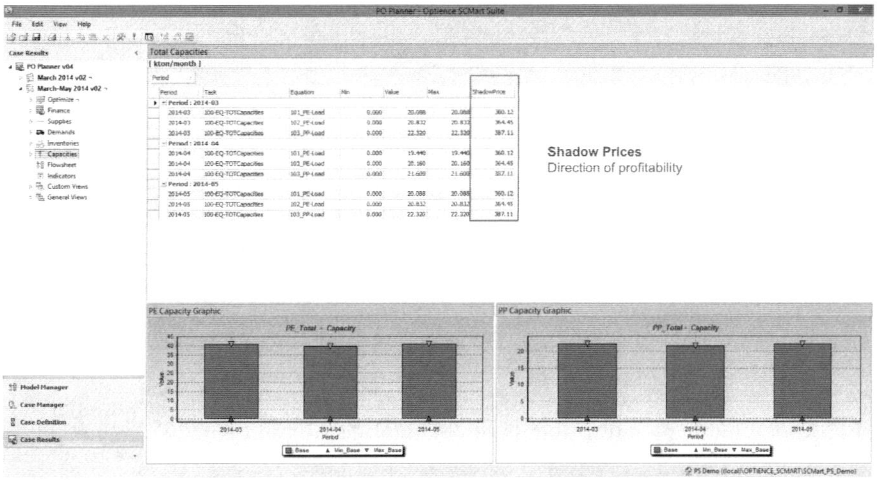

OPTIENCE

Strategic Solutions through Optimization Science

Optimization Results – Demands

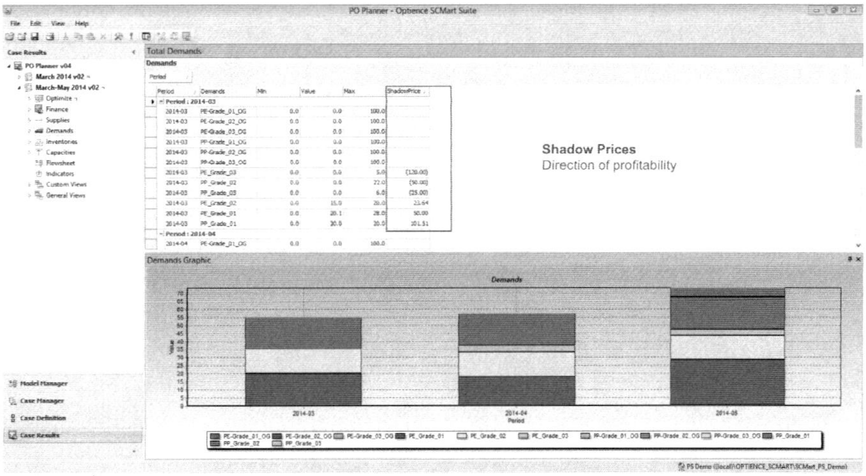

OPTIENCE

Strategic Solutions through Optimization Science

Industrial Practice

What Approach is commonly used in industry now?

What is the most challenging issue to adopt more advanced tool?

OPTIENCE

Strategic Solutions through Optimization Science

Thank You!

OPTIENCE

Strategic Solutions through Optimization Science

Tridiagonal Solutions
compute | innovate | develop

Solving Granular Flow Problems Using Discrete Element Methods

Mothivel Mummudi Ph.D.
Principal Engineer
Tridiagonal Solutions Inc.

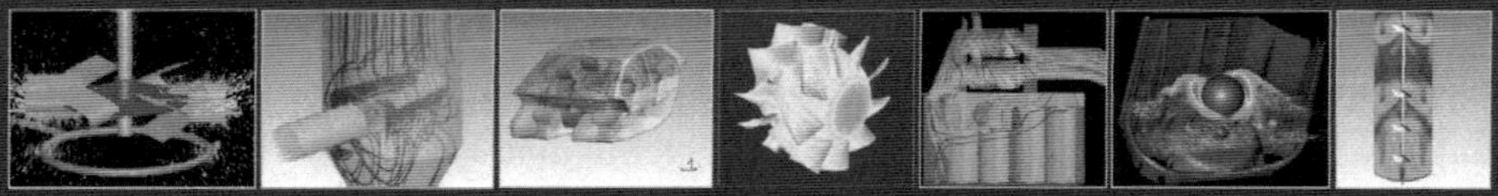

www.tridiagonal.com
info@tridiagonal.com

Tri
Sol
compute | inno

How do we know that creations of worlds are not determined by falling grains of sand ?

Victor Hugo
Les Miserables

- Why study granular flows ?
- How to analyze granular flows ?
- What is the Discrete Element Method ?
- How can the DEM be used effectively to understand granular flows?
  - Applications
  - Extensions

Industrial Granular Flows

Granular materials and flows abound in industry

- Granular flows are used extensively in industrial processes
  - fundamental understanding of processes is important for better process design
  - computational modeling offers a viable means to analyze such flows

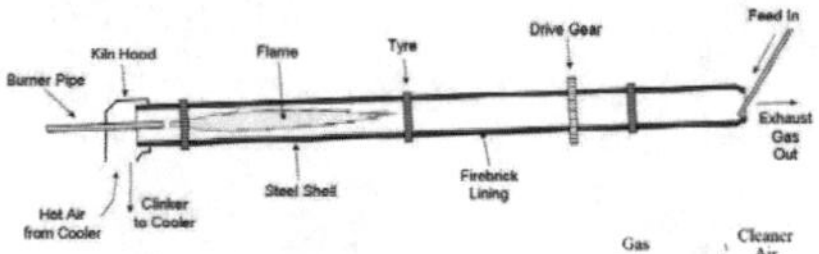

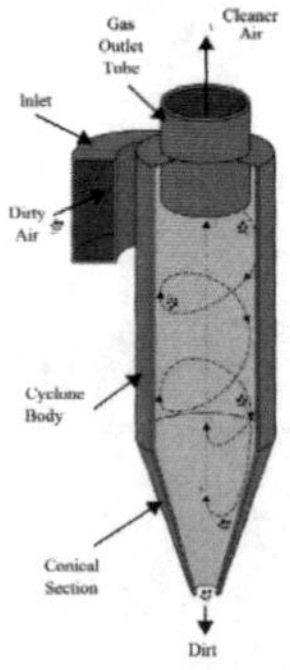

- Computational modeling
  - CFD and FEA are continuum-based techniques
  - Discrete Element Modeling is a fully discrete technique

- Advantages of DEM
  - high level of detail on dynamic behaviour of particles
  - many particle-level phenomena of engineering interest can be described/studied (agglomeration, segregation etc.)
  - interaction between fluid flow and particle/granular flow can be analyzed

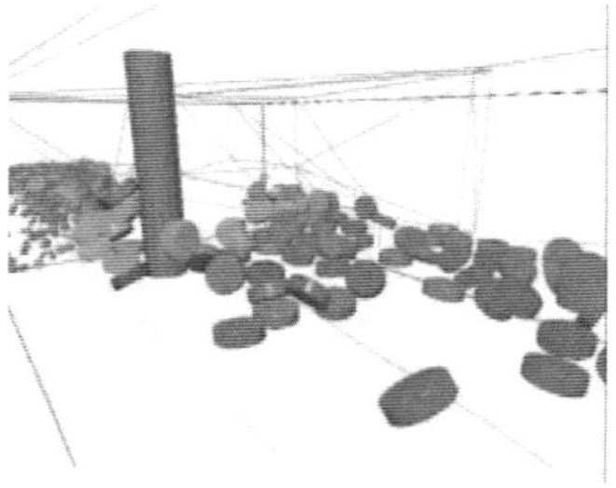

Discrete Element Method
Fundamental principles

Discrete element models describe the motion of individual particles in a collection - as well as their interactions with their neighbours and the environment.

Particle motion is controlled by Newton's second law

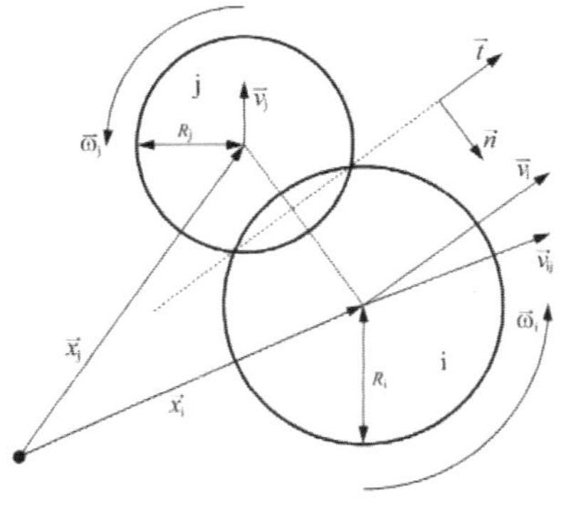

Translational

$$m_i \frac{\mathrm{d}\mathbf{v}_i}{\mathrm{d}t} = \mathbf{F}_i^C + m_i \mathbf{g} + \mathbf{F}_i^{\mathrm{fp}}$$

Rotational

$$\mathbf{I}_i \cdot \frac{\mathrm{d}\omega_i}{\mathrm{d}t} - (\mathbf{I}_i \cdot \omega_i) \times \omega_i = \mathbf{T}_i,$$

Contact point ("neighbors") tracking and contact force models are an important consideration in DEM models

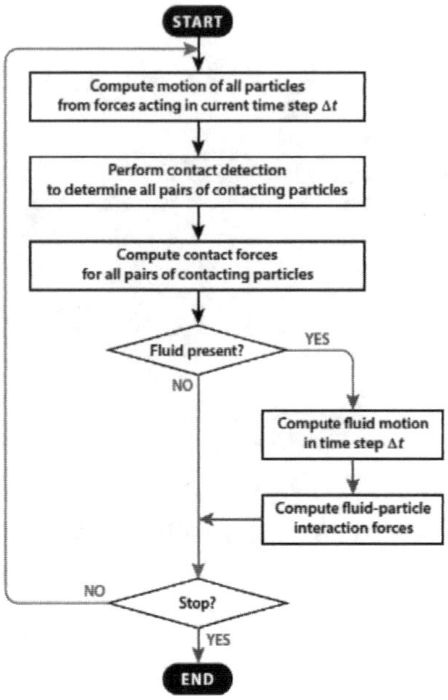

Contact detection can be and is computationally intensive and expensive

Particles can be modeled as "hard" or "soft" spheres

The "Hertz-Mindlin" model is typically used for contact force estimation - other models are available

Typical DEM models involve a "characterization" step to experimentally estimate flow properties such as rolling resistance, cohesiveness, coeff. of restitution etc.

Discrete Element Method

The technique is very flexible and extensible

Tridiagonal Solutions
compute | innovate | develop

Using DEM, a number of different aspects of granular flow can be analyzed

Increasingly complex multiphysics problems can be handled

Large Equipment

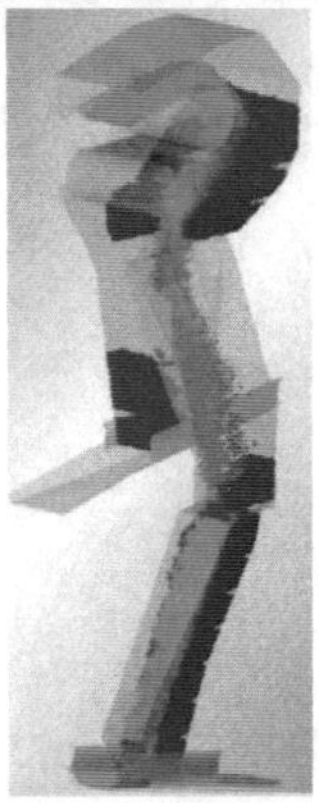

Non-spherical Particles

Complex Heat/Mass Transfer

DEM Applications

A wide range of applications can be treated

Tridiagonal Solutions
compute | innovate | develop

Feeding/Discharging

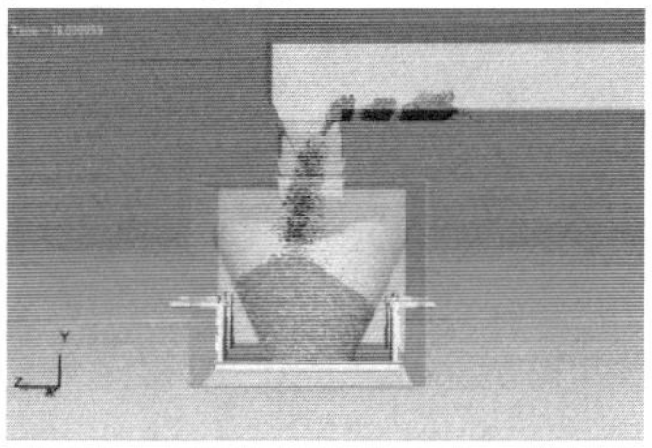

Continuous mixing

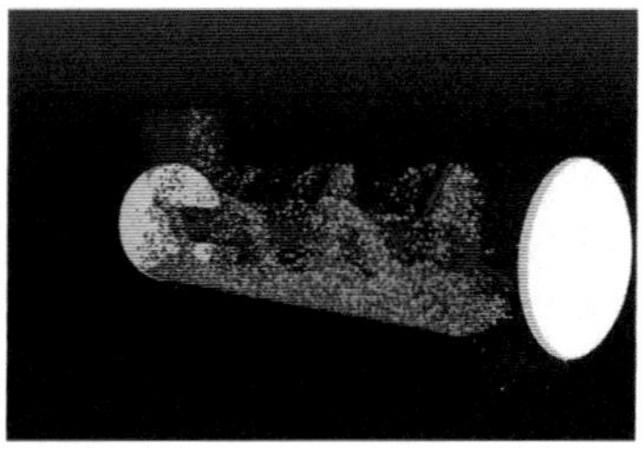

Milling

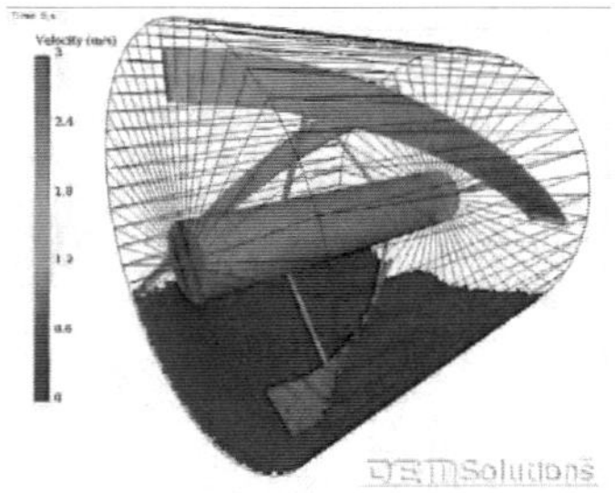

Spray coating

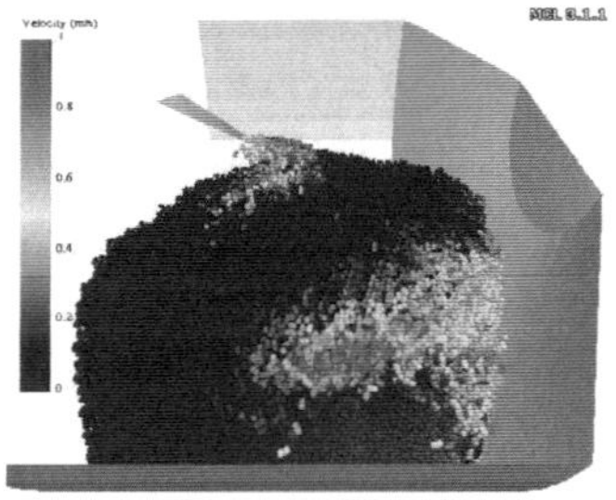

Spray Coating - DEM In Action

Optimization of coating processes via DEM

Tridiagonal Solutions
compute | innovate | develop

Spray coating has applications across the process industries

DEM modeling of spray coating allows optimization of

- Spray coverage
- Mixing element design
- Spraying and drying rates

Used to optimize

- Mixing times
- Spray coverage
    - contact between spray and particles

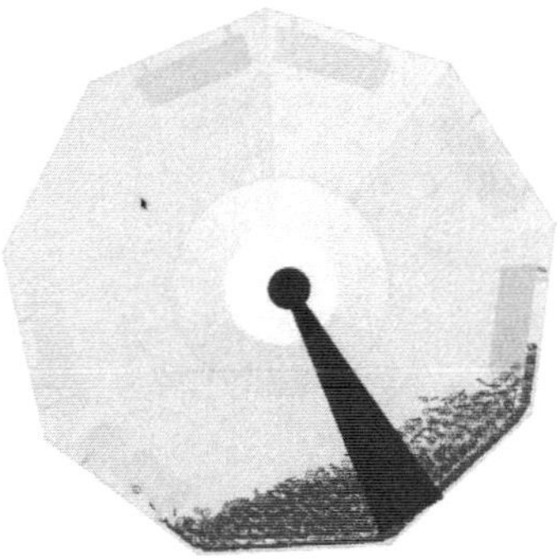

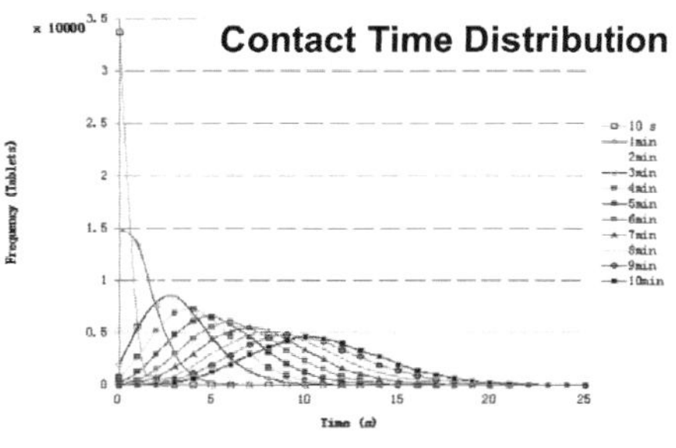

Interaction between particles/granular material and interstitial fluid can be described effectively using coupled EDEM-CFD models

Transfer of momentum, heat and mass between particles and fluid can be captured to a high degree of accuracy.

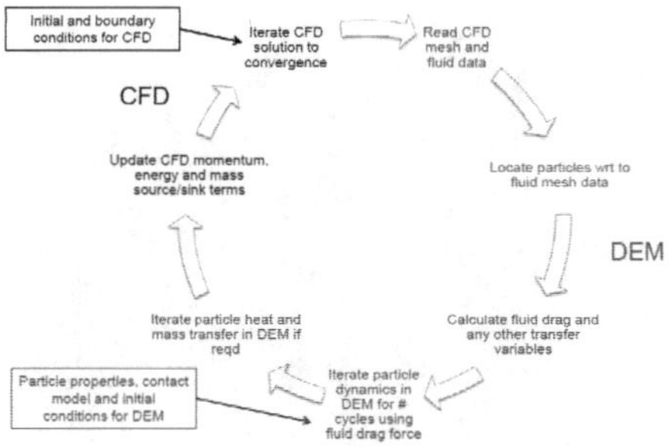

Pneumatic Conveying

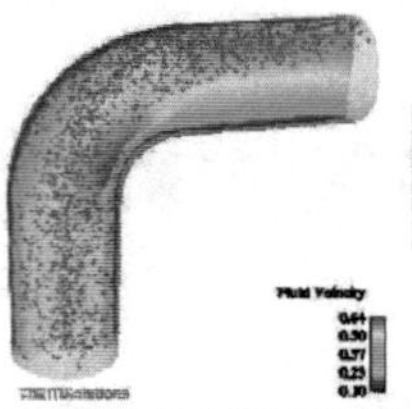

Fluidized Bed

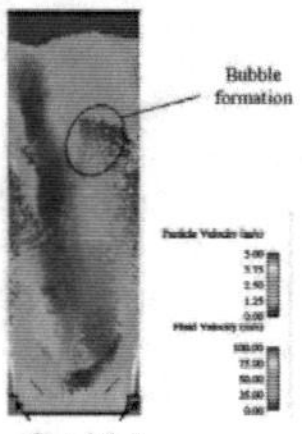

Rotary Kiln

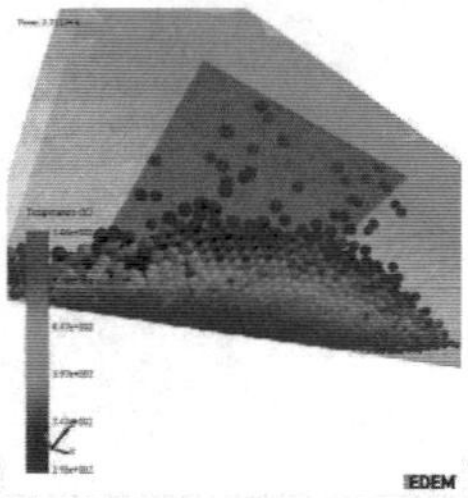

Summary

- Granular flows of industrial importance can be effectively analyzed using the discrete element method (DEM)
  - range of lengthscales
  - non-spherical particles and granules etc.

- DEM models track the flow of individual particles in a given process equipment
  - detailed information on position, velocity, forces/stresses
  - enable optimization of process and product attributes

- Hybrid CFD-DEM models can be used to describe the exchange of momentum, mass and heat between particles/granules and interstitial fluid flow.

- "Granular matter examples" by At_children's_level.jpg: Julie Kertesz from Paris neighbourhood, FranceGravel_small_stones.jpg: Martin Olsson (mnemo on en/sv wikipedia and commons, martin@minimum.se).Lenticchie_z01.jpg: ZyanceSesame_seeds.JPG: Jitender Kumarderivative work: Gsrdzl - At_children's_level.jpgGravel_small_stones.jpgLenticchie_z01.jpgSesame_seeds.JPG. Licensed under CC BY-SA 3.0 via Commons.
- "PELLETIZER" by Feeco International - http://feeco.com/disc-pelletizers/. Licensed under CC BY-SA 3.0 via Commons
- "Nurdles 01 gentlemanrook" by gentlemanrook - originally posted to Flickr as Plastic Pellets - "Nurdles". Licensed under CC BY 2.0 via Commons

https://commons.wikimedia.org

Attributions For Images in Slide 1

- "Cyclone separator" by en:User:Cburnett - Own work with Inkscape based on visual inspection of this image which can be found on on this page. Licensed under CC BY-SA 3.0 via Commons -
- "CemKilnKiln" by LinguisticDemographer - Originally from en.wikipedia; description page. Licensed under Public Domain via Commons.
- "Fluid Catalytic Cracker" by Valero Energy Corporation/TX - http://www.secinfo.com/. Licensed under Public Domain via Wikipedia - https://en.wikipedia.org/wiki/File:Fluid_Catalytic_Cracker.gif
- "Alto Horno, Puerto de Sagunto, España, 2015-01-04, DD 91" by Diego Delso. Licensed under CC BY-SA 4.0 via Commons.

https://commons.wikimedia.org

Outlier Detection and Analysis in Batch and Continuous Processes

Peter J. Ryan, Ph.D., P.E.,
Response Process Consulting, LLC
October 2, 2015

Outlier Detection and Analysis in Continuous and Batch Processes

Contents

- Archived Process Data Overview
  - Missing Data in the Design and Quality Arrays
- Unsupervised Machine Learning – PCA
- Supervised Machine Learning – PLS
- Continuous Process Example
- Batch Process Example
  - Score Projection Plots
  - Contribution Plots – Finding the Relationship between KPI's and Upstream PV's
- Summary

Archived Process Data

- Almost Every Process (Batch or Continuous) is Controlled by a Distributed Control System (DCS) or a PLC Control System
- Almost Every DCS or PLC System is Supported by a Process Data Archiving System
  - ✓ Typically all analog and some discrete process variables (PV's) are recorded (archived) ~ every 2 - 5 seconds
- This Data:
  - ✓ Typically contains a lot of missing data (transmitter failure, off-scan, archiving system not responding, ...)
  - ✓ Is voluminous (almost untenable)
  - ✓ Typically is very correlated and has a low signal-to-noise ratio (noisy data)
  - ✓ Typically lacks a certain degree of accuracy (data reconciliation is usually not preformed on the data before it is "fed" to the DCS or PLC)

Outlier Detection and Analysis in Continuos and Batch Processes
7th Southwest Process Technology Conference, 10/02/15
Page 3

Response Process Consulting, LLC

Archived Process Data

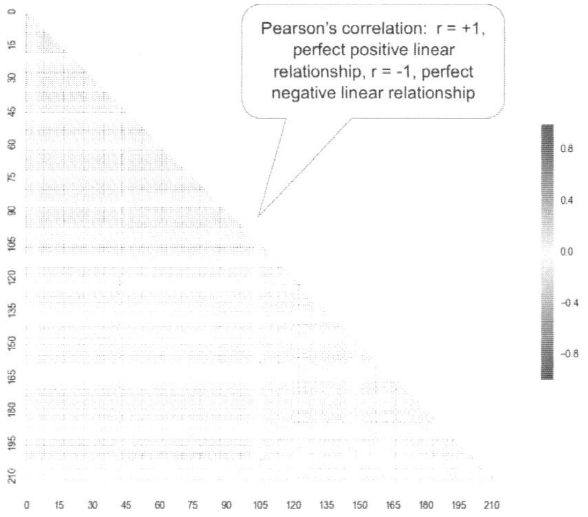

Outlier Detection and Analysis in Continuos and Batch Processes
7th Southwest Process Technology Conference, 10/02/15
Page 4

Response Process Consulting, LLC

Why Use This Resource?

- ***Consider First Principal Models:***
  - ✓ Continuous processes often have a steady-state first-principal model available for analysis and optimization (Aspen Plus, HySys, UniSym, ChemCad, ...)
    - Very good for mass & energy balances, debottlenecking, optimization, but often lacks the granularity needed to describe process quality variables (color, relative viscosity, consistency, taste, ...)
  - ✓ Batch Processes are Dynamic and:
    - Dynamic models are more expensive to create and support,
    - Models are often not available for process analysis,
    - Again, if a dynamic model is available, it often lacks the granularity needed to describe process quality variables

But ...

- ***Archived Process Data can be used to Model Continuous and Batch Processes, and Contains the Granularity to Describe Process Quality Variables:***
  - ✓ We only need methods and tools to sort through this data to produce effective models than can:
    - work with correlated data
    - work with the low signal-to-noise ratio
  - ✓ These methods are latent variable methods Principal Component Analysis (PCA) and Partial Least Squares (PLS)

Consider Latent Variable (LV) Methods – PCA and PLS

- ***Principal Component Analysis (PCA) and Partial Least Squares (PLS)***
  Construct linear combinations of the variables consisting of Scores (T) and Loadings (P) that rotates the data cloud along a new set of axis's:

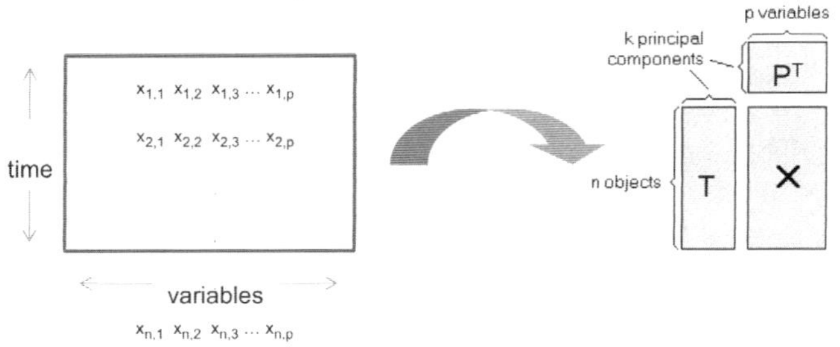

$$X = TP^T + E$$

Outlier Detection and Analysis in Continuos and Batch Processes
7th Southwest Process Technology Conference, 10/02/15
Page 7

Response Process Consulting, LLC

Latent Variable Methods – PCA and PLS

- ***Such That:***

  The first new axis (Principal Component) passes through the data cloud along the path of greatest variability, the second new axis passes through the data cloud along the path of next greatest variability and is orthogonal to the first PC, etc.

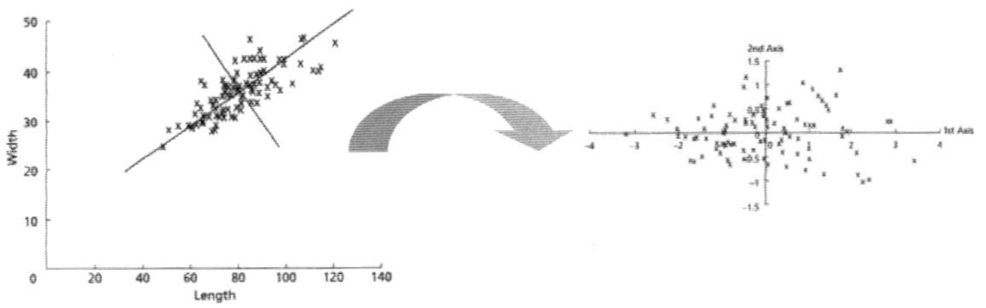

Swan, A.R.H., and M. Sandilands, 1995. Introduction to Geological Data Analysis. Blackwell Science: Oxford, 446 p.

Outlier Detection and Analysis in Continuos and Batch Processes
7th Southwest Process Technology Conference, 10/02/15
Page 8

Response Process Consulting, LLC

Why Latent Variables Models?

LV Models such as PCA and PLS can link observable variables with each other and with Key Process Indicators (KPI's), providing causal structures to the model.

- This is an advantage over other modeling methods such as MVR, SVR, kNNR, EM, NN, etc, all of which deliver a "black box" type of process model.
- LV models are also used to reduce the dimensionality of the process data space, capturing most of the variability in the process data with the first three to four principal components. This allows the model to focus on those PV's that actually drive the process (recall correlated data issues, where EM, SVD and other algorithms fail to converge with too much correlation).

Continuous Process Example

Consider a continuous process consisting of reaction, fractionation and product recovery. One important product specification is the product color.

- Lets develop a model of this process using archived process data from one year of operation. The data contains 1.4% missing data.
- A PLS model is developed, using three principal components.
- Four missing data methods were used to "fill in" the design and quality arrays:
  1) Delete rows with missing values
  2) Column-mean imputation
  3) k-Nearest Neighbors (k = 10)
  4) Iterative Robust Model Imputation (VIM MI)

Handling Missing Data

- If rows with missing data are deleted, 33% of the data set would be lost (for this given example)
- Of the three calculated missing data methods, sampling of performance:

Actual	Mean-impute	Abs % error	knn BioC	Abs % error	VIM MI	Abs % error	
237662.4	231850	2.45	237120	0.23	237300	0.15	
3412.828	1318	61.38	3487.5	2.19	3526.3	3.32	
3335.576	1318	60.49	3487.5	4.55	3356.6	0.63	
28	26.217	6.37	24.444	12.70	26.591	5.03	
3278.256	1318	59.80	3487.5	6.38	3160.2	3.60	
28	26.217	6.37	24.444	12.70	27.678	1.15	
28	26.217	6.37	28.167	0.60	29.725	6.16	
28	26.217	6.37	33.625	20.09	32.129	14.75	
28	26.217	6.37	37.333	33.33	31.996	14.27	
28	26.217	6.37	29.167	4.17	29.552	5.54	
417.8098	419.66	0.44	417.88	0.02	417.96	0.04	
28	26.217	6.37	28	0.00	35.217	25.78	
417.5565	419.66	0.50	416.52	0.25	417.66	0.02	
32	26.217	18.07	33.333	4.17	39.01	21.91	
416.6301	419.66	0.73	418.89	0.54	416.86	0.06	
155.0606	100.98	34.88	82.313	46.92	160.24	3.34	These missing values are going sideways -
25.07543	98.056	291.04	109.19	335.45	16.56	33.96	difficult for knn to impute
1107463	1148600	3.71	1290500	16.53	1104300	0.29	
59676.8	60203	0.88	60064	0.65	59371	0.51	
19986.86	21776	8.95	22864	14.40	20403	2.08	
Maximum error in 20 cells		291.04		335.45		33.96	
Mean error in 20 cells		29.39		25.79		7.13	

Choosing the Number of PC's to Calculate

PCA Variability-Capture Results:

Impute Method	% Variability Captured in PC1	% Variability Captured in PC2	% Variability Captured in PC3	Total % Variability Captured
Mean-impute	52.4	7.2	2.8	62.4
VIM MI	55.4	6.7	2.8	64.9

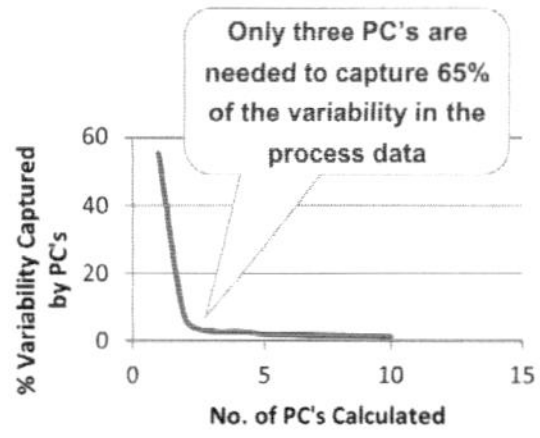

Now, calculate Hotelling T^2 and SPE plots for an overall look at the data, and then examine score plots (t1 v. t2) and look for clusters

- Scores are projected onto the plane formed by the PC's

Two Error Types

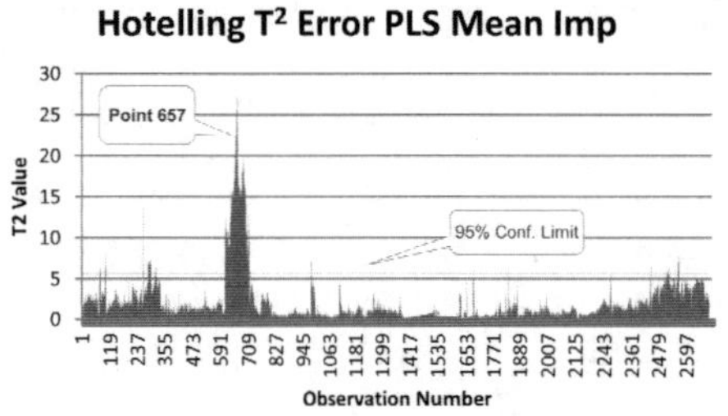

Hotelling T² Error:
Explains deviation of points that are on the model plane but far from the model center-of-mass (n = 3).

$$T^2 = \sum_{i=1}^{n} t_i^2/\lambda_i = \sum_{i=1}^{n} t_i^2/s_i^2$$

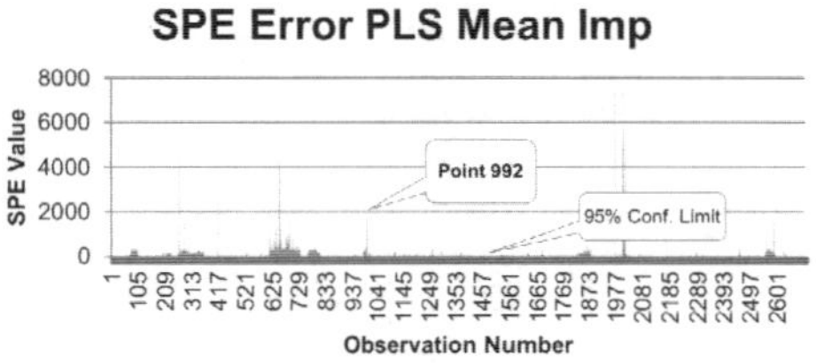

SPE Error:
Explains deviation of points that *may* be close to the model center-of-mass (projection-wise), but are far away from the model plane.

$$SPE_x = \sum_{i=1}^{k} (x_{new,i} - \hat{x}_{new,i})^2$$

Score Plots – Continuous Example

Blue = Good Color
Green = Marginal
Red = Out-of-Spec

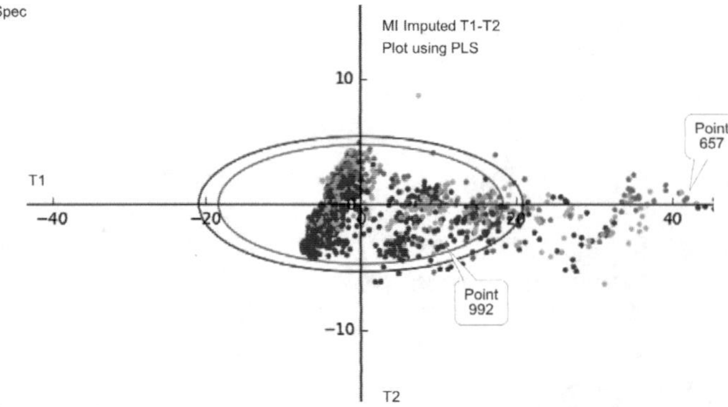

Contribution Plots – Continuous Example

There are small differences between these two contribution plots because the PLS model is maximizing the covariance between the X array and the y (quality) vector. These differences show how the upstream process variables more specifically affect the color KPI (supervised learning) v. the PCA model (unsupervised learning).

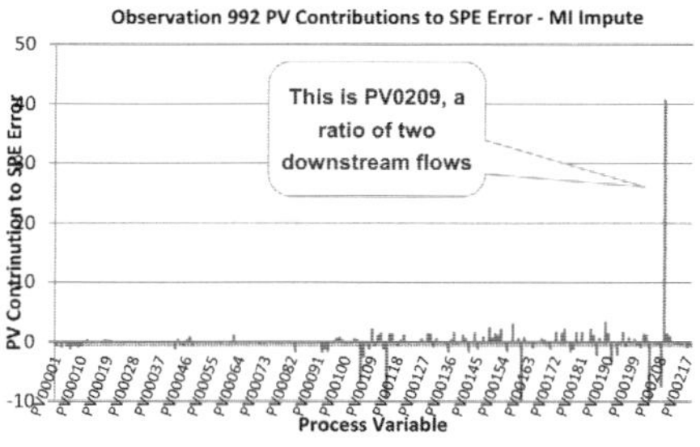

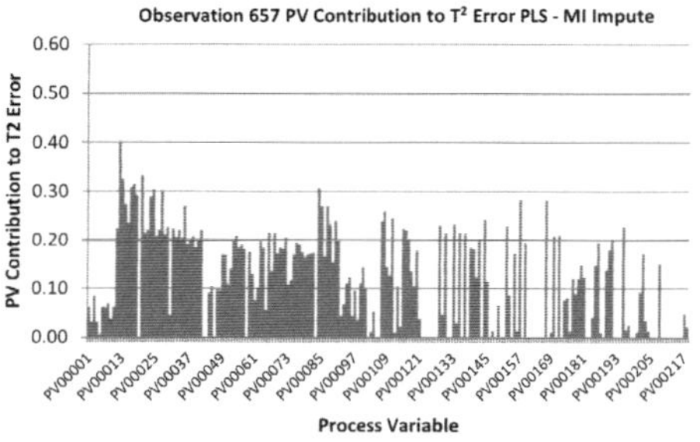

Summary of Continuous Example

A supervised (PLS) model was developed from the archived process data.

- Even though the data only contained 1.4% missing data, eliminating all rows of data that contain missing data would have reduced the data set by 33% - too much information would be lost handling missing data in this way.
- Missing data can be imputed. With both unsupervised and supervised modeling, an imputation method that regresses across rows is best suited for estimating missing data values and preserving information in the data set.
- Model refinement will produce better granularity in the relationship between upstream PV's and KPI's.

Summary of Continuous Example

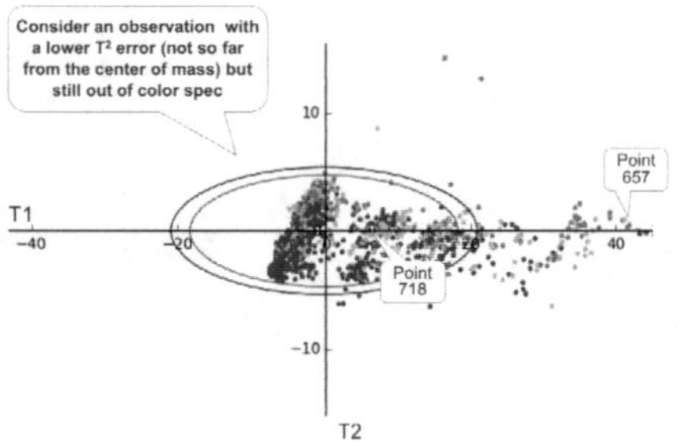

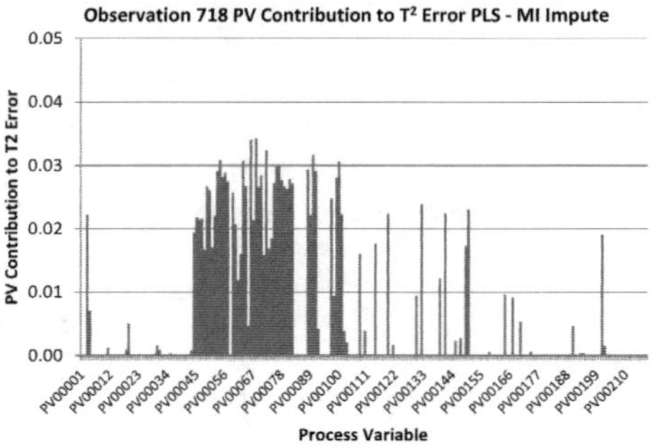

Here strong contributions from PV00043 through PV00081, PV00088 through PV00091, PV00097 through PV00102, and other individual PV's remain – even as we draw closer to the model's center of mass. This suggests a possible differentiation between what PV's are responsible for loss of color spec, whether the observation's T1-T2 projection is close to or farther away from the model's center of mass.

Batch Process Example – Nylon Production

- 1993 (M. Piovoso, K. Dahl, K. Hoo)
- Several batches produced out-of-spec product, but causes were not apparent in either the trajectory data or the measured quality vector (relative viscosity)
- A PCA model was develop to identify and analyze outliers using archived process data

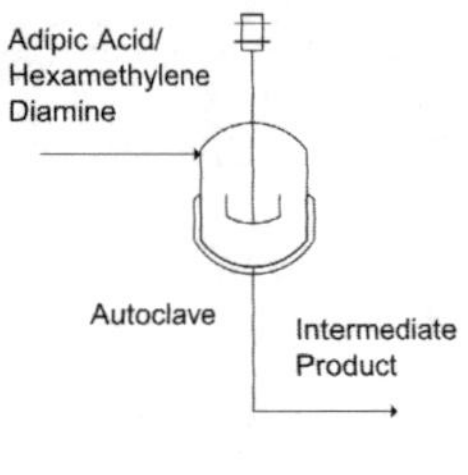

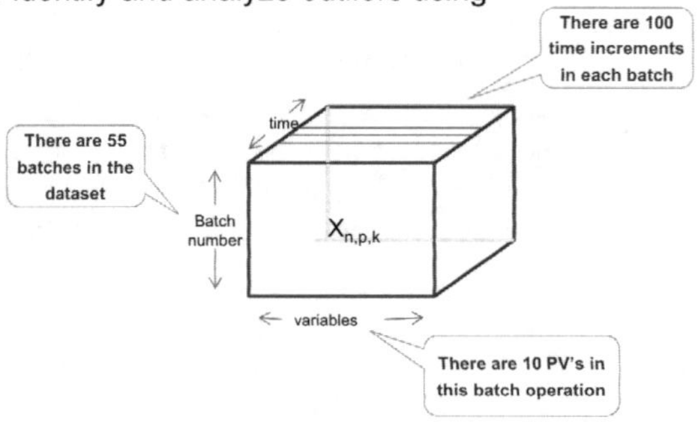

Batch Process Example – Handling Batch Data

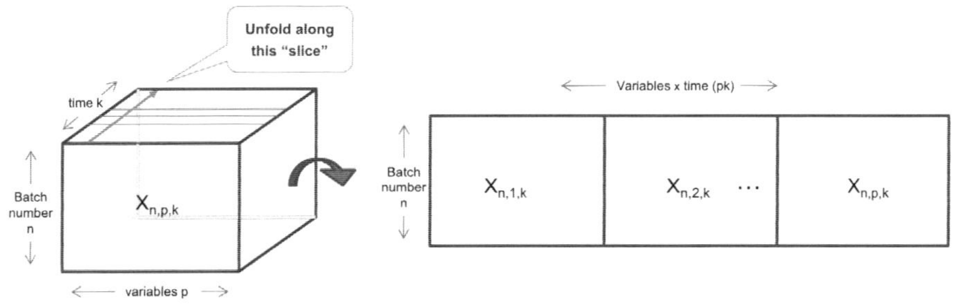

The batch process data must first be "unfolded" before data reduction methods can be applied. Preferred unfolding method is Batch-wise-unfolding (BWU) as shown. One other option is Observation-wise-unfolding (OWU).

The "unfolded" tensor now yields an *n x pk matrix* that can be analyzed just like the n x p matrix containing continuous data. Now PCA or PLS methods can be used to track and analyze batch process performance.

SPE and T^2 Error Plots – Batch Example

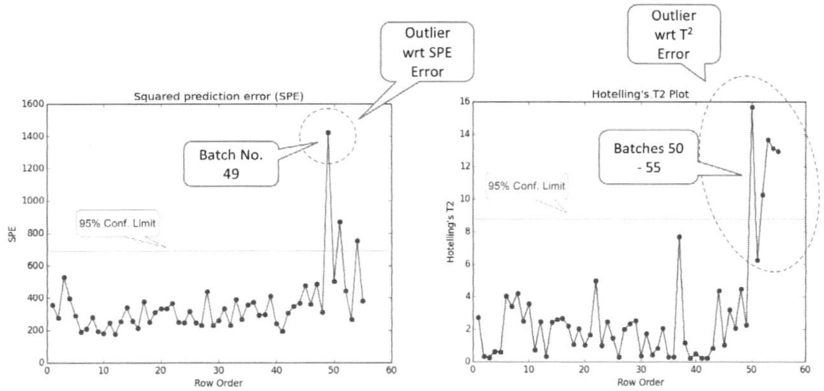

Score Plots – Batch Example

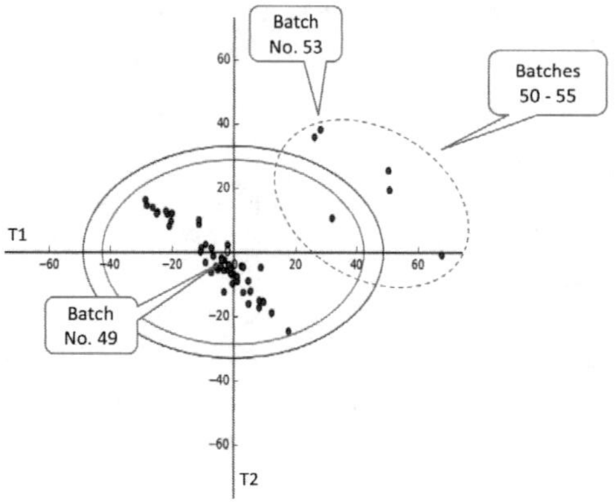

The T1 v. T2 plot captures 55.9% of the variability in the archived process data. Each point on this plot represents 55.9% of the variability of each batch in the data set. The outliers are batch outliers; i.e., batches 50 - 50 are outlying batches.

Score Plots – Batch Example

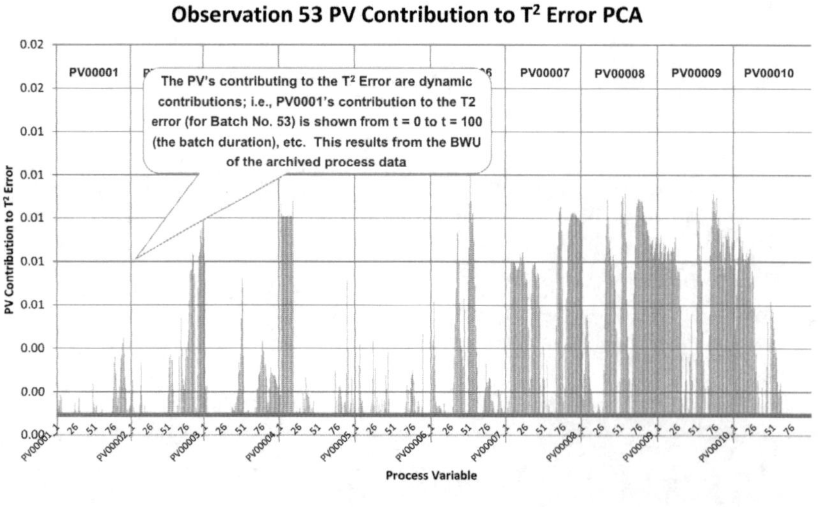

PV Contribution to SPE Plot – Batch Example

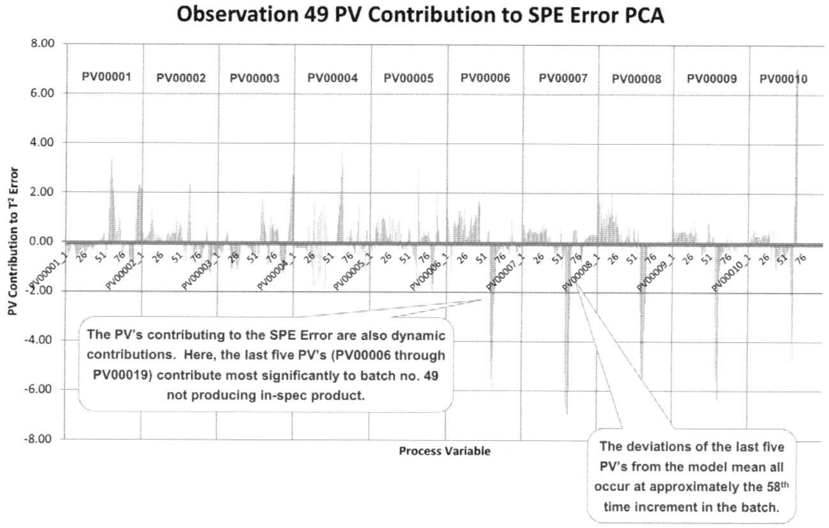

Compare Batch 49 with Other Trajectories

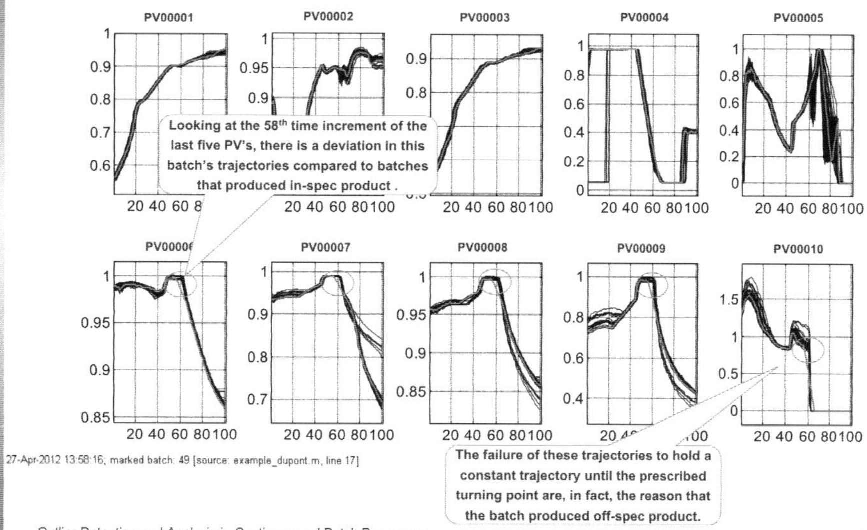

Summary of Batch Example

Only an unsupervised (PCA) was developed from the batch data set, but a supervised model could have been developed if there was a quality vector available.

- Same SPE and T^2 errors were used to review the batches.
- Missing data can be imputed with batch data as well. Here it is especially important to use a regression imputation method because the trajectory data changes as the row is descended (dynamic data).
- The contribution charts now capture the dynamic information of the batch, and can be used to signal the onset of a deviation from "on-spec" batch performance. These methods have been used in on-line applications

Peter J. Ryan, Ph.D., P.E.
(281) 220-7272
peter.ryan@.ResponsePCcom

Development of Numerical Model to Assess Internal CO2 Corrosion of Oil and Gas Pipelines

Srinivasan, V., ANSYS

Originally presented on: 10/2/2015 8:30:00 - 8:54:00

A comprehensive numerical model, developed for computing the corrosion rates generated by CO_2 saturated aqueous solutions, is used to study top-of-the-line corrosion in wet gas pipelines. The fundamental fluid flow is characterized by numerical solutions to Navier-Stokes conservation equations for mass, momentum and energy, accompanied by models to capture turbulence generated by the flow. The corrosion rates are then computed by modeling the electrochemical processes occurring at the metal substrate such as the cathodic reduction of carbonic acid, hydronium ions and anodic oxidation of the metal component. The developed model is first validated against experimental data for CO_2 saturated single phase aqueous solutions and is later extended to model corrosion behavior in gas dominated multiphase flows across various flow network configurations. Elaborate details of the multiphase flow behavior such as time dependent distribution of liquid/gas phasic fractions are provided. The effects of modulated superficial velocity of gas and liquid components as well as the temperature on the overall corrosion rates are monitored and satisfactory trends against available experimental data is observed.

Simulation and Automation Technologies for the Topsides Process Design

Authors
Neda Felorzabihi (Presenter)
Satish Baliga
Jeffrey Feng
Phaneendra Kondapi

KBR
TECHNOLOGY

AIChE
Southwest Process Technology Conference

7th AIChE Southwest Process Technology Conference | October 1-2, 2015 | Galveston, TX 1

Abstract

This presentation describes proven practices, work processes and latest trends in applying simulation and automation technologies to enhance the project lifecycle of upstream projects from design to construction, commissioning and operation.

- Capture the design expertise through the entire project lifecycle
- Validate design to ensure operational excellence
- Leverage design expertise to support construction & commissioning
- Knowledge transfer during operations and revamp

KBR
TECHNOLOGY

AIChE
Southwest Process Technology Conference

2

Why Advanced Technologies for Upstream?

Offshore facilities are typically:

- Capital intensive
- Weight sensitive (over-design must be minimized)
- Characterized by high pressure
- Wide range of throughput
- Variable operating environment

➡ Advanced technology-based methods are best suited to meet these design challenges

KBR TECHNOLOGY

AIChE Southwest Process Technology Conference

3

Advanced Technologies Applications Summary

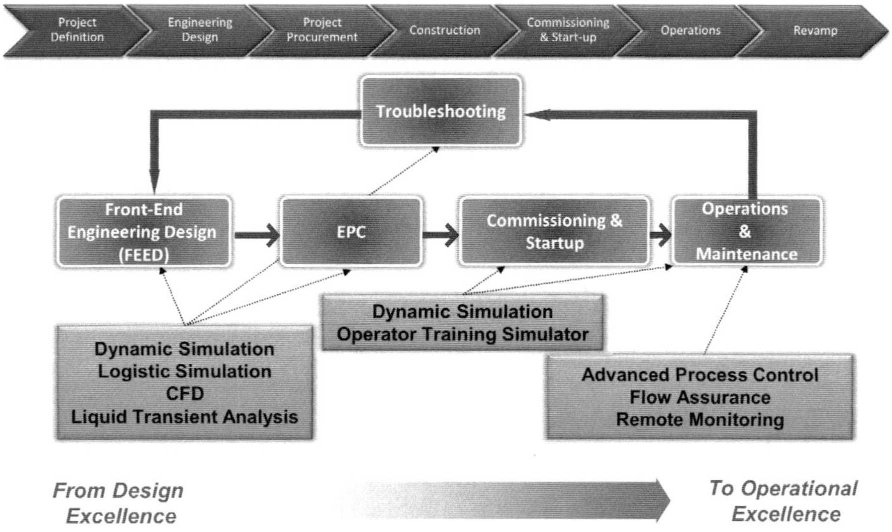

4

Project Lifecycle

Pre-FEED & FEED

EPC

Commissioning & Start up

Operations & Maintenance

KBR TECHNOLOGY

AIChE Southwest Process Technology Conference

5

HIPPS (Case Study)

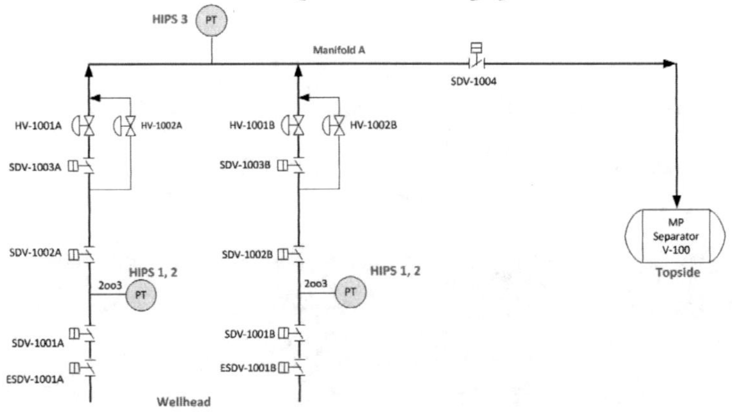

Key Objectives

- Evaluate best over-pressure protection for the system
- Evaluate the benefit of reducing relief load with different combinations of HIPPS strategy and equipment design pressure

KBR TECHNOLOGY

AIChE Southwest Process Technology Conference

6

HIPPS (Case Study)

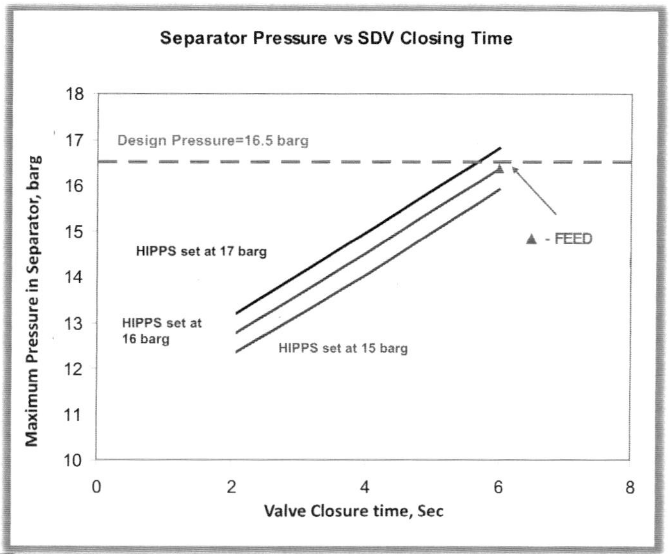

KBR TECHNOLOGY

AIChE Southwest Process Technology Conference

7

CFD (Hot Air Recirculation for Topsides)

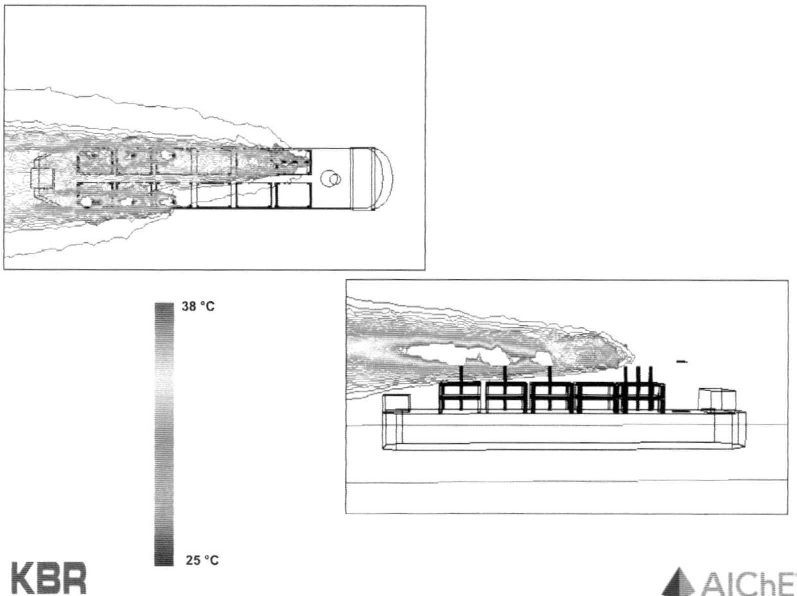

KBR TECHNOLOGY

AIChE Southwest Process Technology Conference

8

Project Lifecycle

Pre-FEED & FEED

EPC

Commissioning & Start up

Operations & Maintenance

KBR
TECHNOLOGY

AIChE
Southwest Process Technology Conference

9

Topsides Process Dynamic Studies

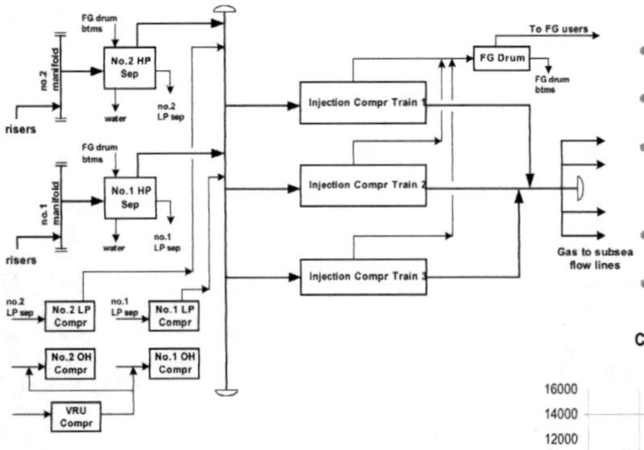

- Adequacy of separator design
- Adequacy of anti-surge systems
- Requirements for decoupling train interactions
- Control philosophy validation
- HIPPS strategy

C-UC-122301B-02 (2nd Stage) Head vs Flow

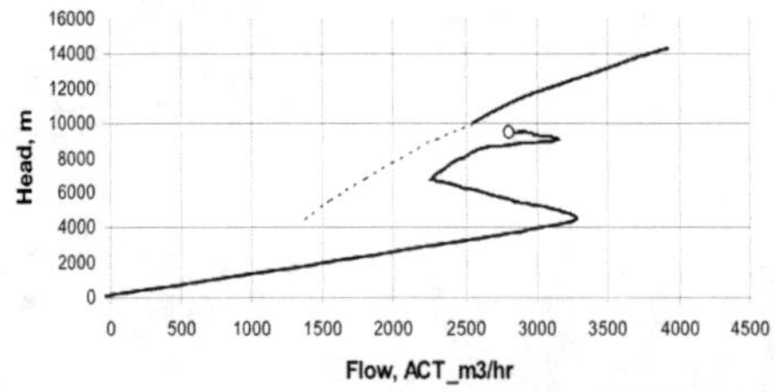

KBR
TECHNOLOGY

AIChE
Southwest Process Technology Conference

10

Compressor Control & Dynamic Simulation

High-fidelity dynamic model linked with CCC emulator

- Rigorous compressor control study
- Pre-tuning controllers prior to commissioning

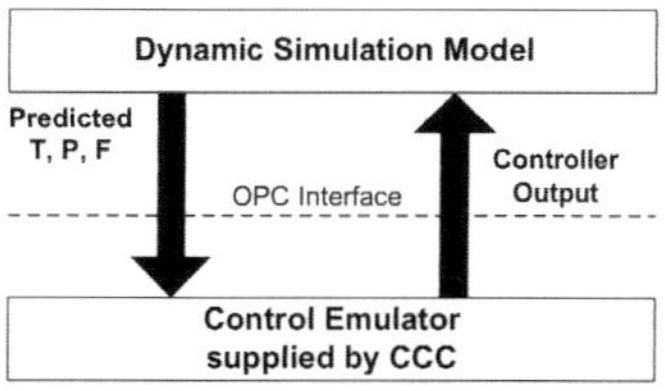

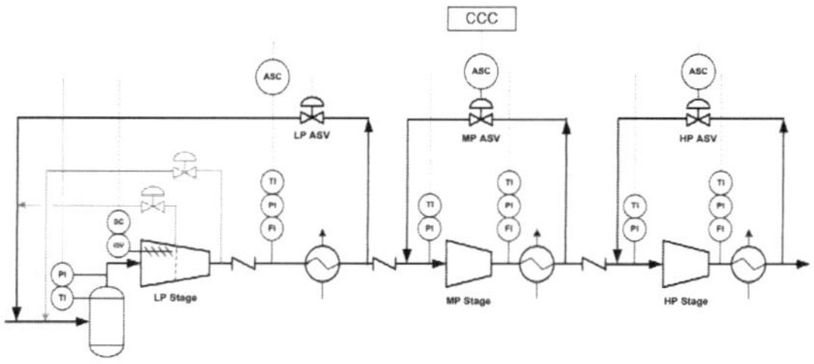

KBR
TECHNOLOGY

AIChE
Southwest Process Technology Conference

11

Liquid Transient Surge Analysis

LNG loading lines on a typical LNG Facility

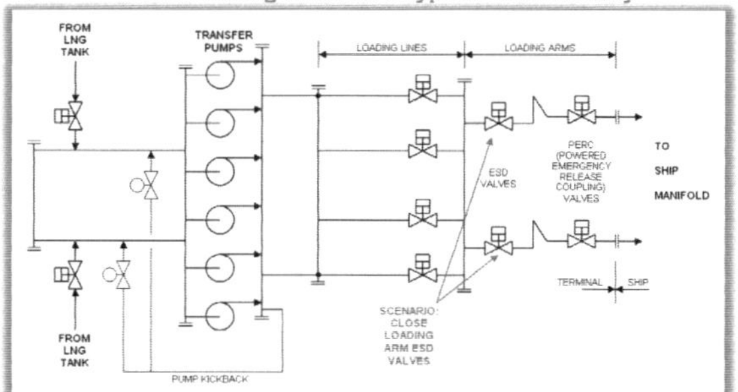

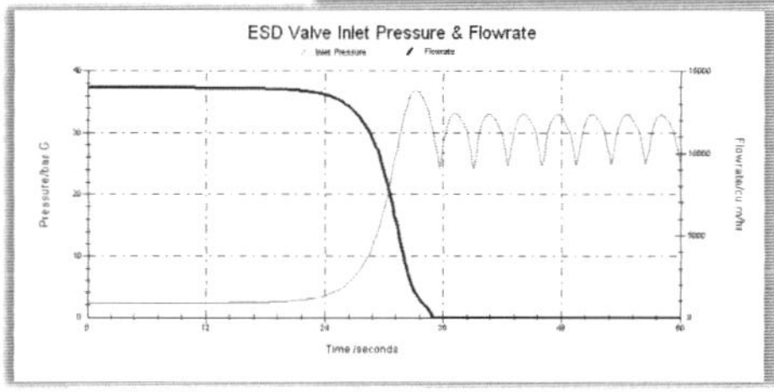

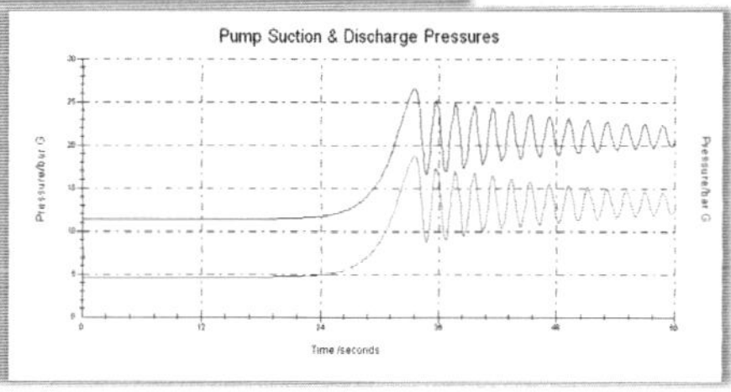

12

Logistics Simulation – Case Studies

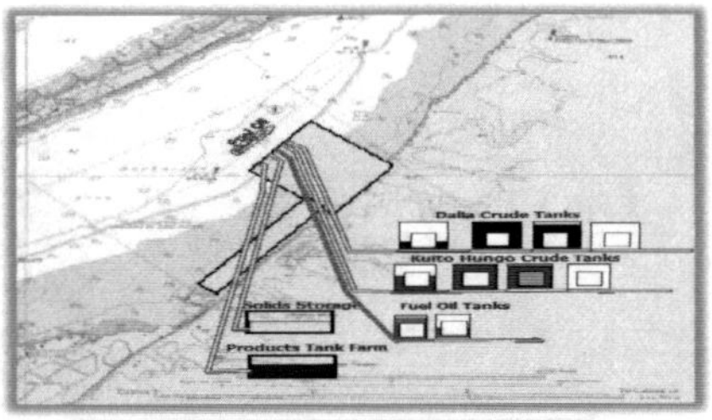

Shipping and Storage Study

- Optimal tank size and marine berth number considering resource utilization, production loss, cycle time, and inventory
- Vessel fleet size taking uncertainty in trip times, currents and weather into account

Construction Logistics

- Match freight delivery plan with construction schedule considering constraints of ports, vessel, barge, rail and trucking fleets
- Ensure civil infrastructure in place can handle the labor and freight traffic loads
- Predict lay down space requirements
- Identify risks up front and contingency plans during the planning process

KBR TECHNOLOGY

AIChE Southwest Process Technology Conference

13

Project Lifecycle

Pre-FEED & FEED

EPC

Commissioning & Start up

Operations & Maintenance

KBR TECHNOLOGY

AIChE Southwest Process Technology Conference

14

Operator Training Simulator (OTS)

- Train operators on start-up, shutdown and emergency response
- Validate operating procedures
- Models used to support start-up teams

FPSO OTS- Case Study

✓ OTS scope includes integration of all systems operated by board operators

- Topsides turret, oil/gas separation and compression, utilities
- Marine systems for storage, ballast, loading & unloading
- Production, gas injection and water injection wells and risers (60 total wells, 17 risers)
- Chemical injection systems

✓ Execution

- Modeling scope – topsides, marine and subsea
- Project kick off October 2013, FAT completed January 2015

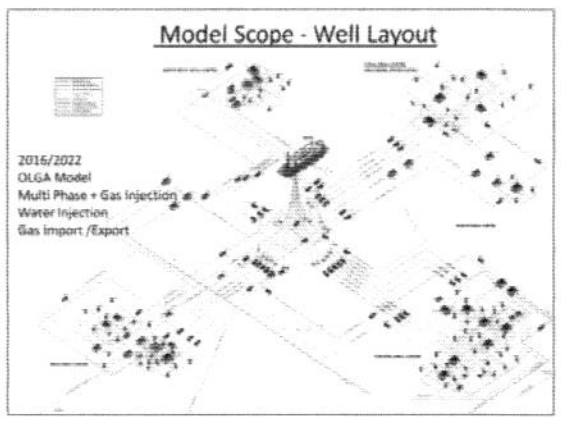

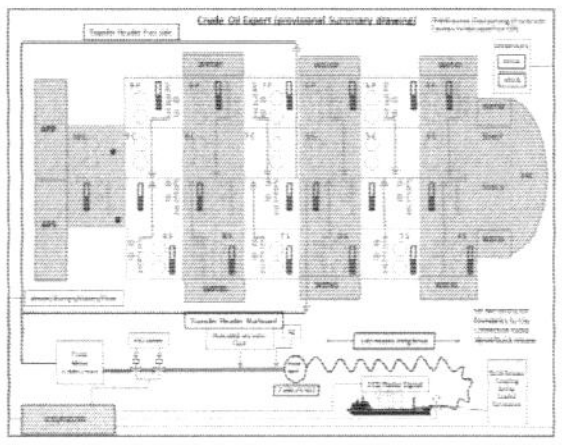

OTS Model Validation

Overall Plant Response During Plant Turndown

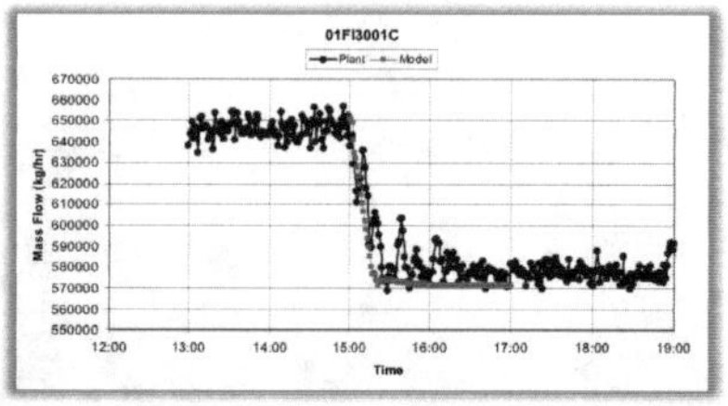

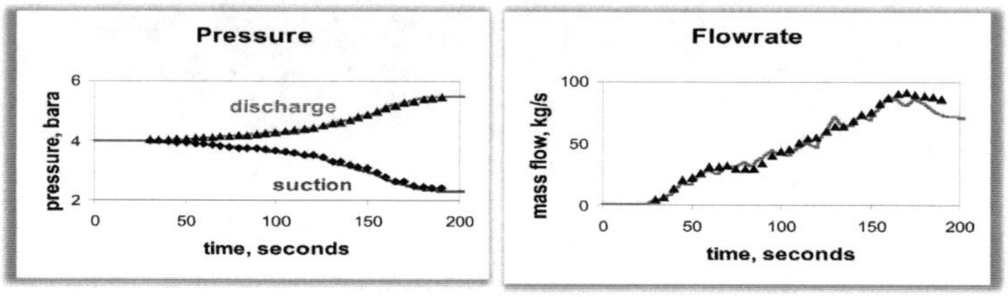

Startup of a 3-Stage Compressor

KBR
TECHNOLOGY

AIChE
Southwest Process Technology Conference

17

Project Lifecycle

Pre-FEED & FEED

EPC

Commissioning & Start up

Operations & Maintenance

KBR
TECHNOLOGY

AIChE
Southwest Process Technology Conference

18

Remote Monitoring for Intelligent Operations

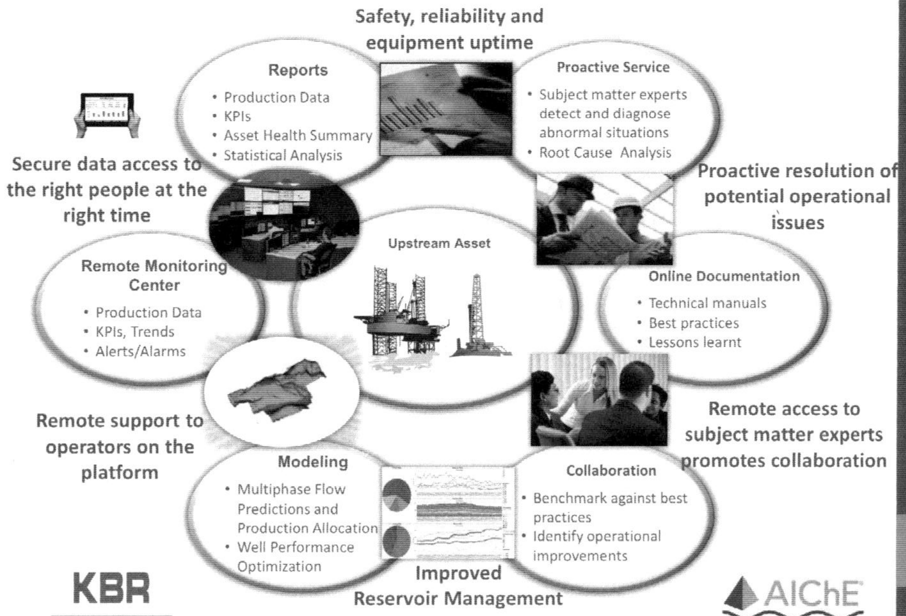

KBR TECHNOLOGY

AIChE Southwest Process Technology Conference

19

Advanced Process Control

✓ Slug Mitigation
  - APC dynamically manipulate topside choke at separator inlet to dampen the impact of upstream pressure fluctuations on the production facility

✓ Slug Management
  - "Ride" the slug and utilize surge capacity of "swing" separators to minimize impact on secondary separator and downstream units
  - Improved stability, less equipment trips and increased production (1-2 %)

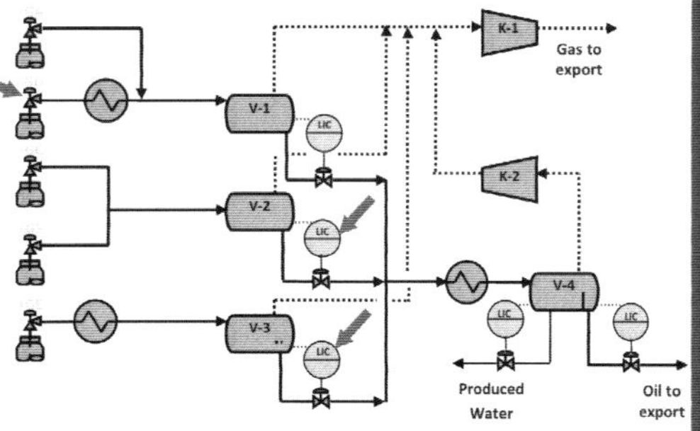

When a liquid slug enters an inlet separator, the APC raises the set points of level controllers for other inlet separators thereby reducing their liquid outlet flow rates and reducing impact on downstream separation and processing.

KBR TECHNOLOGY

AIChE Southwest Process Technology Conference

20

Flow Assurance & Production Allocation

- Multiphase Flow Prediction
  - Online, multiphase, rigorous PVT compositional model-based system for predicting oil, gas and water flows from individual wells in an oil field
  - Allocation of production to different reservoir zones helps improve reservoir management and allows unproductive wells to be shut-in

- Hydrate Management & Control
  - Software to simulate transient cooling of subsea flowlines and to predict time remaining before hydrate formation
  - Minimize chemical injection (methanol) to reduce operating expenses

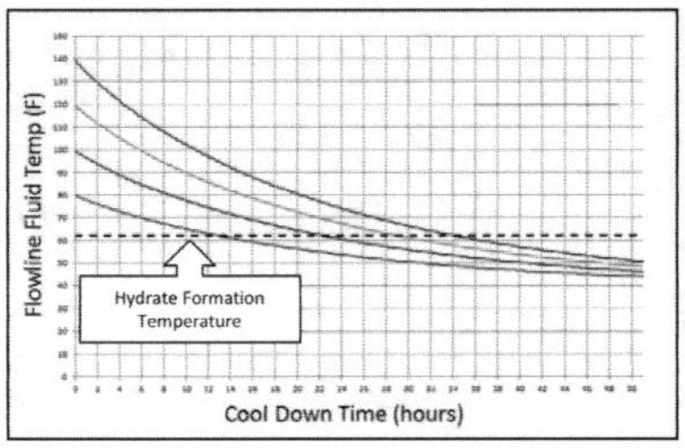

KBR
TECHNOLOGY

AIChE
Southwest Process Technology Conference

21

Summary

Proven practices and work processes in applying Advanced Simulations & Automation Technologies through the lifecycle of upstream projects from front-end design, EPC, commissioning & startup to operations were presented. Case studies with application of different simulation tools for upstream projects were also introduced. Using these tools during the lifecycle of upstream projects enables:

- *Capture the design expertise through the entire project lifecycle*
- *Validate design to ensure operational excellence*
- *Leverage design expertise to support construction & commissioning*
- *Knowledge transfer during operations and revamp*

From Design Excellence → ***To Operational Excellence***

KBR
TECHNOLOGY

AIChE
Southwest Process Technology Conference

22

Questions

KBR
TECHNOLOGY

AIChE
Southwest Process Technology Conference

23

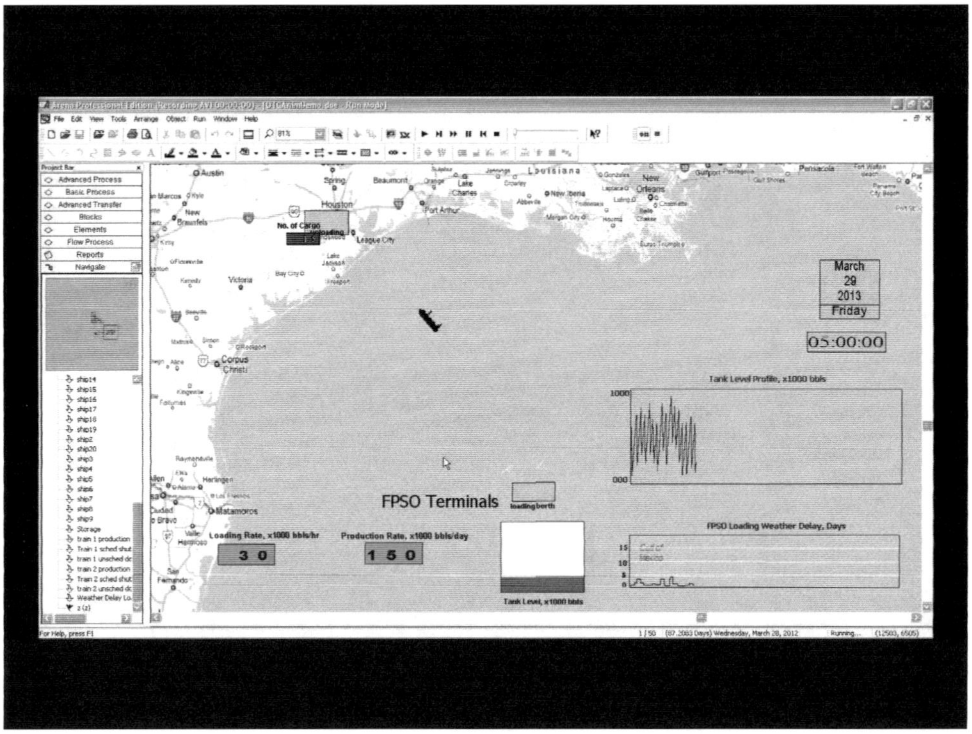

Gas Evolution Rates in Crude Oil for Upstream Applications

Aichele, C. P., Oklahoma State University
Lavenson, D., Chevron Energy Technology Company
Originally presented on: 10/2/2015 9:18:00 - 9:42:00

As gas-liquid separators are reduced in size to accommodate offshore and subsea specifications, ensuring that the units are appropriately sized has become increasingly important. Failure to size separators correctly will hinder both the safety and reliability of the unit, and could compromise processes downstream. While the vessel volume is typically calculated from estimated flow rates and well head conditions, the rate of gas evolution during flow and at high pressures has been largely overlooked. This presentation provides an overview of the research that is relevant to gas evolution with specific emphasis on the impact of flow on the rate of gas evolution. Sources from literature show that the rate of gas evolution can be affected by parameters including initial saturation pressure, liquid temperature, rate of agitation, and the presence of solid particulates. Equilibrium and dynamic modeling is also discussed in the presentation.

Application of Microfluidics for Emulsion Characterization

Presented by
Subarna Kole
Oklahoma State University

AIChE
Southwest Process Technology Conference

7th AIChE Southwest Process Technology Conference | October 1-2, 2015 | Galveston, TX

Introduction to Emulsions in Petroleum Industries

- Emulsification and demulsification processes have a wide range of applications in various fields such as food, pharmaceuticals and petroleum industries.
- The crude oil produced from the well contains significant amount of water and this water has to be separated before any further downstream processing.

AIChE
Southwest Process Technology Conference

7th AIChE Southwest Process Technology Conference | October 1-2, 2015 | Galveston, TX

Microfluidic Facility

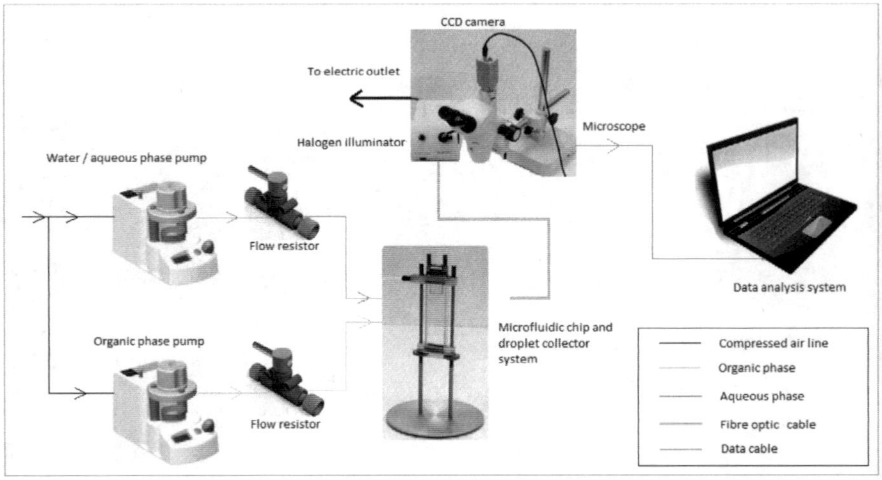

Flow diagram

Photograph of the Experimental Facility

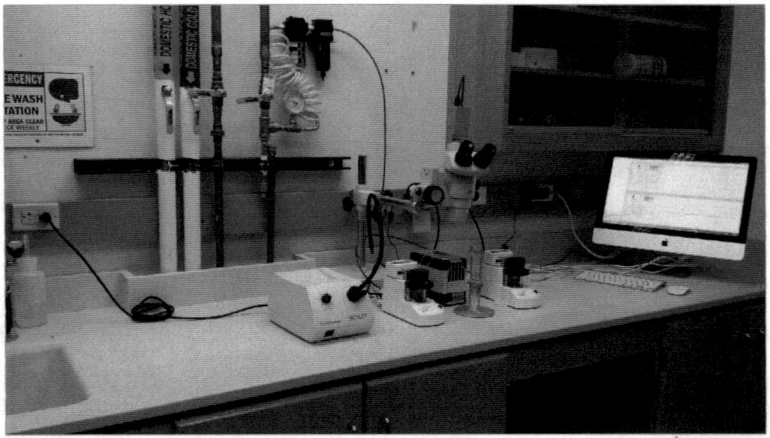

Details of Components used in Experimental Facility

Components	Specifications	Functions
Droplet junction chip	100 µm etch depth, hydrophilic	Monodispersed oil-in-water emulsion formation
Droplet junction chip	100 µm etch depth, hydrophobic	Monodispersed water-in-oil emulsion formation
Mitos P-Pump Basic	Up to 10 bar pressure	Compatible with N_2 gas bottle or compressor, ideal for droplet formation
Microscope	7.5x zoom range, up to 50x magnification	Three-dimensional visualization of the sample
High speed camera	1.3 MP, 1000 fps	Still images and high quality videos
Flow rate sensors	0.2-5 ml/min, 70-1500 nl/min	Fast and accurate measurements of ultra-low liquid rates

AIChE Southwest Process Technology Conference

Droplet Generation Chip

100 µm Droplet Generation Chip

AIChE Southwest Process Technology Conference

7th AIChE Southwest Process Technology Conference | October 1-2, 2015 | Galveston, TX

Droplet formation at T-junction

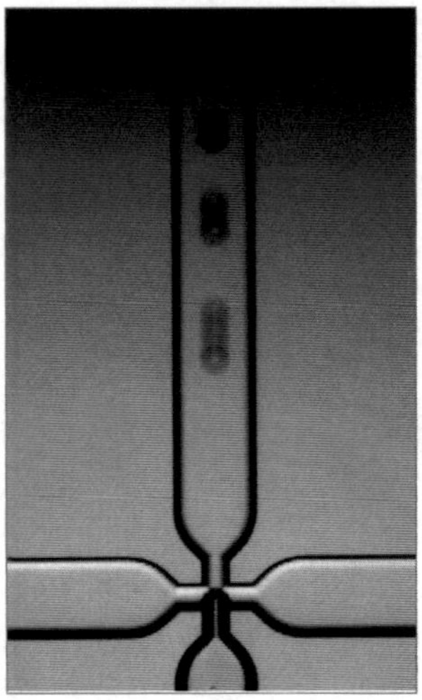

Monodispersed Stream of Droplets Formed at T-junction

- Interaction of droplets at a T-junction gives rise to various primary responses: no collision, coalescence, slipping and splitting.
- The above responses depend on relative velocity of the approaching particles and initial droplet size.

AIChE Southwest Process Technology Conference

7th AIChE Southwest Process Technology Conference | October 1-2, 2015 | Galveston, TX

Effect of Process Parameters on Droplet Size

AIChE Southwest Process Technology Conference

7th AIChE Southwest Process Technology Conference | October 1-2, 2015 | Galveston, TX

Effect of Organic Phase on Oil Droplet Size

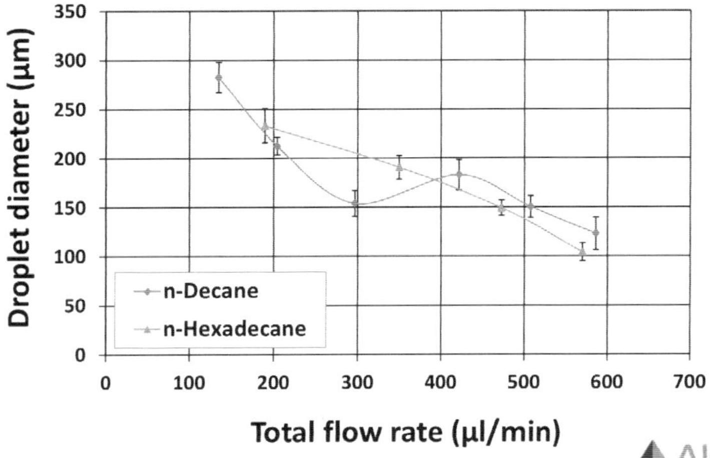

❖ *100 μm hydrophilic Chip*

Effect of Organic Phase on Water Droplet Size

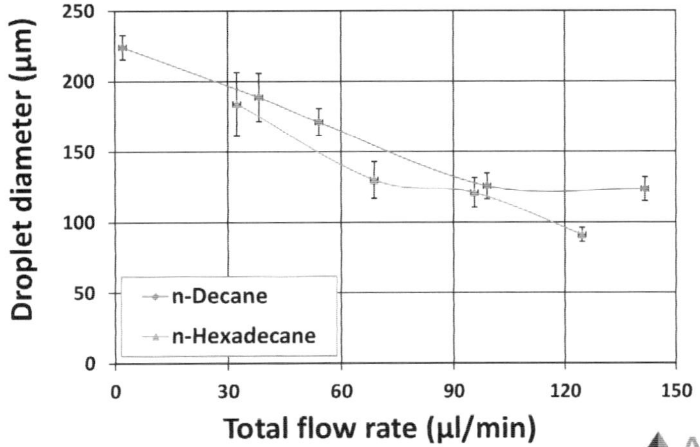

❖ *100 μm hydrophobic chip*

Effect of Salt Concentration and Salt Type on Oil Droplet Size

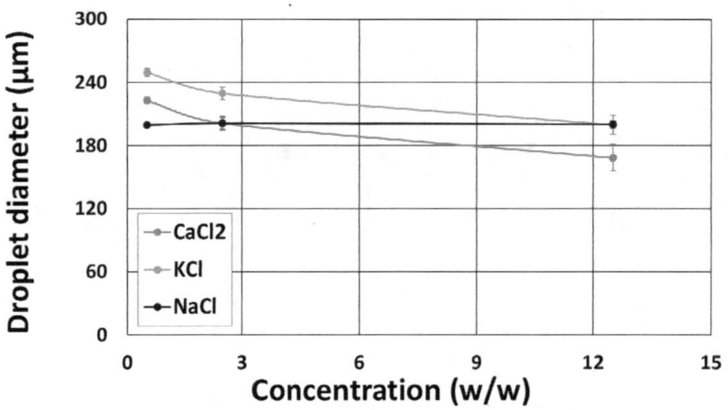

- *100 μm hydrophilic chip; water flow rate = 100 μl/min, oil (hexadecane) flow rate = 1 μl/min*

AIChE Southwest Process Technology Conference

7th AIChE Southwest Process Technology Conference | October 1-2, 2015 | Galveston, TX

Effect of Salt Concentration and Salt Type on Water Droplet Size

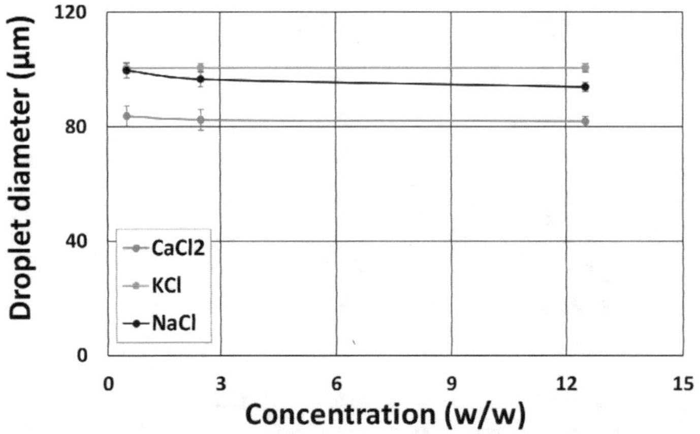

- *100 micron hydrophobic chip; water flow rate = 1 μl/min, oil (hexadecane) flow rate = 100 μl/min*

AIChE Southwest Process Technology Conference

7th AIChE Southwest Process Technology Conference | October 1-2, 2015 | Galveston, TX

Effect of Total Flow Rate on Water and Oil Droplet Size

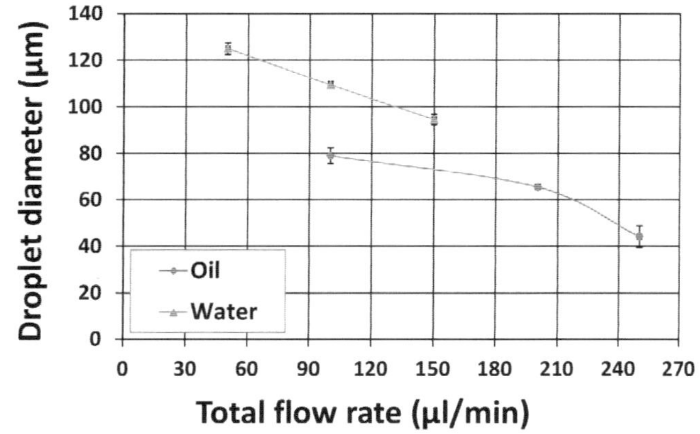

- *Aqueous phase: 3.5 g/l SDS solution; oil phase: hexadecane; water-cut = 1% & 99.98% for 100 µm hydrophobic & hydrophilic chips respectively*

AIChE Southwest Process Technology Conference

7th AIChE Southwest Process Technology Conference | October 1-2, 2015 | Galveston, TX

Effect of Water-cut on Water Droplet Size

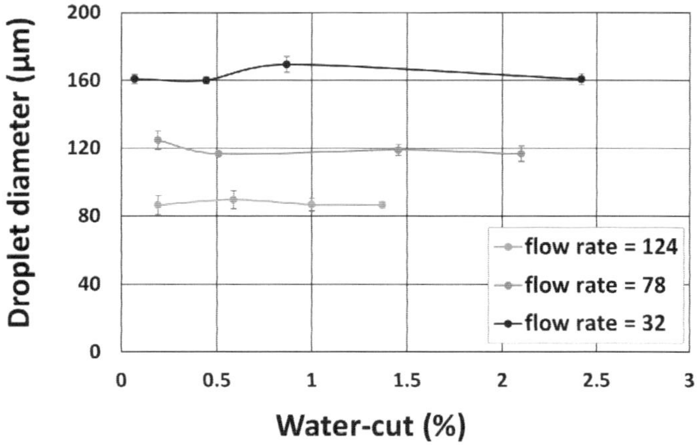

- *100 µm hydrophobic chip; aqueous phase: deionized water; organic phase: hexadecane*

AIChE Southwest Process Technology Conference

7th AIChE Southwest Process Technology Conference | October 1-2, 2015 | Galveston, TX

Stability Analysis (O/W Emulsion)

- 33.3 % hexadecane and 66.67% SDS-water solution after emulsion is stable
- Rest of the water separated in few seconds

Stability Analysis (W/O Emulsion)

- Water-cut = 11% (3.5 g/l SDS solution) and oil percent = 89 % (n-hexadecane) after stable emulsion is formed

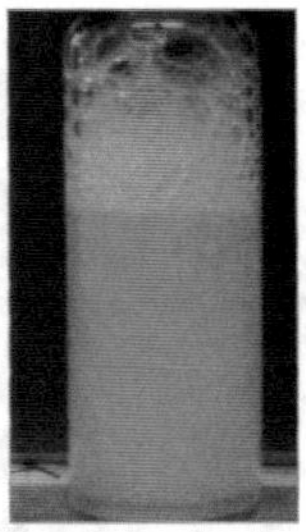

At 0 minute

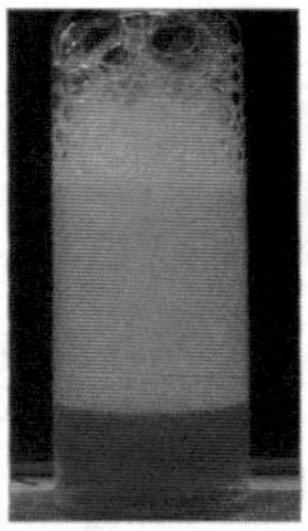

After 1 minute

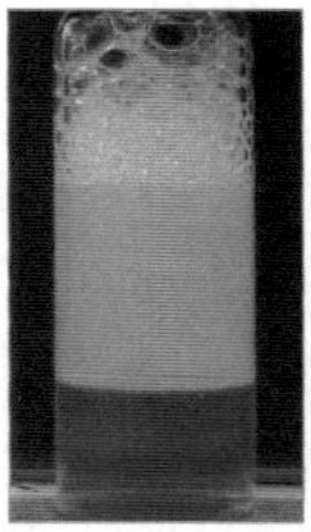

After 5 minutes

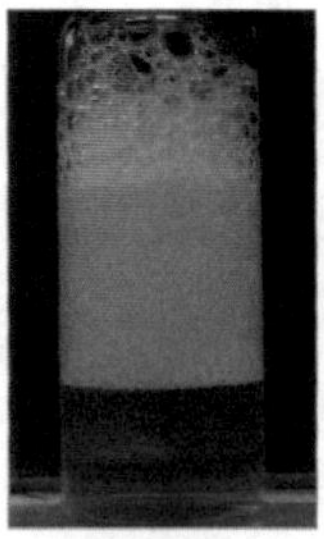

After 30 min

Conclusions

- Smaller water droplets are formed in case of hexadecane as compared to decane for hydrophobic chip
- In case of hydrophilic chip no such effect of organic phase is observed
- For hydrophilic chip, NaCl concentration did not affect droplet size while KCl and $CaCl_2$ resulted in a reduction of droplet size at high concentration
- For hydrophobic chip, increase in NaCl concentration reduced droplet size while KCl and $CaCl_2$ concentrations did not have any effect
- Increase in total flow rate led to a decrease in droplet size for both hydrophobic and hydrophilic chips

AIChE Southwest Process Technology Conference

7th AIChE Southwest Process Technology Conference | October 1-2, 2015 | Galveston, TX

Conclusions

- For a constant total flow rate, water fraction did not seem to affect droplet size
- In case of oil in water emulsion (3.5 g/l SDS solution and hexadecane), minimum oil fraction (v/v) for a stable emulsion was 33.3 %
- In case of water in oil emulsion (3.5 g/l SDS solution and hexadecane), minimum water fraction (v/v) for a stable emulsion was 11 %

AIChE Southwest Process Technology Conference

7th AIChE Southwest Process Technology Conference | October 1-2, 2015 | Galveston, TX

Thank you!

Questions?

AIChE
Southwest Process Technology Conference

Tridiagonal Solutions

compute | innovate | develop

Fine Particles Erosion -Experiments and CFD Analysis

7th AIChE Southwest Process Technology Conference

Praveen Gonuguntla
Swaroop Dhavale
Tridiagonal Solutions Inc.

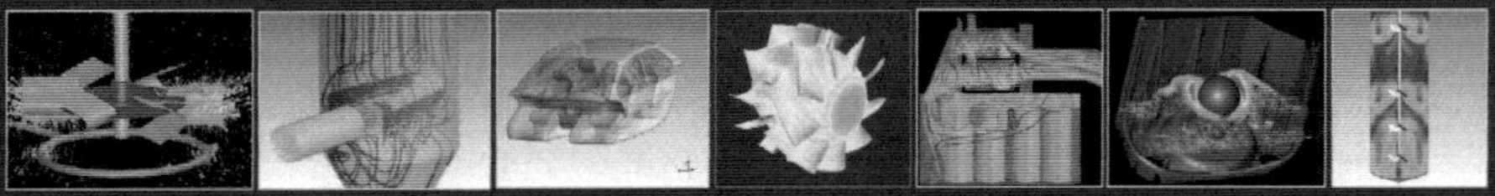

www.tridiagonal.com
info@tridiagonal.com

Outline

- Why to study fine sand erosion
- Experiments of fine sand erosion in 2'' loop
  - Experimental setup
  - Measurements
  - Erosion patterns
- CFD analysis
  - Problem approach
  - CFD results
- Comparison of CFD results with the experiments
- Summary

- Erosion due to fine particles (2 to 50 microns) occurs in oil sands, off shore oil &gas and refinery operations.
- In spite of sand control techniques implemented down hole, fine sand often finds it way in the pipelines which causes erosion
- The above phenomenon has not been studied extensively
- Reliable methods to predict the fines erosion is not readily available
- Common assumption is that fines may not cause erosion since the particle sizes and hence the mass is very small

Study performed

- Performed experiments with average sand size of ~28 microns
- Validation of the fine sand experimental data with CFD

Experimental setup

Tri
Sol
compute | inno

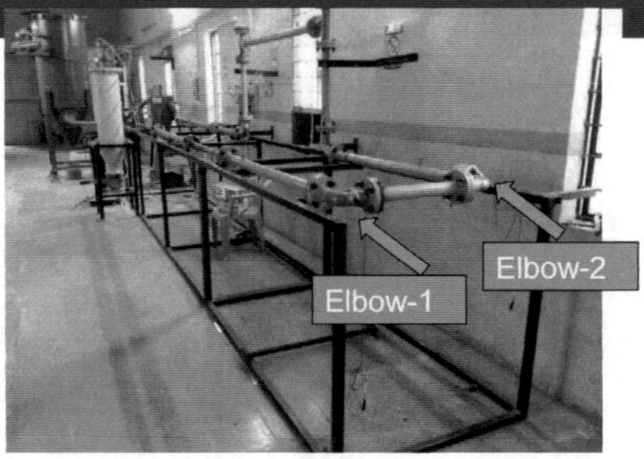

- Erosion experiments are performed in 2" pipe loop
  - Sand fines with average particle size of 28 microns is used (AGSCO #325 sand)
  - Carbon steel-SA106 Grade B, Schedule-80 pipes are considered for the erosion study
  - Elbow-1&2 are the short radius elbow

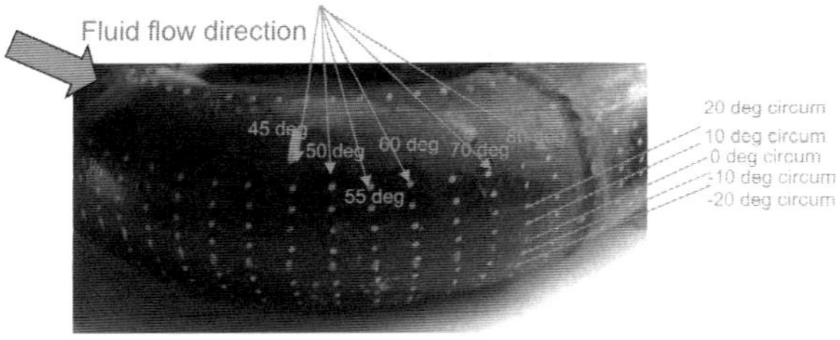

Markings, where measurements are made

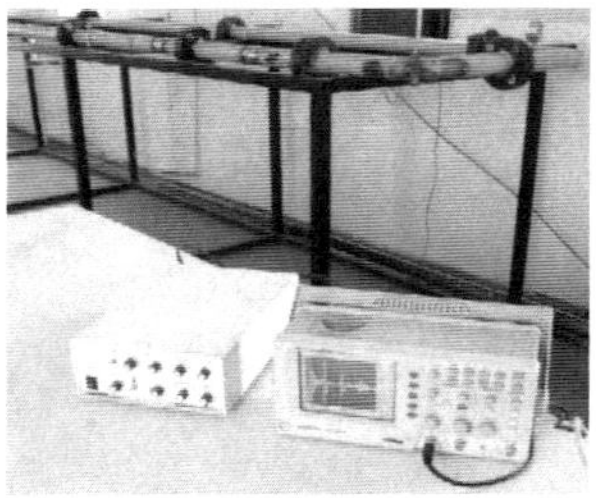

Ultrasound thickness measurement device

- Thickness measurements are made before and after the experiments and thickness loss is reported
- Ultrasound device able to measure the thickness loss even for 10 microns (for fixed transducer)

Erosion patterns

Tridiagonal Solutions
compute | innovate | develop

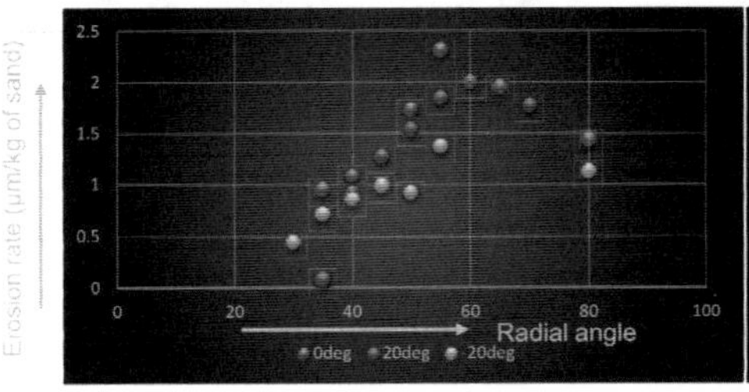

Elbow-1

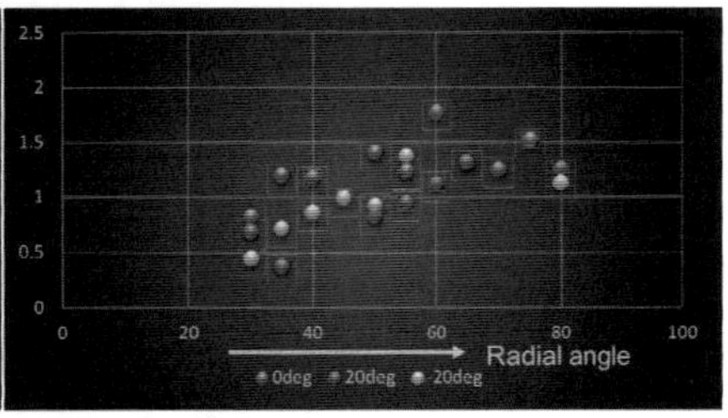

Elbow-2

Gas velocities	34 m/s
Sand flow rate	16 kg/hr
Total sand passed	234 kg
Run Time	14 hrs 07 min

- Elbow-1 experiences more erosion with peak value close to 2.31 microns/kg of sand, where as elbow-2 peak value is 1.78 microns/kg of sand
- Erosion in elbow-2 is relatively uniform along the surface at different circumferential angle unlike elbow-1
  - Elbow-1 has more straight length where particle attains momentum before reaching elbow-1

Erosion spots in elbow-1&2

Tridiagonal Solutions
compute | innovate | develop

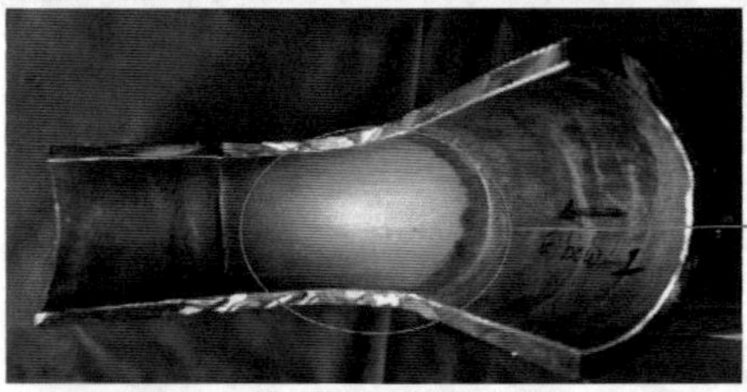

Elbow-1

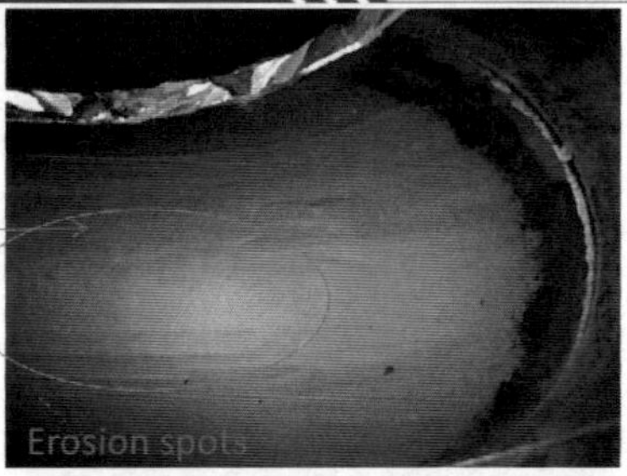

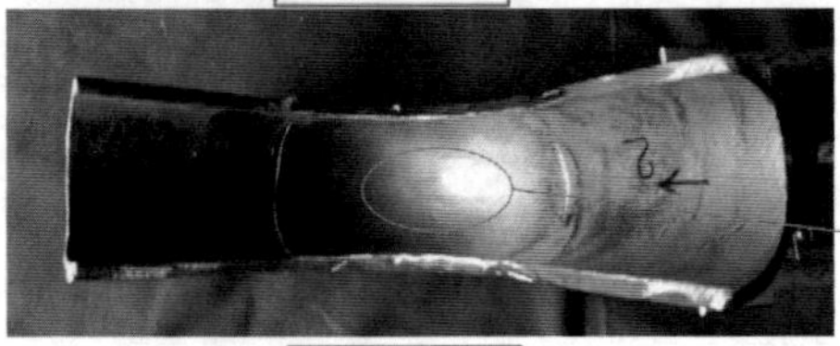

Elbow-2

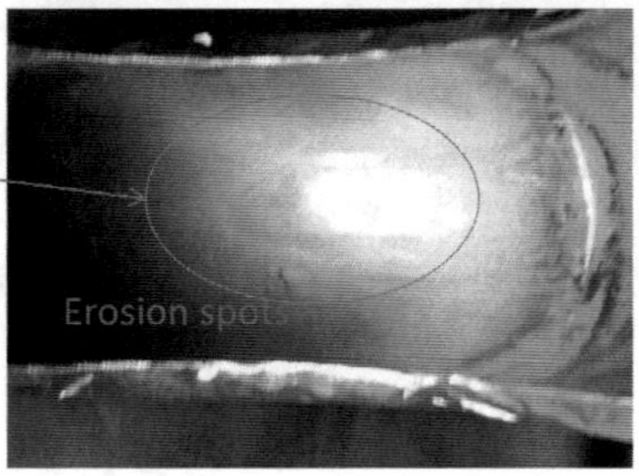

- Elbows are sectioned with intrados and extrados, images shown above are for extrados section
- Erosion spots are also identified by visual inspection

CFD analysis

Tridiagonal Solutions
compute | innovate | develop

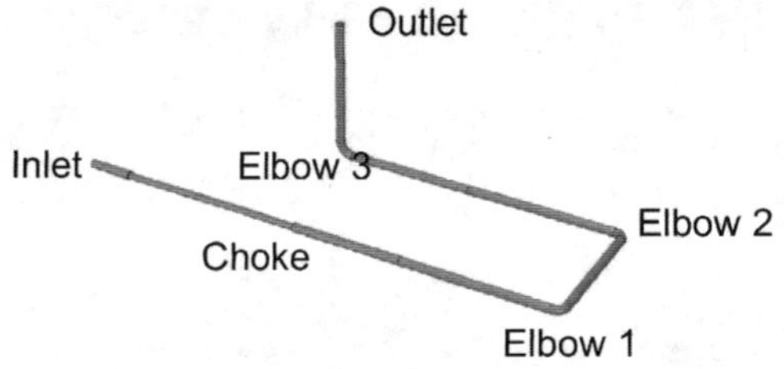

- CFD analysis is performed to predict the erosion in elbow-1&2
- CFD domain is considered from the straight pipe before the choke to the downstream of the elbow-3

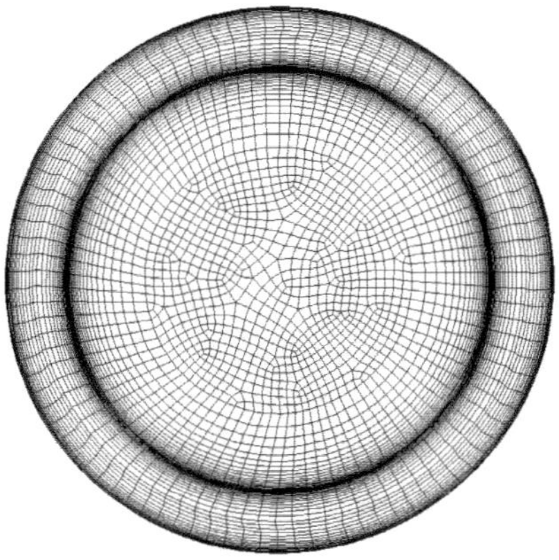

Mesh at Inlet

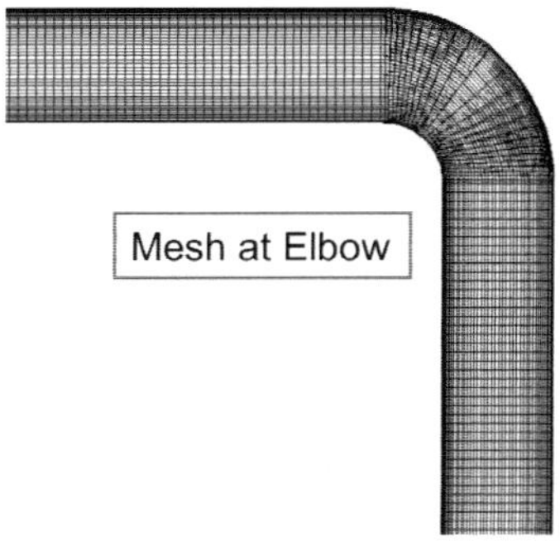

Mesh at Elbow

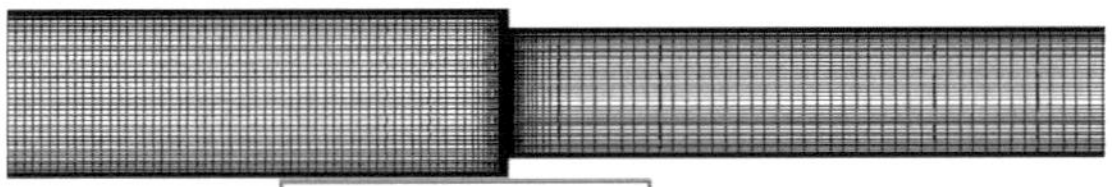

Mesh at Choke

Mesh Details

Total mesh : 11 million

Boundary layer is developed in the elbows and pipe section

Problem approach

- Steady state approach
- Realizable k-epsilon model is used to model turbulence with enhanced wall treatment
  - Wall y+ of 2 is maintained
- To track the particles path and to study the erosion, lagrangian framework is used.
  - DPM (Discrete phase model) is used to track the sand particles in the gas phase
  - Two way coupling is used i.e. gas to particle and particle to gas interactions are accounted
  - Oka et al. model is used to predict the erosion rate.
    - UDF (User defined function) is used to incorporate this model in FLUENT

Oka Erosion Model equation

$$E_V = f(\alpha)E_{90}$$

$$E_{90} = K(H_{v,w})^{k1}\left(\frac{u_p}{u^*}\right)^{k2}\left(\frac{d_p}{d^*}\right)^{k3}$$

$$f(\alpha) = [\sin(\alpha)]^{n_1}\{1 + H_V[1 - \sin(\alpha)]\}^{n_2}$$

* Above equations are from the Oka et al. literature data

- Many models are available in the literature, of which Oka model is quite popular, since it accounts for erodent material properties
- This model accounts for angle of impingement, material hardness, particle impingement velocities and particle diameter function
- Oka model reports erosion in mm3/kg of sand

CFD results

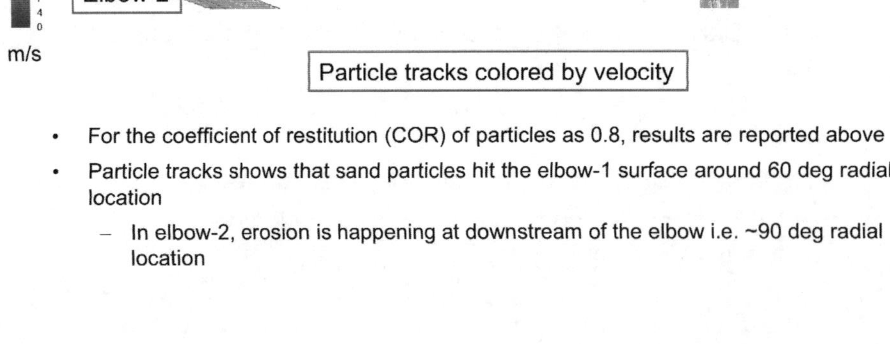

Particle tracks colored by velocity

- For the coefficient of restitution (COR) of particles as 0.8, results are reported above
- Particle tracks shows that sand particles hit the elbow-1 surface around 60 deg radial location
  - In elbow-2, erosion is happening at downstream of the elbow i.e. ~90 deg radial location

CFD vs. Experiments

Tridiagonal Solutions
compute | innovate | develop

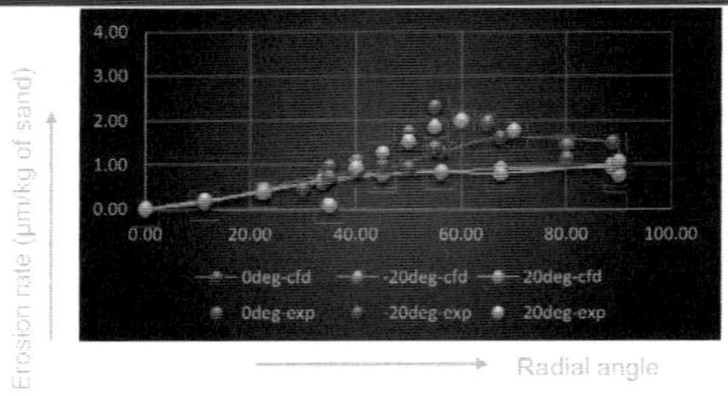

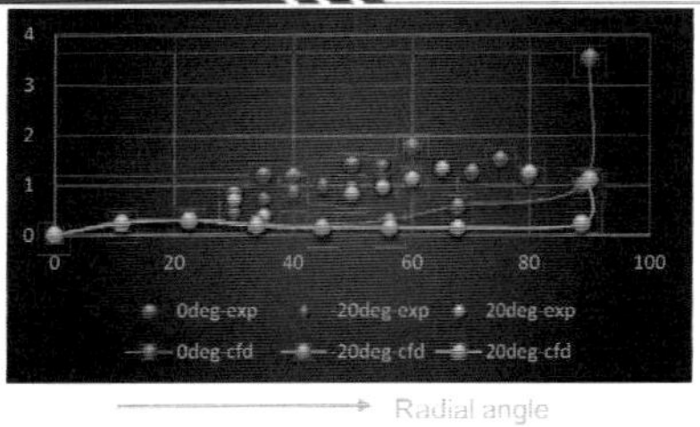

Elbow-1

Elbow-2

- Elbow-1 experiences more erosion with peak value close to 2.31 microns/kg of sand, where as elbow-2 peak value is 1.78 microns/kg of sand
- Erosion in elbow-2 is more uniform along the surface at different circumferential angle unlike elbow-1, this is due to more straight length where particle attains momentum before reaching elbow-1
- Predictions of the erosion location and values from CFD is reasonable for the elbow-1
  - However capturing the erosion location in elbow-2 becomes challenging in the CFD

Effect of coefficient of restitution

Tridiagonal Solutions
compute | innovate | develop

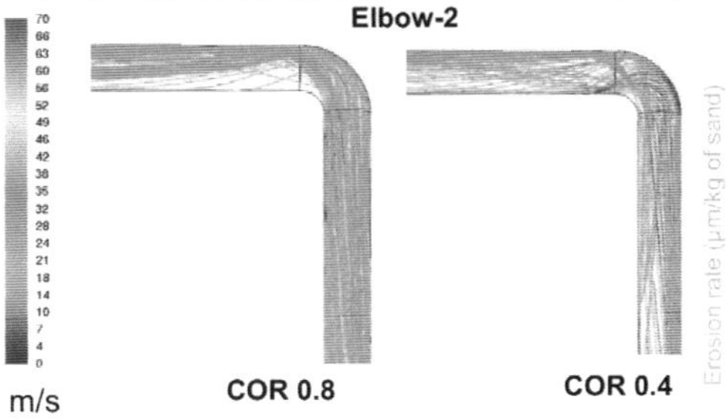

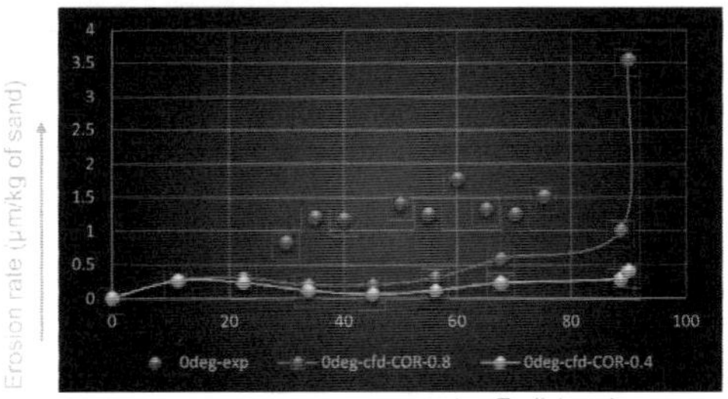

Comparison of erosion rates in elbow-2

- Attempts were made to capture the erosion location appropriately in elbow-2, by reducing the COR to 0.4
  - i.e. assuming particle looses 60% of the energy after hitting the walls surface
- Low COR is studied to avoid the rebound of the particles from the elbow-1 that reaches the intrados section of the elbow-2
- Erosion is further under predicted for the low COR

- High erosion rates are noticed in the elbows for the fine sand
  - About 2.31 microns/kg of sand for 34 m/s air velocity
  - Elbow-2 is having relatively low erosion compared to elbow-1 due to momentum loss after reaching elbow-1
- Common assumption of no/minimum erosion for fine sand is proved to be false
- CFD better predicts the erosion location for the elbow-1
  - Oka model under predicts the erosion
  - Effect of particles loosing energy after it hits the wall surface is investigated
  - However, erosion effects of back to back elbows needs to be further investigated in CFD

Effective Interviewing

Greg Yeo
ExxonMobil Chemical Company

AIChE
Southwest Process Technology Conference

Agenda

- Before the Interview
  - Preparing as if it's an important class
    - Understand the industry and company: What are they looking for?
    - Be prepared to tell your story in a relevant way
- At the Interview
    - Be authentic, relevant and motivated
- After the Interview
    - Followup, build a network, build a career

AIChE
Southwest Process Technology Conference

Interviewer Interaction

- What's the interviewer looking for? *The "right" balance of:*
  - **Technical Aptitude/Capability**
  - **Demonstrated Leadership**
  - **A person they can work with**
    - *People Aptitude*
    - *Enthusiasm, Interest*
    - *Flexibility*
    - *Communication Skills*
  - ***"Grit"***
- What does the interviewer have to go on?
  - **Academic record**
  - **Resume**
  - **Face to face interview**
  - *Internship (maybe?)*
  - *Campus interview (maybe?)*
  - *Phone interview (maybe?)*

"Does this person make my team better? Is this an engineer that I want on my next project?"

7th AIChE Southwest Process Technology Conference | October 1-2, 2015 | Galveston, TX

Prepare

- Research the company in advance: *Most candidates do not!*
  - Understand the company's business
  - What they are looking for?
- Prepare an interesting resume: *Most resumes look alike!*
  - Emphasize *your* experience and skills
  - *Make it a platform to tell your story*
- Analyze Yourself and ***prepare***: *Most students do not prepare!*
  - Know your strengths and weaknesses
  - What do they need? Do your skills fit? Do you know what you want? Are you flexible?
  - Would ***you*** hire you for this position?
- Prepare examples, stories, questions
  - This is a future colleague. Be interesting, memorable, able to tell an engaging story.
  - ***I remember engaging stories from energized, capable candidates!***

7th AIChE Southwest Process Technology Conference | October 1-2, 2015 | Galveston, TX

If You Have A Lunch / Dinner Meeting

- This may be part of the interview process
  - Be yourself! This person is trying to get to know you.
  - Are you the kind of person they would want to work with?
- Ask lots of questions
  - Show genuine interest. Let this meeting be part of your prep
  - This person got a job – learn from them
- Be energetic, interested, professional

Before I interview a student,
I always talk to the person who took them to dinner.

AIChE
Southwest Process Technology Conference

7th AIChE Southwest Process Technology Conference | October 1-2, 2015 | Galveston, TX

At the Interview

- Dress for Success
  - If you don't know, ask or overdress. And be comfortable with it.
- Silence your phone in advance
- Use positive body language
- Be enthusiastic, engaging, and authentic.
  - Make eye contact. Show interest. And mean it.
- Create a two way conversation
  - You are interested in this interviewer and company
  - Ask questions. Take the lead when it's offered.
  - ***Be able to tell an <u>interesting story that relates to the position/company.</u>***
    - Leadership
    - Working with / connecting with people
    - Technical depth
    - Ability to stick with tough challenges

AIChE
Southwest Process Technology Conference

7th AIChE Southwest Process Technology Conference | October 1-2, 2015 | Galveston, TX

At the Interview

- Direct the conversation about your qualifications, previous accomplishments, and interest in the job
  - Highlight how previous experiences prepared you for this position
  - *If you understand the position, you can tell the right story about yourself*
- Don't be afraid to highlight areas that company may not ask about
  - Language skills
  - Interest in travel – if it's a component of the position
  - Long term career goals that are consistent with the company
  - Your innate skills and energy:
    1. Technical excellence
    2. Leadership
    3. Team success

AIChE Southwest Process Technology Conference

Make it a Conversation!!

- What are the primary sources of revenue?
- Ask/talk about the key products or services
- How many of your senior leaders are engineers?
- What is a typical career path for a ChE like me?
- Do you see my experience and skills as a good fit for this position?
- What opportunities are there for training, travel, continuing education?
- How do you help employees with work-life balance?
- Is the company growing? What drives growth? Are there global operations?

AIChE Southwest Process Technology Conference

7th AIChE Southwest Process Technology Conference | October 1-2, 2015 | Galveston, TX

After the Interview

- Before you leave, ask, "What's the next step?"
  - When can you expect to hear from them?
- Thank everyone that you talk to
- Send a follow-up note or email
- Analyze your performance
  - What should I do differently next time?
- Follow up

AIChE
Southwest Process Technology Conference

Be Careful, Be Thoughtful, Be Professional

- Reputation and ethics matter
- Be careful with social media – LinkedIn, Facebook, Twitter
- Build bridges – don't burn them
- Learn about industries, companies, cultures interviewing experience
- Find ways to stay current

AIChE
Southwest Process Technology Conference

7th AIChE Southwest Process Technology Conference | October 1-2, 2015 | Galveston, TX

Questions?

Or conversation later today

7th AIChE Southwest Process Technology Conference | October 1-2, 2015 | Galveston, TX

11

Chemical Engineers in Oil and Gas

Dmitry Galushko

Josh Bird

About – Josh Bird

- 2013 UH ChemE Alumni
- ConocoPhillips Facilities Engineer
  - Eagle Ford Development (June 2013 - Current)

Upstream Industry

- Commonly known as the Exploration & Production(E&P) sector
- Two major sectors

Offshore:

Onshore:

Onshore Oil and Gas

- Concentrated on different types of reservoirs

Conventional:

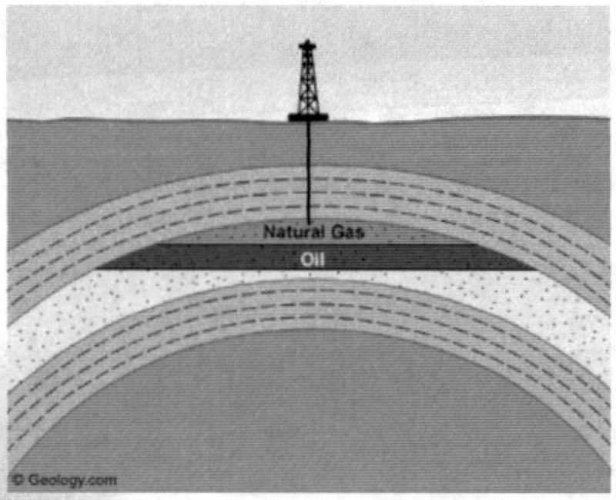

Unconventional:

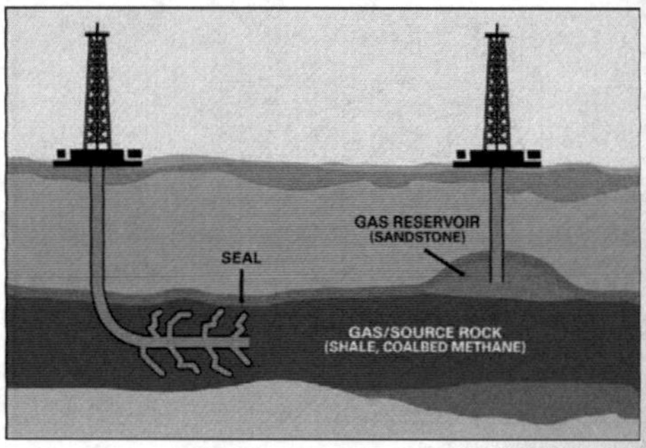

Unconventional Reservoirs

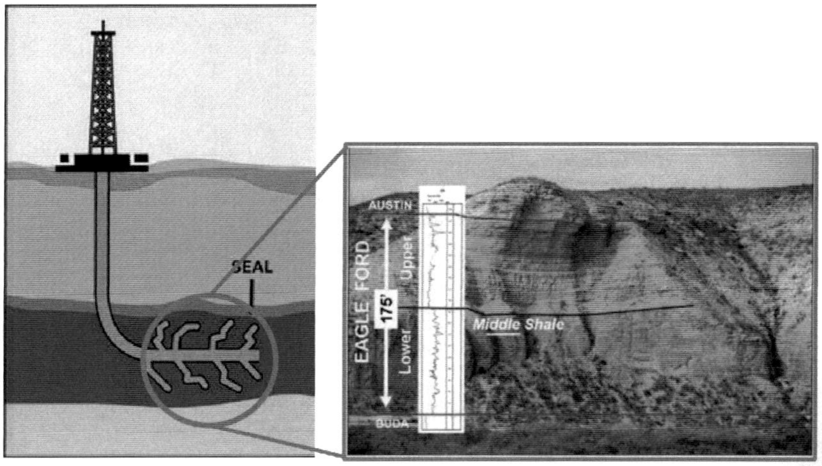

Unconventional Revolution

- Unconventional reservoirs have revolutionized the oil & gas industry
  - Hydraulic Fracturing
  - Horizontal Drilling

- Three factors leading to the boom in Upstream
  - Unconventional reservoirs (Eagle Ford Shale)

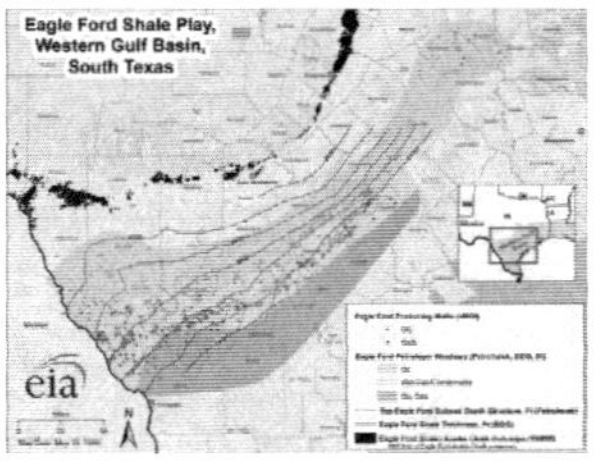

  - High oil prices and low transportation costs
  - Aging workforce

Unconventional Revolution

- Unique factors about onshore unconventional operations
  - Hundreds and even thousands of wells and small facilities
  - Have only recovered a fraction of oil in place
    - Recovery close to 5%
  - Still has a lot of technological advancement to go

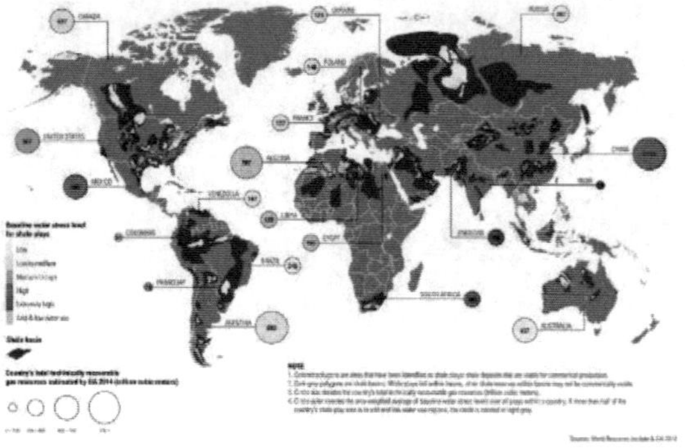

Job Tasks

- Facilities Engineer in Eagle Ford Development
- Currently in charge of the design of all well facilities across the Eagle Ford field
  - Working with engineering, construction, operations, and supply chain to optimize facilities and lower costs

Job Task

- Reviewing current designs and optimizing for constructability and costs

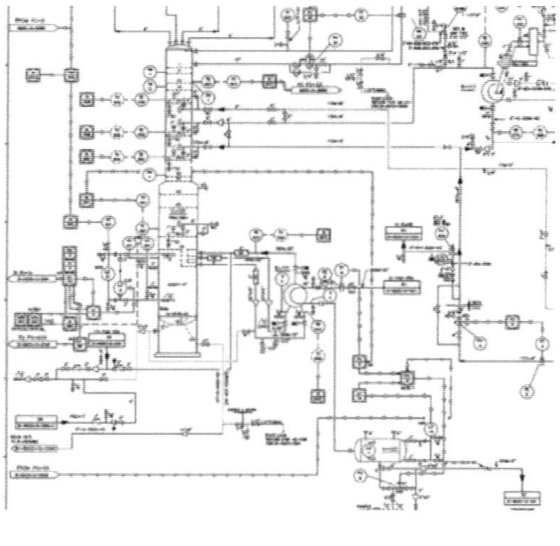

Typical Tasks

- Looking forward into the future and challenging the status quo to make creative and more efficient designs

Typical Tasks

- Working on multifunctional teams to fundamentally change the way we do business to achieve our targets

- People skills are KEY!

Career Options

- Chemical Engineers are very versatile
- Engineering Positions
  - Drilling Engineering
  - Reservoir Engineering
  - Completions Engineering
  - Production Engineering
  - Facilities Engineering
  - Field Engineering (contractor companies mostly)
- Chemical Engineers typically focus in Reservoir, Production, Facilities, and Field Engineering

Resume Tips

- Now is the best time to begin!
- Want to focus on three things
  1. Good GPA
  2. Solid Work Experience
  3. Leadership opportunities
- Combine this with a good attitude and you are on your way!

Transitioning To Professional Life

- Go to sleep early, wake up early
- Spend time researching how to handle money the right way
  - Do not go into debt to buy cars and houses in your first years at work
- Spend time creating personal and professional goals
  - Where do you want to be in 10 years? 5 years? 1 year?
  - What small actions can you take on a consistent basis to reach those goals?
- Have a passion for your work
  - If you can't find passion for it, find another job
- Enjoy your newly found free time!
  - Pick up hobbies
  - Rebuild your social life that is now in shambles
  - NETWORK!

Career Options

- Post Graduate Degrees may or may not align with your career goals
- Most job functions do not require advanced degrees
  - Including management, both business and technical
- Some job functions that advanced degrees will benefit
  - Subject matter experts (SME's)
  - Research and Development
- MBA's
  - Can assist in getting into upper level management
  - But so can hard work, dedication, and a passion for your work

Questions

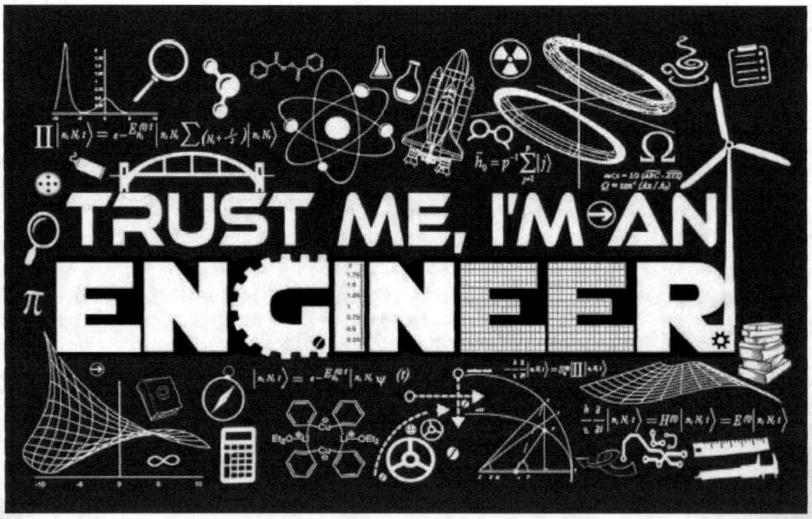

A DAY IN THE LIFE OF A CHEMICAL ENGINEER

Jaime Robles, Senior Process Engineer, Fluor Corporation
AIChE SPTC 2015 – Student Session

Best advice for ChemE's…

"The Chemical Engineer needs to be flexible and adaptive to change. Ask questions and challenge everything."

AIChE Southwest Process Technology Conference
7th AIChE Southwest Process Technology Conference | October 1-2, 2015 | Galveston, TX

Transitioning into the real world...

- So you've graduated (or close to)...now what?
  - Wait, what does an engineer do anyway?
    - Solve problems safely and efficiently to improve lives.
    - More importantly, we turn ideas into $$$$$$!
  - Time to get paid for that brain!
    - Time off? Vacation? Job searching? Starting a job?
    - Operating, Design, Consulting, R&D
  - Life in general
    - Steady, more predictable schedule.
    - You may miss the "college atmosphere".
    - Start thinking of other "grown-up' things like retirement, insurance.

AIChE
Southwest Process Technology Conference

7th AIChE Southwest Process Technology Conference | October 1-2, 2015 | Galveston, TX

Day-to-day for Process

- Significant role throughout the course of a project
  - Conceptual, design, troubleshoot, go-to team
- Perform calculations...lots of calculations...
  - ✓ Process flow and optimization
  - ✓ Process simulation
  - ✓ Heat and material balances
  - ✓ Process flow diagrams
  - ✓ Piping and instrumentation diagrams
  - ✓ Equipment sizing and specification
  - ✓ Line sizing and hydraulics
  - ✓ Instrumentation specification
  - ✓ Process safety reviews
  - ✓ Relief scenario analysis
  - ✓ Utility and effluent summaries
  - ✓ Technical support for other disciplines
  - ✓ Client interaction
  - ✓ "Data detective"
- Travel, initiatives and other fun activities

AIChE
Southwest Process Technology Conference

7th AIChE Southwest Process Technology Conference | October 1-2, 2015 | Galveston, TX

"Engineer At Work"

AIChE
Southwest Process Technology Conference

7th AIChE Southwest Process Technology Conference | October 1-2, 2015 | Galveston, TX

It starts with an idea…

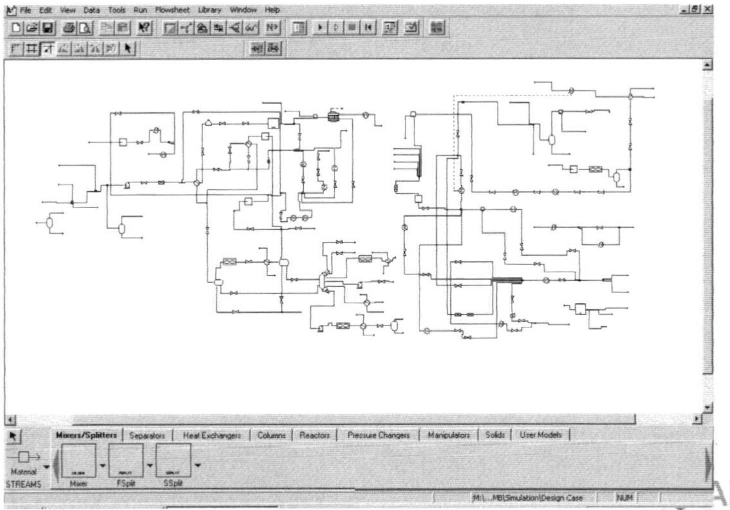

Process Simulation

AIChE
Southwest Process Technology Conference

7th AIChE Southwest Process Technology Conference | October 1-2, 2015 | Galveston, TX

…and it starts taking shape…

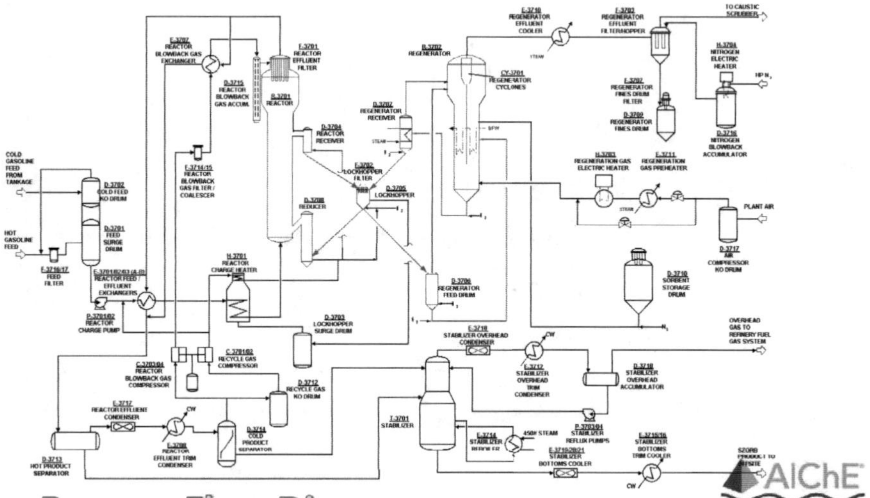

Process Flow Diagram

AIChE Southwest Process Technology Conference

7th AIChE Southwest Process Technology Conference | October 1-2, 2015 | Galveston, TX

…the details are designed…

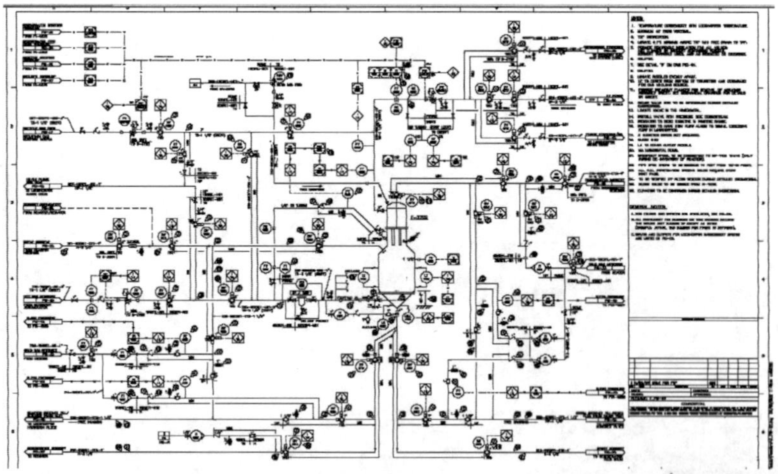

Piping & Instrumentation Diagram

AIChE Southwest Process Technology Conference

7th AIChE Southwest Process Technology Conference | October 1-2, 2015 | Galveston, TX

3D Model

Southwest Process Technology Conference

7th AIChE Southwest Process Technology Conference | October 1-2, 2015 | Galveston, TX

But you can also have a little fun…for your sanity!

- Stay active and involved, both inside and outside of work!

BPMS150

Fluor Cares

CrossFit
Forging Elite Fitness

AIChE
Southwest Process Technology Conference

Career paths for a Chemical Engineer

- Honestly, your options are almost endless!

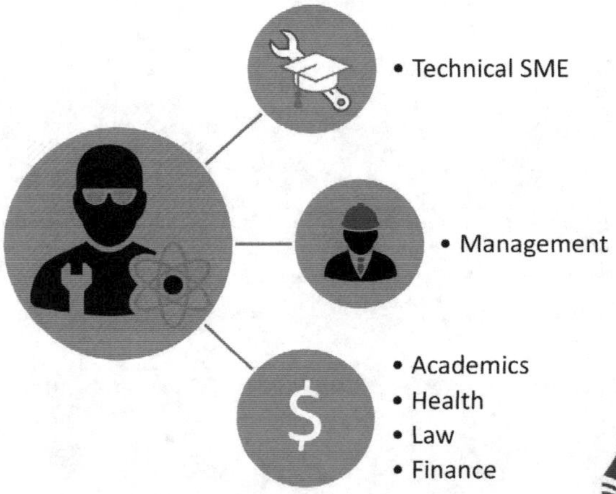

AIChE
Southwest Process Technology Conference

Survival tips & things I wish I knew...

- Be organized and disciplined.
- Stay mindful of the future position you want.
- If you run into a problem, always offer a solution.
- Hone communication skills.
- No inhibitions – ask about designs and decisions made.
- Adapt to change; be flexible; there's not just "one" answer.
- Step up to challenges...and menial tasks.
- Make your supervisors look good.
- Learn something else besides Chemical Engineering.
- See the forest **and** the trees.
- Be humble, respectful, and responsible.
- Provide value to your client.
- Take vacations, find a hobby, save for rainy days.

7th AIChE Southwest Process Technology Conference | October 1-2, 2015 | Galveston, TX

Thank you!

7th AIChE Southwest Process Technology Conference | October 1-2, 2015 | Galveston, TX

PSM – a regulation or culture ?

Swarup Bade, P.E
Process Safety Consultant

Lloyd's Register Energy Americas, Inc.
1330 Enclave Parkway, Houston, TX 77077, USA
M +1 916 220 2594, E swarup.bade@lr.org

Lloyd's Register Energy

Working together
for a safer world

PSM

Bhopal, India (December 1984)

- Pesticide production facility
- Water introduced into methyl isocyanate storage
- MIC toxic vapor release from vent system
- 3000 early fatalities;
- ~200,000 injuries
- Plant shut down; Union Carbide eventually sold
- PSM prompted

Introduction

29 CFR 1910 OSHA

As required by the U.S. Department of Labor, Occupational Safety and Health Administration

GENERAL INDUSTRY REGULATIONS

Process Safety Management

Process Safety Management is the application of management principles and systems to the identification, understanding and control of process hazards to protect employees, facility assets and the environment.

PSM 14 Elements

1. Employee Participation
2. Process Safety Information
3. Process Hazard Analysis
4. Operating Procedures
5. Training
6. Mechanical Integrity
7. Contractors
8. Pre-Startup Safety Review
9. Work Permits
10. Management of Change
11. Incident Investigation
12. Emergency Response Plan
13. Compliance Audits
14. Trade Secrets

1. Employee Participation

- Requires employers to:

a) Develop a written Employee Participation Plan for Employee Involvement
b) Consult with employees on the conduct of the development of PSM Elements
c) Provide access to PSM information

2. Process Safety Information (PSI)

OSHA groups PSI in 3 categories:

a. Hazards of the chemicals and flammables in the process

b. Information related to the technology of the process

c. Information pertaining to the equipment in the process.

Lloyd's Register Energy

3. Process Hazard Analysis (PHA)

- What if
- Check list
- What if / Check list
- HAZOP (Hazard and Operability study)
- FMEA (Failure Mode Effect Analysis)
- FTA (Fault Tree Analysis)

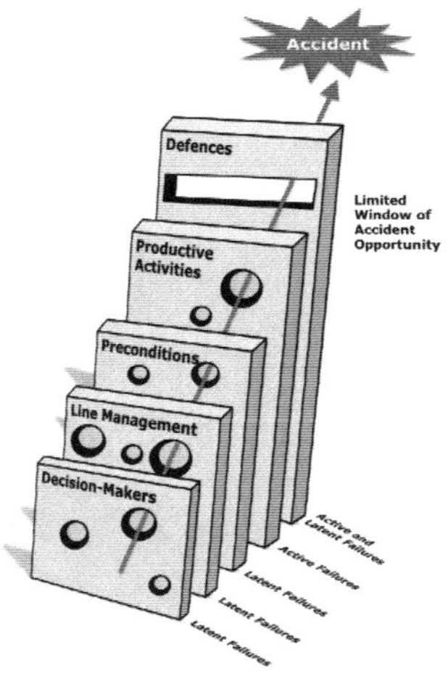

Lloyd's Register Energy

4. Operating Procedures

- Initial startup; Normal, temporary and emergency operations.
- Normal and emergency shutdown.
- Operating limits; Consequences and corrections of deviation
- Engineering & administrative controls and personal protective equipment
- Safety systems and their functions

Lloyd's Register Energy

5. Training

- Process Overview and operating procedures
- Safety & health hazards, emergency procedures
- Refresher Training
- Record of the training

Lloyd's Register Energy

6. Mechanical Integrity:

Establish and implement written procedures (MIP: Mechanical Integrity Program) to maintain the on-going integrity of facility equipment.

This includes:

a) Test & Inspections (T&Is)
b) Documentation of T&Is
c) Equipment deficiencies
d) New Equipment.
e) Material Control.

Lloyd's Register Energy

7. Contractors

Requirements include:

- Obtain and evaluate safety performance and programs
- Provisions of the emergency action plan
- Develop and implement safe work practices
- Contract employee log

Lloyd's Register Energy

8. Pre-Startup Safety Review

- Construction and equipment is in accordance with design specifications
- Safety, operating, maintenance, and emergency procedures
- Modified facilities meet the requirements contained in Management of Change
- Training of each employee involved in operating a process

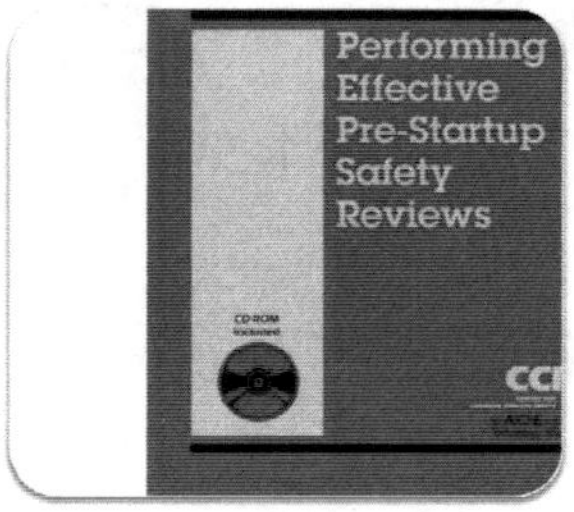

Lloyd's Register Energy

9. Work Permits

- Fire protection and prevention to be implemented (29 CFR 1910.252(a))
- Dates and identity of the object on which hot work is performed

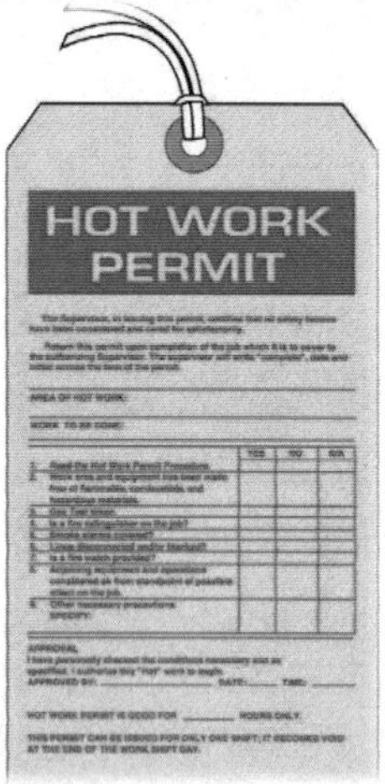

Lloyd's Register Energy

10. Management of Change (MOC):

Establish and implement written procedures to manage changes (except for "replacements in kind") to process chemicals, technology, equipment, and procedures; and, changes to facilities that affect a covered process.

Lloyd's Register Energy

11. Incident Investigation

- Investigate each incident
- Establish an incident investigation team
- Incident Reports
- Corrective Actions
- Report Review.

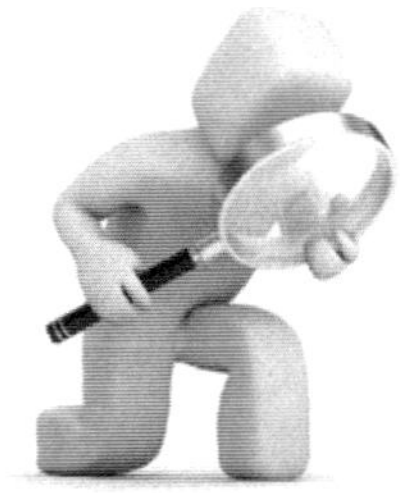

Lloyd's Register Energy

12. Emergency Planning & Response

- Emergency action plan for the entire plant
- Procedures for handling small releases

Lloyd's Register Energy

13. Compliance Audits

Certify compliance with the provisions of the PSM Standard at least every three years to verify that the procedures and practices developed under the standard are adequate and are being followed.

Lloyd's Register Energy

14. Trade Secrets

Standard 1910.1200, employees and their designated representatives shall have access to trade secret information contained within the process hazard analysis and other documents required to be developed by this standard.

Lloyd's Register Energy

Lloyd's Register Energy

Texas City, Texas (March 2005)

- Refinery isomerization unit
- One valve not opened during unit restart
- Release of hot flammable material from blowdown
- Ignition and vapor cloud explosion
- 15 fatalities, 170+ injuries; BP losses and impacts

Lloyd's Register Energy

National Emphasis Programs (NEP)

National Emphasis Programs (NEP):

- Provide for planned inspections in high hazard industries
- Focus efforts on specific hazards
- Refinery NEP focuses on implementation of PSM in Refineries
- Chemical NEP focuses on implementation of PSM in all other PSM covered facilities

Lloyd's Register Energy

OSHA NEP Program

Comparison of Refinery and Chemical NEP's (as per OSHA 2012 report)

Type of PSM Inspection Program	Citation Per Inspection	Penalty Per Inspection
Refinery NEP	11.2	$ 76,800
Chemical NEP	8.4	$ 31,600

Top 1910 Standards violated (as per OSHA 2012 report)

Standard	Description	Number	%
1910.119	Process Safety Management	891	59.9 %
1910.147	Lockout / Tagout	55	3.7 %
1910.120	Hazardous Waste & Emergency Response	47	3.2 %
1910.134	Respiratory Protection	36	2.4 %

Lloyd's Register Energy

OSHA NEP Citations

Top Refinery NEP Cited PSM Elements
(as per OSHA 2012 report)

Element	Description	Number	%
J	Mechanical Integrity	198	19.5 %
d	Process Safety Information	177	17.4 %
f	Operating Procedures	174	17.1 %
e	Process Hazard Analysis	168	16.5 %
I	Management of Change	92	9 %

Top Chemical NEP Citations by PSM Elements
(as per OSHA 2012 report)

Element	Description	Number	%
j	Mechanical Integrity	207	23.2 %
d	Process Safety Information	186	20.9 %
e	Process Hazard Analysis	141	15.8 %
f	Operating Procedures	125	14 %
I	Management of Change	49	5.5 %

Lloyd's Register Energy

Questions

Lloyd's Register Energy

An Introduction to Low Pressure Storage Tanks

Johnson, M., Foster Wheeler

Originally presented on: 10/2/2015 14:00:00 - 14:30:00

Storage tanks are an overlooked part of any refinery or chemical plant. They don't make boatloads of money, but they are an integral part of keeping a process running smoothly and provide a buffer to help minimize lost production and shutdown time. Adequately sizing raw feed, intermediate product, and final product storage tanks should include a number of scenarios which are often overlooked by "rule-of-thumb" type considerations. This presentation is an introduction to the various design considerations for storage tanks including types, sizing, breathing calculations, and safety considerations encountered when evaluating tanks in a manufacturing facility.

Introduction to Pressure Relief Devices

Emily Jackson-Crosby
Process Engineer

Overview

- Importance of Relief Valves
- Codes and Standards
- Definitions
- Types of Relief Devices
- Scenario Analysis
- Additional Analysis Steps
- Things to Consider

7th AIChE Southwest Process Technology Conference | October 1-2, 2015 | Galveston, TX

Importance of Relief Valves

- Safety
  - Employees
  - Community
- Process Containment
- Mechanical Integrity
- Process Hazard Analysis
- Compliance
- Environmental

Codes and Standards

- ASME Section I - Boilers
- ASME Section VIII – Pressure Vessels, etc.
- ASME B31.3 – Piping
- API 520 Pt I and Pt II – Sizing, Selection, and Installation of Pressure-relieving Devices
- API 521 – Pressure-relieving and Depressuring Systems
- API 2000 – Atmospheric Storage Tanks
- API 620 – Large, Welded, Low-pressure Storage Tanks
- API 650 – Welded Tanks
- ASHRAE 15 – Refrigeration Systems
- Company Standards

Definitions

- Pressure Relief Device
  - actuated by inlet static pressure and designed to open during emergency or abnormal conditions
  - prevents a rise of internal fluid pressure in excess of a specified design value and may be designed to prevent excessive internal vacuum
- Maximum Allowable Working Pressure (MAWP)
  - maximum gauge pressure permissible at the top of a vessel in its normal operating position at the designated coincident temperature

AIChE Southwest Process Technology Conference

Definitions

- Accumulation
  - % pressure increase above MAWP allowed during overpressure
- Overpressure
  - % pressure above device set point
- Set Pressure
  - The inlet gauge pressure at which the pressure relief device is set to open under service conditions.
- Relieving Pressure
  - Valve set pressure plus overpressure or accumulation.
- Orifice Area
  - Required – based on required relieving rate
  - Available – installed orifice size

AIChE Southwest Process Technology Conference

7th AIChE Southwest Process Technology Conference | October 1-2, 2015 | Galveston, TX

Definitions

- Blowdown
  - The difference between the set pressure and the closing pressure of a PRV (typically 7% but can range higher)
- Simmer
  - The audible or visible escape of compressible fluid between the seat and disc just below set pressure

Types of Relief Devices

- Conventional Relief Valve

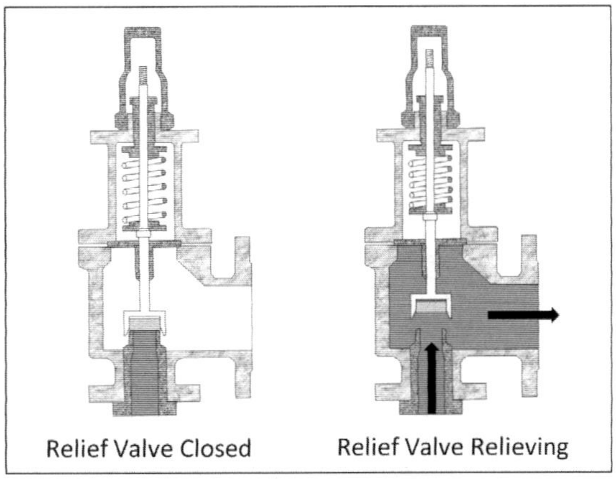

www.mycheme.com

Types of Relief Devices

- Balanced Bellows Relief Valve

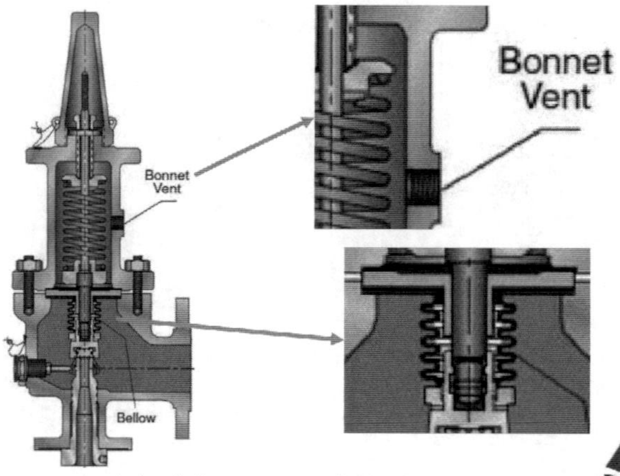

http://www.wermac.org/valves/valves_pressure_relief.html

Types of Relief Devices

- Pilot Operated Relief Valve

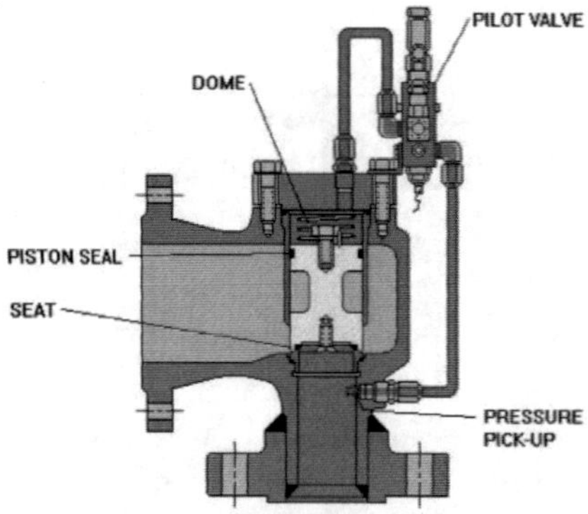

https://controls.engin.umich.edu/wiki/index.php/ValveTypesSelection

Types of Relief Devices

- Rupture Disk

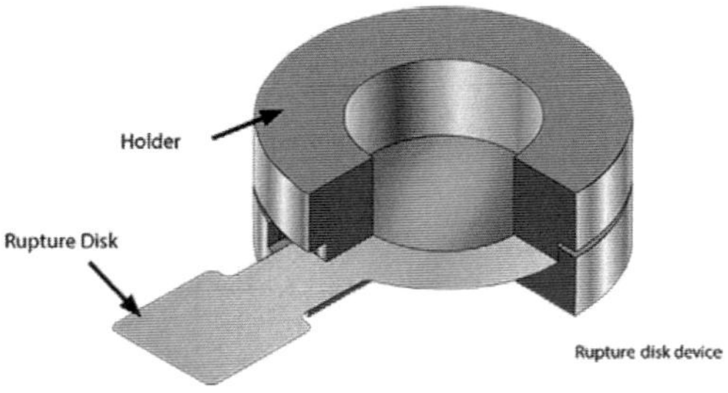

http://www.nationalboard.org/Index.aspx?pageID=164&ID=381

Types of Relief Devices

- Conservation Vent/Emergency Vent

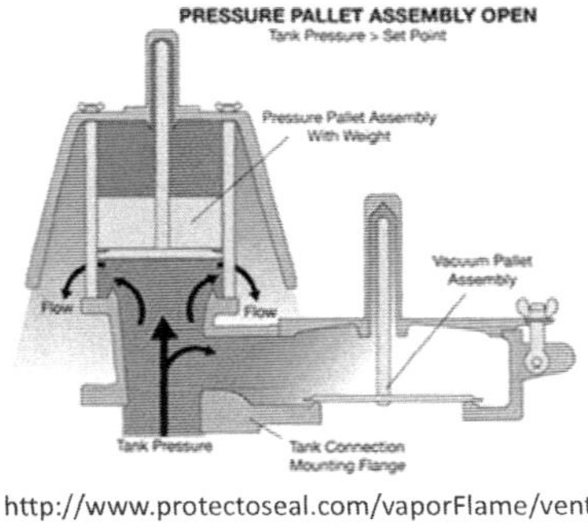

http://www.protectoseal.com/vaporFlame/venting_safety.cfm

http://mx.valves.pentair.com/en/valves/products/pressure_relief_valves/tank_protection/?id=tcm:204-5792&catid=tcm:204-4115-1024

Types of Relief Devices

- Rupture Pin

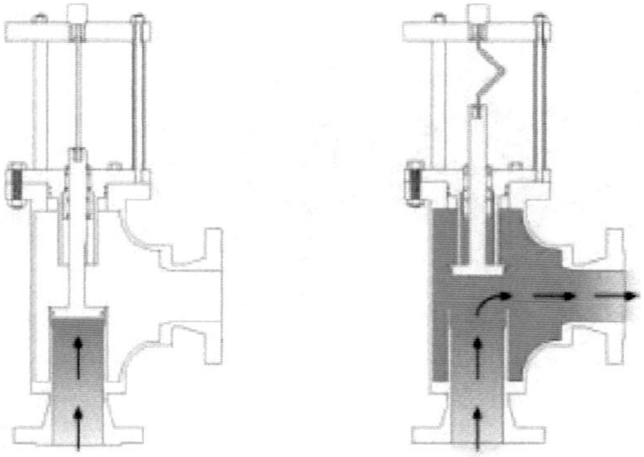

http://www.kingsenergy.com/products/valves/rupture-pin-technology/

AIChE Southwest Process Technology Conference

Type of Relief Devices

Type of Device	Design Considerations
Conventional Relief Valve	Most common, less expensive, Differential pressure device
Balanced Bellows Relief Valve	Minimize effect of higher outlet pressure drops, can be installed on some conventional relief valves, spring is isolated from process fluid
Pilot Operated Relief Valve	Major relieving device is combined with and controlled by a self-actuated auxiliary relief valve (pilot), can help mitigate high inlet and outlet pressure drops
Rupture Disk	Non-reclosing, lower cost, actuated by static differential pressure between inlet and outlet
Conservation Vent/Emergency Vent	API Tanks, Low pressure service, can vent to atmosphere or header system
Rupture Pin	Non-reclosing, can be reset without having to clear up the system

Scenario Analysis

- Define your system boundaries
- Determine the MAWP of the protected system
- Evaluate potential overpressure sources
  - Lists provided in API
  - Need to understand the process
  - Need to understand start-up/shutdown procedures
- Drives required relief rate calculations
- Take this time to learn the process!

AIChE
Southwest Process Technology Conference

Additional Analysis Steps

- Determine if installed orifice area is adequate
- Calculate inlet and outlet pressure drops
- Verify if installation is aligned with code
- Determine if device discharges to a safe location
- Calculate noise limit at discharge and safe distances
- Determine reaction forces to ensure adequate support

AIChE
Southwest Process Technology Conference

Things to Consider

- Chatter and Water Hammer
  - When the pressure at the valve inlet falls below the blowdown setting, the valve will shut.
  - Flow stops, so inlet dP goes to zero, and pressure at the valve inlet forces it open again.
  - Flow resumes, and the valve shuts again.
  - Rapid repetition will damage the valve and limit flow.
- Steam Relief
- High Viscosity Fluids
- Multiple Relief Valves
- Protecting Multiple Pieces of Equipment
- High Operating Ratios

Questions?

7th AIChE Southwest Process Technology Conference | October 1-2, 2015 | Galveston, TX

Thank you!

AIChE

Southwest Process Technology Conference

7th AIChE Southwest Process Technology Conference | October 1-2, 2015 | Galveston, TX

Challenges Facing Young Professionals

CAROL SCHMIDT, P.E. (TX)

OCT 2015

The whole enchilada…

- Myths of graduation
- Resumes/Interviews
- Adjusting to work life
- Career development
  - General
  - PE
  - Graduate degree
  - MBA
  - Volunteering
- Work and family
- When to change jobs
- When/how to ask for a raise
- Recommended reading

Obligatory Engineering Joke

Newton asked, "How do you write 4 in between 5?"

- Doctor: "What!"
- Scientist: "Impossible!"
- Manager: "Can't find it on the internet"
- Engineer: "F(IV)E"

Myths before graduation

- Degree = job
  - A degree is a tool, not a guarantee
- Employers want you
  - They want profit
- You will find a job on campus
  - Maybe, but it isn't the only avenue
- Some people are just naturally better than others
  - You are what you practice

Resumes

- Develop a self description
- Make every word count
- Zero mistakes

RAMIT SETHI

- Consistency – no, Seriously
- Include accomplishments
  - Be quantitative
  - Link to projects
- Test it out

Interviewing

It's not just what you know, it's how what you know will help them

- Research and practice
- Show up
- Story telling
- Know your industry
- Be honest
- Ask questions
- Be thankful

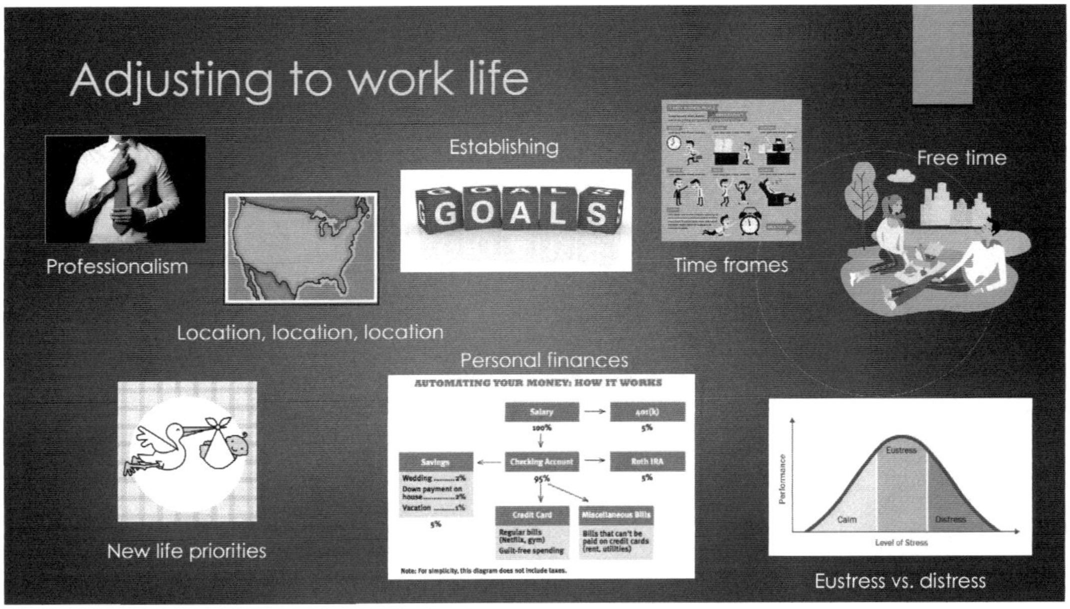
Adjusting to work life
Professionalism
Location, location, location
Establishing
GOALS
Time frames
Free time
Personal finances
AUTOMATING YOUR MONEY: HOW IT WORKS
Salary
401(k)
Savings
Checking Account
Roth IRA
Credit Card
Miscellaneous Bills
New life priorities
Eustress
Calm
Distress
Performance
Level of Stress
Eustress vs. distress

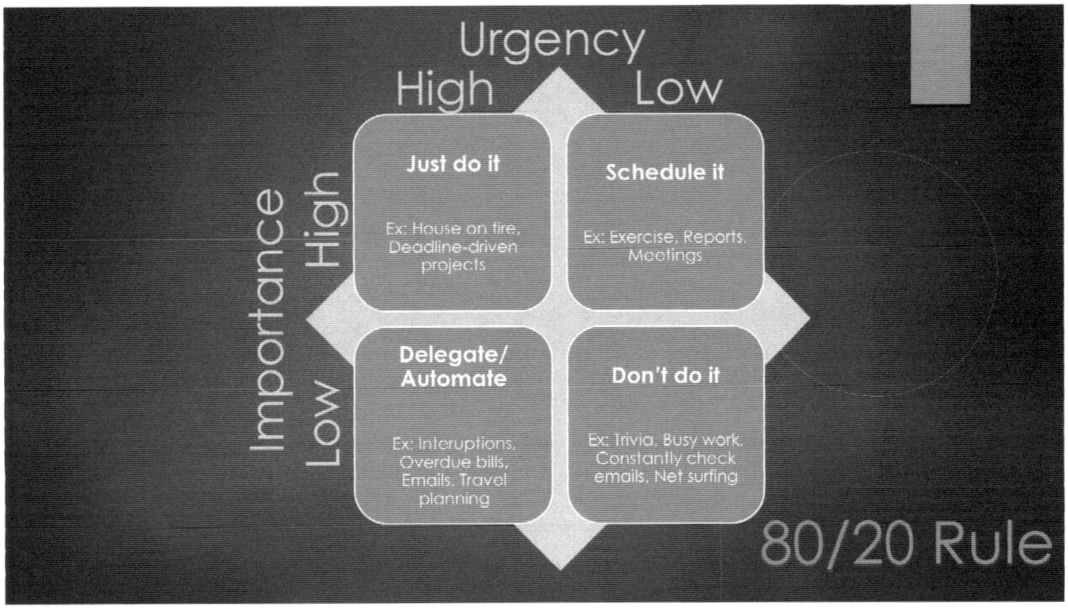
Urgency
High
Low
Importance
High
Low
Just do it
Ex: House on fire, Deadline-driven projects
Schedule it
Ex: Exercise, Reports, Meetings
Delegate/ Automate
Ex: Interuptions, Overdue bills, Emails, Travel planning
Don't do it
Ex: Trivia, Busy work, Constantly check emails, Net surfing
80/20 Rule

Career development - General

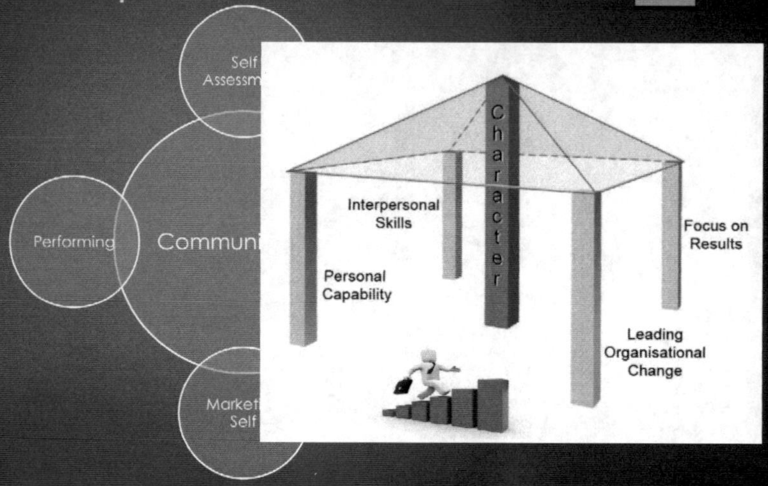

Career development - PE

- PE = Technical proficiency = "engineer"
- Benefits – Least time consuming/cheapest, can represent yourself, no work time lost (1 day + study time)
- Downside – Continual upkeep, potential misuse, still a salary ceiling
- Every state is different, many have reciprocity
  - TBPE
    - FE – Junior yr
    - PE – Grad + 4yrs

Texas PE Application Deadlines and Exam Dates

Exam Year	PE Application Deadline	PE Exam Registration Deadline	PE and SE Vertical Forces Exam	SE Lateral Forces Exam
2015	December 12, 2014	February 19, 2015	April 17, 2015	April 18, 2015
	July 10, 2015	September 3, 2015	October 30, 2015	October 31, 2015
2016	December 11, 2015	February 18, 2016	April 15, 2016	April 16, 2016
	July 15, 2016	September 1, 2016	October 28, 2016	October 29, 2016
2017	December 9, 2016	To Be Determined	April 21, 2017	April 22, 2017
	July 14, 2017	To Be Determined	October 27, 2017	October 28, 2017
2018	December 8, 2017	To Be Determined	April 13, 2018	April 14, 2018
	July 13, 2018	To Be Determined	October 26, 2018	October 27, 2018
2019	December 7, 2018	To Be Determined	April 5, 2019	April 6, 2019
	July 12, 2019	To Be Determined	October 25, 2019	October 26, 2019
2020	December 6, 2019	To Be Determined	April 17, 2020	April 18, 2020
	July 10, 2020	To Be Determined	October 23, 2020	October 24, 2020

https://www.egr.uh.edu/professionals/fe-pe/pepp-review-course

Career development – Graduate degree

- Graduate degree = Highly specialized

- Benefits – Highest pay upon graduation, smaller employee pool
- Downsides – Harder to qualify for non-niche jobs, more susceptible to market fluctuations, difficult to work and go to school (4-6+ years)

- Different graduate schools
  - Chemical/Petroleum/Subsea
  - Law
  - Medical
  - Business

Career development - MBA

- MBA = Management track.

- Benefits – Possible fast track, learn to start your own business, learn while working (2-4 yrs)
- Downsides – Not a guarantee, dime a dozen, expensive

- Different MBAs for different sectors
  - Executive MBA (not as highly valued)
  - Energy
  - Finance
  - Leadership
  - Entrepreneurship
- Masters in Engineering Management

- Most programs 2-4 yrs, many have night programs, some online

Career Development - Volunteering

- Different sectors
- Low stakes
- Minimal time commitment
- Resume builder
- Shows interest/initiative
- Peers get your jokes
- Competition: professional vs. college
- Confidence builder

Work and Family

You can do more than you think you can

~But~

You cannot do everything

When to change jobs

- Better opportunity
  - $$$!= happiness
- Undervalued
- Downturn

Reverse engineer

When to ask for a raise

- 90 days before annual review
- Start of big project
- New huge responsibility
- Taking on another persons workload
- Boss acknowledges your contribution

NOT when:

- Emotionally negative
- Not producing
- Desperate

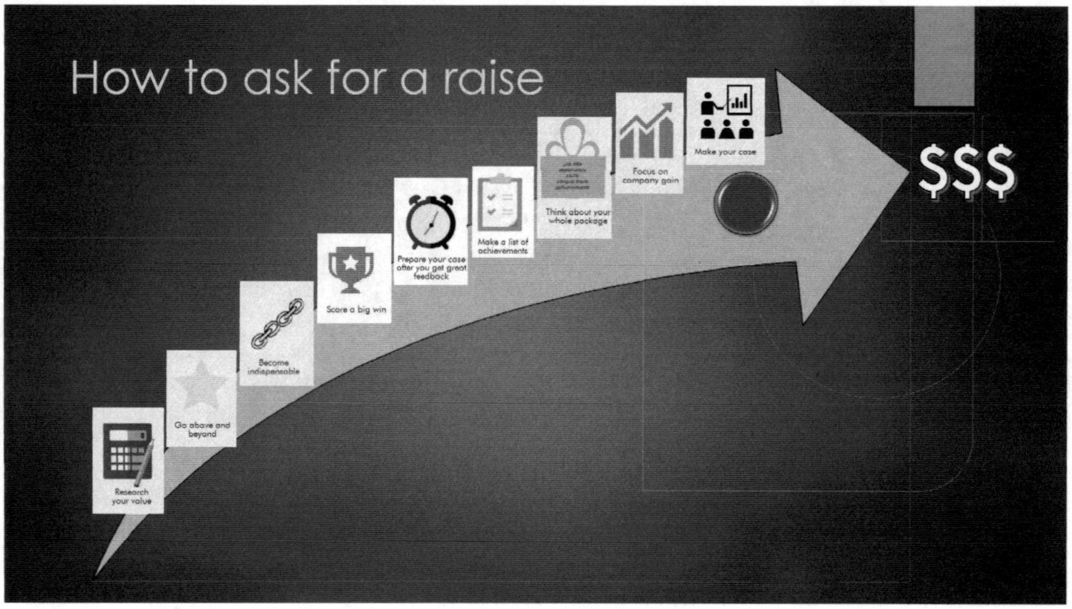

Recommended reading

- I will teach you to be rich – Ramit Sethi
- The 4 hr work week – Tim Ferris
- Quiet – Susan Cain
- Outliers – Malcolm Gladwell
- Leadership and self deception – The Arbinger Institute
- The richest man in Babylon – George Clason
- The millionaire next door – Thomas Stanley
- Play it away – Charlie Hoehn

From Micro to Macro without Migraines

Bennett, E.,

Originally presented on: 10/2/2015 13:30:00 - 14:00:00

The design of industrial chemical processes typically starts at a lab scale and passes through a battery of analyses on the road to production scale volumes. This session presents a case study of a batch reaction scale-up for a multimillion dollar specialty chemical project and approaches taken to address the challenges encountered along the way. Tools that will be discussed as part of the analysis include VisiMix and Statgraphics.

Options to reduce cooling water consumption in fertilizer plants

Discussion of challenges, tools and case studies

Edgar Amaro-Ronces, Eberhard Lucke – CH2M

7th AIChE Southwest Process Technology Conference

October 2, 2015

Options to reduce cooling water consumption in fertilizer plants

- Introduction
- Challenge
- Methodologies
  - Models
- Case Studies
- Conclusion

Introduction

7th AIChE Southwest Process Technology Conference

3

Introduction

- Key theme for many projects: water sustainability
  - Maintain • Support • Endure
- Cost of treatment for supply (water sources) and discharge (water sinks)
- Future trends in industry and facility design
  - Adopt efficient systems that minimize use of fresh water
- Challenge exists across all major industries
- Large cooling towers
  - require significant volumes of make-up water
- Pitfalls of recycling water
- Environmental regulations

4

Challenge

7th AIChE Southwest Process Technology Conference

AIChE

Southwest Process Technology Conference

ch2m.

5

Challenge

- How to deal with the complexity of the issues around cooling water?
- Broad range of applicable methods to model/describe the problem
  - Each with specific strengths and weaknesses
- Factors like complexity of the heat exchanger network (HEN), capital cost, operating and maintenance cost, safety, required treatment
- Constraints like limited sources, limited discharge options, financial limitations, existing hardware systems, water versus air cooling
- Complexity of problem challenges the application of tools that allow finding the optimum solution to the problem

6

7th AIChE Southwest Process Technology Conference - Edgar Amaro-Ronces, Eberhard Lucke - CH2M, October 2, 2015

Level of Optimization

- Heat/energy optimization
  - Local optimization – single heat exchangers/heat exchange services
  - Global optimization – HEN; overall system – multiple services
- Source optimization
  - Fresh water sources – rivers; aquifers; others
  - Water reuse – external/internal; treated waste water
  - Water recycling
- Discharge optimization
  - Waste water treatment – requirements; outlets; reuse and recycle options
- Cost optimization
  - CAPEX – use of existing equipment and facilities; new equipment and facilities
  - OPEX – minimize consumption; fresh water treatment; waste water treatment
  - Other cost factors – environmental impact; competing with other users (priorities); seasonal fluctuations (e.g. dry season)
- More.......site specific • project specific • technology specific..........

Methodologies

7th AIChE Southwest Process Technology Conference

Methodologies

- Technical models
  - Spreadsheets (e.g. water balance); **Pinch Analysis**; Process Simulation (static, dynamic)
- Financial models
  - CAPEX models; OPEX models; **Combined models**; other cost factors
- Mathematical models
  - Genetic Algorithm (GA); Ant Colony Optimization (ACO); Particle Swarm Optimization (PSO); **Cat Swarm Optimization (CSO)**
- Engineers
  - Knowledge; Expertise; Judgement

Technical Models

- Example: Pinch Analysis
  - Aka Process Integration, Heat Integration, Energy Integration or Pinch Technology
  - Methodology used to minimize energy consumption by calculating thermodynamically feasible targets and achieving these targets by optimizing heat recovery systems, energy supply methods and process operating conditions
- Application
  - Originally designed to support energy saving initiatives and energy conservation
  - Later found to be applicable for improving effluent quality, reducing emissions, increasing product yields, debottlenecking, improving process flexibility and process safety

Water Pinch Analysis [1]

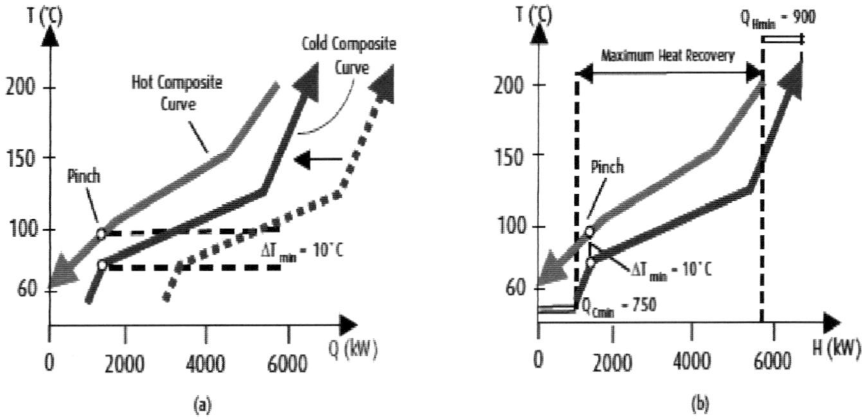

11

Water Pinch Analysis

- Strengths
  - Well known methodology with proven success record
  - Several software options available on the market
  - Has been further developed to expand range of application
    - Grid Diagram Table (combination with Problem Table Algorithm)
    - CPA (Chemical Pinch Analysis)
    - WCA (Water Cascade Analysis)
    - MENs (Mass Exchange Networks)
    - WPA (Water Pinch Analysis)
    - ......others

12

Water Pinch Analysis

- Weaknesses
  - Provides a technical solution only
  - Financial considerations need to be factored in via separate model
  - Tend to create very complex heat exchanger networks that lack practical application
  - Limited to application within the process itself
  - Very difficult to apply to solving multi-component systems
  - Solution difficult for "Threshold Pinch" cases
    - Example: Ammonia process = no hot utility required

13

7th AIChE Southwest Process Technology Conference - Edgar Amaro-Ronces, Eberhard Lucke - CH2M, October 2, 2015

Financial Models [2]

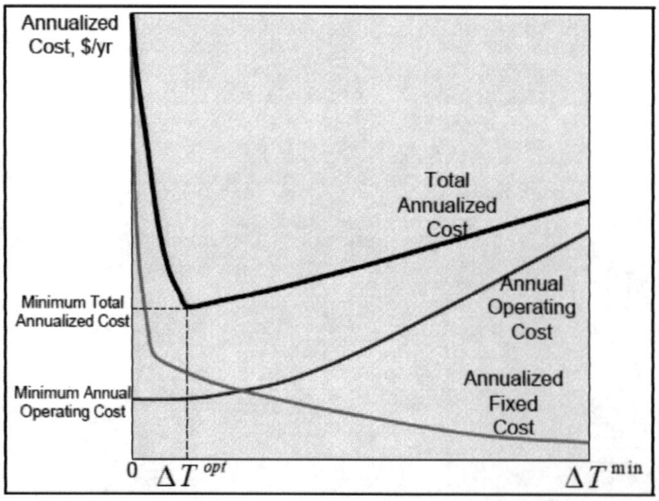

14

7th AIChE Southwest Process Technology Conference - Edgar Amaro-Ronces, Eberhard Lucke - CH2M, October 2, 2015

Financial Models [3]

Most common formulation for minimizing total annual cost of a heat exchanger network:

$$MinJ = \sum_{i=1}^{HP} C_{CU} Q_{CU,i} + \sum_{j=1}^{CP} C_{HU} Q_{HU,j} + \sum_{k=1}^{NE} (C_0 + C_1 A_P^m) + \sum_{i=1}^{HP} (C_2 + C_3 A_i^m) + \sum_{j=1}^{CP} (C_2 + C_3 A_j^m)$$

Financial Models [3]

Most common formulation for minimizing total annual cost of a heat exchanger network:

$$MinJ = \sum_{i=1}^{HP} C_{CU} Q_{CU,i} + \sum_{j=1}^{CP} C_{HU} Q_{HU,j} + \sum_{k=1}^{NE} (C_0 + C_1 A_P^m) + \sum_{i=1}^{HP} (C_2 + C_3 A_i^m) + \sum_{j=1}^{CP} (C_2 + C_3 A_j^m)$$

Objective Function

Financial Models [3]

Most common formulation for minimizing total annual cost of a heat exchanger network:

$$MinJ = \sum_{i=1}^{HP} C_{CU} Q_{CU,i} + \sum_{j=1}^{CP} C_{HU} Q_{HU,j} + \sum_{k=1}^{NE} (C_0 + C_1 A_P^m) + \sum_{i=1}^{HP} (C_2 + C_3 A_i^m) + \sum_{j=1}^{CP} (C_2 + C_3 A_j^m)$$

Cold Utility Cost

Financial Models [3]

Most common formulation for minimizing total annual cost of a heat exchanger network:

$$MinJ = \sum_{i=1}^{HP} C_{CU} Q_{CU,i} + \sum_{j=1}^{CP} C_{HU} Q_{HU,j} + \sum_{k=1}^{NE} (C_0 + C_1 A_P^m) + \sum_{i=1}^{HP} (C_2 + C_3 A_i^m) + \sum_{j=1}^{CP} (C_2 + C_3 A_j^m)$$

Hot Utility Cost

Financial Models [3]

Most common formulation for minimizing total annual cost of a heat exchanger network:

$$MinJ = \sum_{i=1}^{HP} C_{CU} Q_{CU,i} + \sum_{j=1}^{CP} C_{HU} Q_{HU,j} + \sum_{k=1}^{NE} (C_0 + C_1 A_P^m) + \sum_{i=1}^{HP} (C_2 + C_3 A_i^m) + \sum_{j=1}^{CP} (C_2 + C_3 A_j^m)$$

Capital Cost for Process Heat Exchangers

Financial Models [3]

Most common formulation for minimizing total annual cost of a heat exchanger network:

$$MinJ = \sum_{i=1}^{HP} C_{CU} Q_{CU,i} + \sum_{j=1}^{CP} C_{HU} Q_{HU,j} + \sum_{k=1}^{NE} (C_0 + C_1 A_P^m) + \sum_{i=1}^{HP} (C_2 + C_3 A_i^m) + \sum_{j=1}^{CP} (C_2 + C_3 A_j^m)$$

Capital Cost for Heaters

Financial Models [3]

Most common formulation for minimizing total annual cost of a heat exchanger network:

$$MinJ = \sum_{i=1}^{HP} C_{CU} Q_{CU,i} + \sum_{j=1}^{CP} C_{HU} Q_{HU,j} + \sum_{k=1}^{NE} (C_0 + C_1 A_P^m) +$$
$$\sum_{i=1}^{HP} (C_2 + C_3 A_i^m) + \sum_{j=1}^{CP} (C_2 + C_3 A_j^m)$$

Capital Cost for Coolers

21

Financial Models [3]

- Constraints
  - Required to solve the problem = minimize objective function
  - Overall heat balance for each stream must be solved
  - Minimum temperature approach limit must be met
  - Non-negativity for Energy Streams must be maintained
  - Mass conservation in branches must be met

22

Mathematical Models and Algorithms [3]

- Example: Cat Swarm Optimization
  - Developed from Particle Swarm Optimization and Ant Colony Optimization
  - Models the behavior of CATS to solve the optimization problem
  - CSO Algorithm has two sub-models:
    - Seeking mode: moves carefully and slowly

    - Tracing mode: runs after target

Cat Swarm Optimization [3]

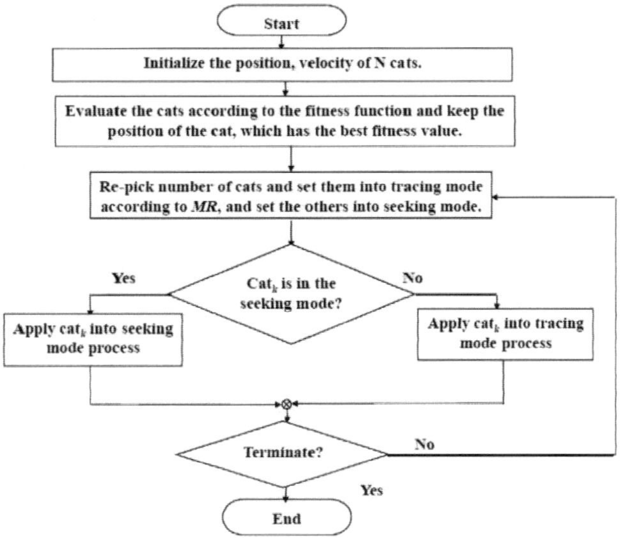

Cat Swarm Optimization [3]

- Strengths
  - Powerful algorithm that finds better solutions than most other approaches
- Weaknesses
  - Not a "typical" engineering tool
  - Requires detailed mathematical knowledge
  - Requires very high computation time

Example: VOYAGE™

- Created in 2004, developed by CH2M
- Analyzes Resource Systems with Multiple Components
- Powerful planning tool
  - Integrated analysis of whole systems
  - Time dependent capital improvements and operations
  - Evaluate alternative "what if" scenarios
- Advanced optimization
  - Analyze numerous alternatives to meet goals
    - Economic
    - Environmental
    - Social
- Great communication tool
  - Stakeholder understanding
  - Demonstrate impact of decisions

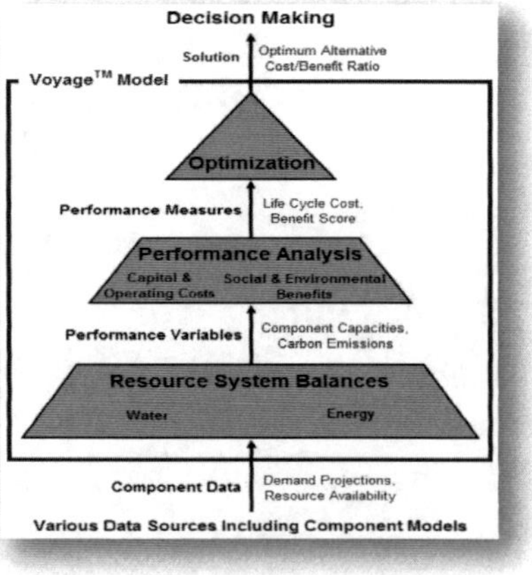

Facilitates defensible decision making

Example: VOYAGE[TM]

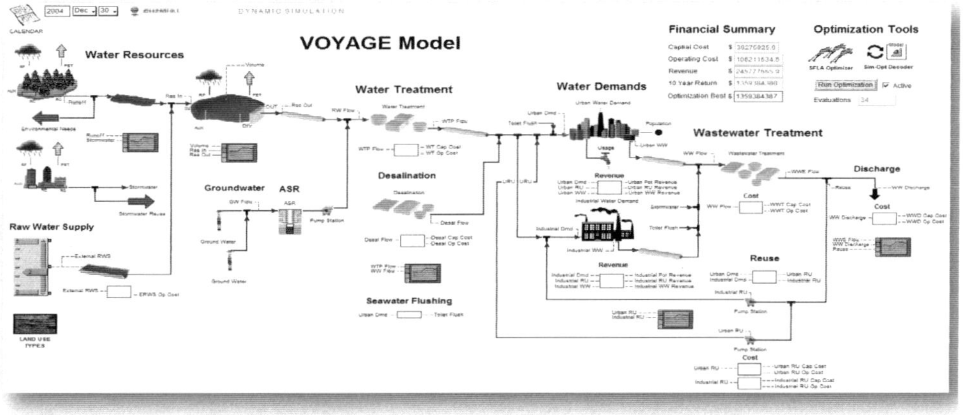

27

Example: VOYAGE[TM]

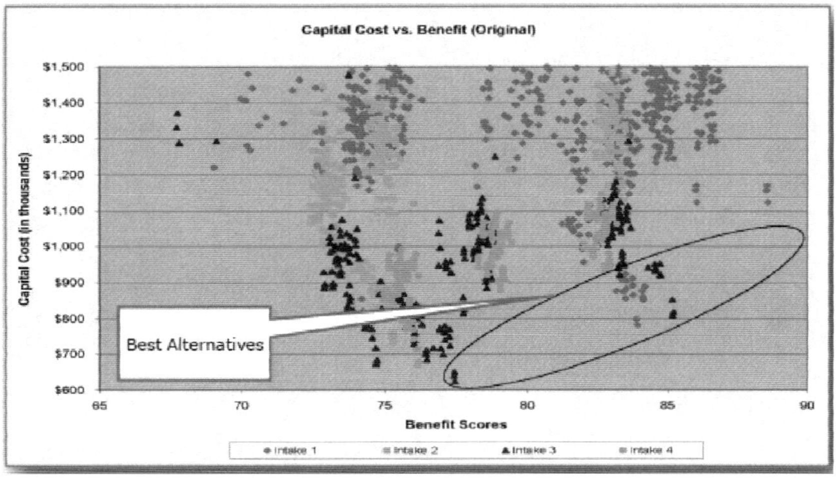

28

My Favorite "Tool" – Process Engineer

29

7th AIChE Southwest Process Technology Conference - Edgar Amaro-Ronces, Eberhard Lucke - CH2M, October 2, 2015

Case Studies

7th AIChE Southwest Process Technology Conference

AIChE

Southwest Process Technology Conference

ch2m

30

Case Study 1 [4]

- Basis of optimization study:
  - Ammonia Plant with 1,360 tpd production capacity
  - Cooling provided by open evaporative cooling tower system – 10 cells – induced draft – cross flow
  - Make-up water requirement up to 600m^3/h
  - Owner suspected significant savings are achievable by applying better technologies and reuse of water
- Economical Driver: savings of over $2,000 per year per m^3/h fresh water conserved

Case Study 1 [4]

- Idea #1: Reduce Cycle of Concentration (COC)
  - COC is a measure for the build-up of minerals in the cooling water calculated as the ration of the concentration of minerals in the recirculating water to the concentration of minerals in the make-up water
  - COC (design) = 6
  - COC (new) = 2.5
  - Required optimization of chemical cooling water treatment
- Reduction in make-up water = 141 m^3/h
- Savings = $280,000/year

Case Study 1 [4]

- Idea #2: Treat Wastewater and Reuse
  - Required additional cooling water treatment (see idea #1)
- Proven capability to reuse about 78 m^3/h
- Savings = $150,000/year

Case Study 1 [4]

- Idea #3: Use Steam Boiler Blowdown as Make-up Water
  - Required additional cooling water treatment (see idea #1)
- Proven capability to use about 7 m^3/h
- Savings = $14,000/year

Case Study 1 [4]

- Idea #4: Replace Drift Eliminators
  - Estimated savings of about $20,000/year
  - High capital expense to buy and replace existing drift eliminators with newer design
- Decided to wait until existing draft eliminators will have to be replaced

Case Study 1 [4]

- Summary
  - Idea # 1: $280,000/year
  - Idea # 2: $150,000/year
  - Idea #3: $14,000/year
  - Idea #4: no short term savings
- Total savings: $444,000/year
- Reduced make-up water by 228 m^3/h (38% reduction)
- Required optimization of chemical treatment system to give desired flexibility for reduced COC and multiple make-up water streams

Case Study 2

- New ammonia fertilizer complex
- Production capacity of 2400 tpd Ammonia/3,500 tpd Urea
- Cooling water supply was a challenge due to plant location and limited water sourcing options
  - 1st Cooling Tower - Low Chloride Water – 3 cells
  - 2nd Cooling Tower - High Chloride Water - 10 cells
    - Total circulation flow: 155,600 gpm / 35,341 m^3/h
    - Blowdown: 1.75%
    - Evaporation losses: 2,720 gpm / 618 m^3/h

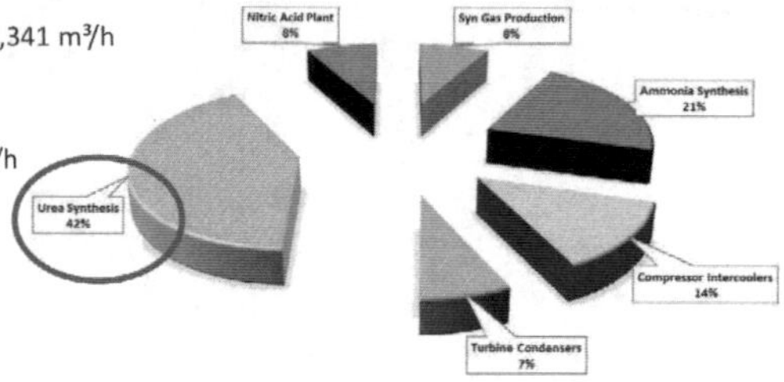

37

Case Study 2

- Simple Engineering Study: replace water cooling with air cooling where feasible
    - Total circulation flow: 123,900 gpm / 28,141 m^3/h
    - Blowdown: 1.75%
    - Evaporation losses: 2,165 gpm / 492 m^3/h
- No reduction in cooling tower size (client decision)
- Water cost reduction
  - Reduced evaporation losses
    - 555 gpm / 126 m^3/h
  - Estimated $250,000/year

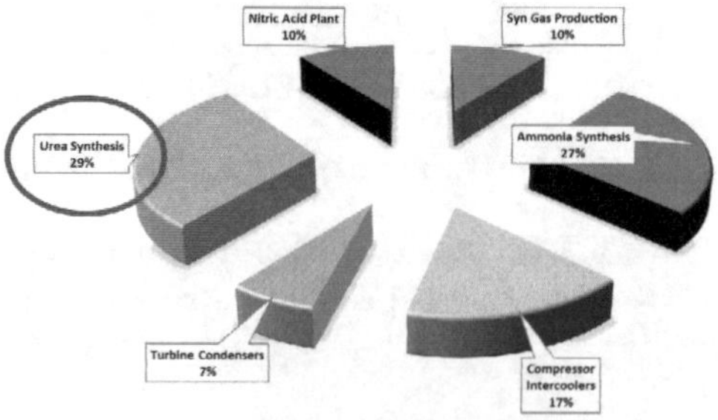

38

Conclusions

7th AIChE Southwest Process Technology Conference

AIChE Southwest Process Technology Conference

ch2m.

39

Conclusions

- Sustainable water sourcing will continue to be an issue
  - Especially in remote areas with limited water sources
  - Also in areas with competing water users
- We have a lot of tools available, we just have to understand their strengths and their limitations
- There is no "universal" approach for all the different challenges
- In most cases an integrated approach using multiple tools will be required
- The Process Engineer is the "glue" that keeps it all together
- Even small and simple studies can yield significant results

40

7th AIChE Southwest Process Technology Conference - Edgar Amaro-Ronces, Eberhard Lucke - CH2M, October 2, 2015

References

1. "Pinch Analysis: For the Efficient Use of Energy, Water & Hydrogen", Natural Resources Canada, 2003
2. "An Energy Integration Approach on UHDE Ammonia Process", Alnouss M. Ahmed, Al-Nuaimi A. Ibrahim, International Scholarly and Scientific Research & Innovation 6 (11), 2012
3. "Optimization of Heat Exchanger Networks using an Evolutionary Method", Abolfazl Ghiasvand, Alireza Fazlali, Tayebeh S. Ghiasi, Mahdi Aliyari-Shoorehdeli, Amir H. Mohammadi, Advances in Energy Research – Volume 18, Chapter 10, 2014
4. "Optimization of Make Up Water Reuse Saves $100,000/yr for Ammonia Plant", Davor Kesner, Tanja Vulinec Losso, GE Water & Process Technologies, 2007

41

7th AIChE Southwest Process Technology Conference - Edgar Amaro-Ronces, Eberhard Lucke - CH2M, October 2, 2015

THANK YOU!

7th AIChE Southwest Process Technology Conference

AIChE

Southwest Process Technology Conference

42

ch2m

What have I learnt from my Process Development & Commercialization Experience?

Anil Mehta
at
7th AIChE Southwest Process Technology Conference
October 2, 2015
Galveston, TX

The Dow Chemical Company

A little bit on my background...

- 30+ years of experience in all phases of process development & commercialization
  - Experience with an EPC company
  - Research on Supercritical fluids at UT, Austin
  - Broad range of process technology experience at Dow Chemical

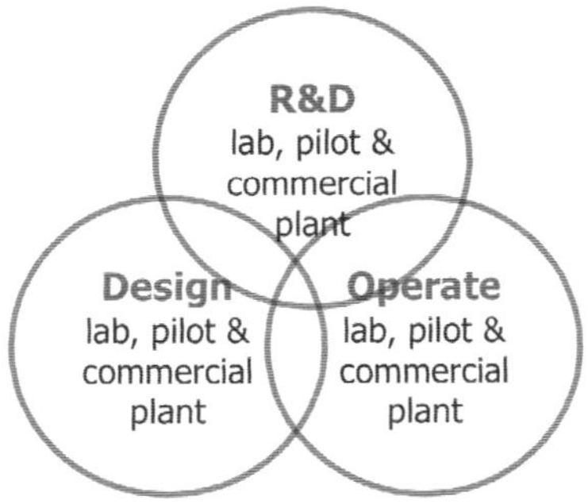

What is Process Development?

- What's a Process?
  - A sequence of actions to accomplish an objective
  - Will restrict the discussion to the context of "manufacturing a chemical or a material"

- What is Process Development?
  - Well defined in the description for this session
  - In concise terms:

  Taking the (material) manufacturing concept to a commercial scale while meeting product performance and economic goals in a safe and environmentally sustainable method

Types of Process Development

- Develop a new process
  - For a new product – tend to have more low volume products
    - Biodiesel (methyl esters of natural fatty acids)
    - Biofuel from algae
  - For an existing product (new synthesis routes)
    - Carbon fibers
    - Ethanol from cellulosic biomass instead of corn
    - Ethanol from carbon dioxide and sunlight – Joule Unlimited, MA
- Improve an existing process significantly
  - Increase product yield
  - Decrease waste production
  - Decrease energy consumption
  - Decrease capital cost

Process Development – Real Examples

Looking back at few of the Process Development & Commercialization projects I was involved with...

Example 1: A cycloaliphatic epoxy resin process development

- Joint research project to replace isocyanates in automotive top-coat application
- Started from a bench scale batch process
- Led scale up to a 2000-gallon reactor scale
- Made 10s of 1000s of lbs product
- Dow product had better properties in all major aspects

Example 1 (cycloaliphatic epoxy) - Results

	Dow	Competitor
Higher Epoxide, %	~27	25
Lower Viscosity, cps	~320	~800
Lower color, APHA	~80	95+
Total chlorides, ppm	~8000	~12000
Hyd. Chlorides, ppm	<5000	<6000
Lower volatiles, %	0.1-0.4	0.9-1

- Development work for a continuous process started
- The project discontinued after a couple of years
  - The targeted product application did not pan out
- Business decided not to pursue the project even though we could make a better product

Example 1 - Learning

- Only a few out of many projects will get commercialized and be successful - it's the nature of research
- Things will not always go our way – do your best always
- Document the work well in scientific reports

Example 2 – Glycerine To Epichlorohydrin (GTE)

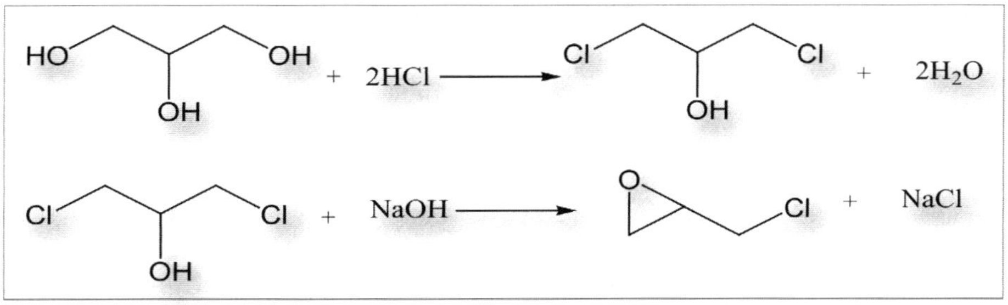

- Basic chemistry known since early 1900s but no commercial process development until the last decade

GTE Process Development

- Goal: Develop and demonstrate the process, generate data for designing a commercial plant to be built in China
- Based on the experimental work done at <1L scale in a semi-batch reactor
- Economy of scale (large volume) favored a continuous process

GTE - The Strategy

- Systematic approach – a text book example of process development
- First, defined the commercial process - a continuous process with a recycle loop, full use of process simulation tools, optimized
- Defined clearly the objectives for the pilot plant and prepared detailed list of the data required for designing a commercial plant
- Ensured subject matter expert input in the details
- Made full use of expertise available within the company
- Established required samples analysis capability

GTE - What scale for the pilot plant?

- "Scale down to scale up"
- Defined the minimum scale required for each unit operation – reaction, distillation etc.
- The smallest available glass lined vessel defined the size of the hydrochlorination pilot plant
- The smallest size required for piloting the epoxidation step was at least an order of magnitude larger

GTE Process Development - Challenges

- Orders of magnitude increase in complexity in going from a semi-batch (bench scale) to a continuous process with multiple unit operations and recycle stream
- Highly corrosive process – handle aqueous HCl at high temperature
- Extremely tight time-line

Example 2 (GTE) - Results

- Designed, built and started up the pilot plant in ~8 months
- A new proprietary continuous GTE process developed and demonstrated in <12 months
- Many patent applications filed and granted
- Commercial plant designed and approved by the company's board
- The 2008 economic crisis followed soon
- The project put on hold – not revived since then

Example 2 – Learning

- Can't be done alone – requires a team working together towards a common goal – the sheer amount of work and the number of skills required is overwhelming for any one person
- Process development is expensive
- Make use of skills and expertise available within and outside the company
- Recognizing people for their contribution goes a long way
- Safety of people must be a top priority – especially when dealing with hazardous chemicals, do not take shortcuts to meet the timeline
- Extract maximum possible amount of information from each experiment since it takes so much investment of effort to build the lab or pilot scale capability
- High quality analysis of the process samples goes a long way – each process sample is a result of a real physical phenomena – even a sample from a process upset or startup and shutdown presents an opportunity to learn more about the process

Example 3: Amines Decolorization

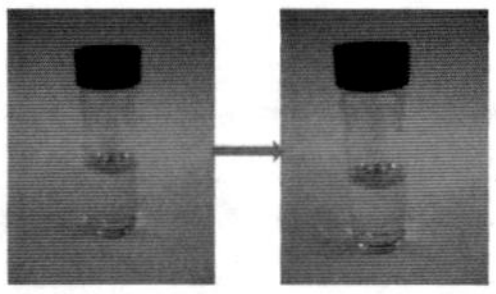

- Decolorizing amines using a weak organic acid (US8853459)
- Previous best: <50% yield of low color amine
- Improved yield of low color amine to ~90%
- Scaled up the improvement from ~100 grams (bench scale) to >1 million lbs – batch process successfully
- How was this achieved? -
  - Studied thoroughly the available knowledge, the work done before, proposed a hypotheses and tested it (experiments)
  - Identified key variables for the scale-up

Example 4: Low-chlorides HOCl

- HOCl (hypochlorous acid) used in the production of chlorohydrins

$$Cl_2 + H_2O \leftrightarrow HOCl + HCl$$

- Benefits of low-chlorides HOCl
  - Reduced formation of byproducts resulting from the presence of chlorides
  - Improved raw material yield
  - Reduced water use and wastewater generation

Low-chlorides HOCl – Raw material

- At least three choices for alkaline raw material
  - Soda ash (Na_2CO_3)
  - Lime (CaO)
  - Caustic soda (NaOH)

Example 4a: HOCl from Soda ash

- Was asked to design, build and start-up a pilot plant for producing HOCl from soda ash
- Spent ~5 months on designing a pilot plant, designing specifying and ordering equipment and starting construction of the pilot plant
- Comparative economic analysis of the process with lime → Raw material cost for soda ash much greater than for lime → Stopped the project quickly → Documented the work in scientific reports
- Learning:
  - Kill the project as soon as possible if it's no longer a good idea
  - Do not get attached emotionally to your project - Be honest to yourself – you can not defeat the mother nature

Example 4b: HOCl from Caustic - Chemistry

HOCl Generation:

$Cl_2(g) \leftrightarrow Cl_2(aq)$

$Cl_2(aq) + 2\ NaOH(aq) + \rightarrow NaOCl\ (aq) + NaCl\ (aq) + H2O$

$Cl_2(aq) + NaOCl(aq) + H_2O \rightarrow 2\ HOCl\ (aq) + NaCl\ (aq)$

HOCl Desorbtion:

$HOCl\ (aq) \leftrightarrow HOCl(g)$

$2HOCl(g) \leftrightarrow Cl_2O(g) + H_2O(g)$

HOCL Decomposition:

$6HOCl(aq) + NaCl(aq) \rightarrow NaCLO_3(aq) + 3Cl_2 + 3H_2O(l)$

HOCl Technology Commercialization

- HOCl recovery maximized by desorbing it before it has time to decompose
- Rotating packed bed reactors offered enhanced mass-transfer capability
- Process piloted and successfully scaled-up to commercial scale
- In operation for >15 years
- Benefits realized from commercialization of the technology:
  - 1/3+ reduction in chlorinated byproducts and wastewater

HOCl - Did we miss anything?

- Yes, missed a few things – problems showed up in the least expected areas
- Solved all of the problems in a couple of years and achieved the desired performance
- The new plant won the Technology center's "Best In Class" award out of 18 total plants – **noteworthy achievement for a technology in commercial operation for only 2 years**

What made HOCl Commercialization Successful?

- Close collaboration between R&D and the commercial plant design team - 2-3 key R&D team members on the commercial plant design team
- Experienced and competent project manager
- Skill level of key people on the project
- Attention paid to details – every key design document and equipment & instrument specification was reviewed
- Stakeholders' support

Example 4 - Learning

- Great pressure when problems manifested! - Not solving some of the problems timely could have resulted in a permanent shutdown of the new plant
  - It's a thin line between success and failure
- The devil is in the details
- Be 'hands-on'
- Proper choice of materials of construction is critical – we were dealing with a very corrosive processes
- Advise from my project manager "do not compromise on safety and quality – no one will remember if the plant starts up a little late but no one will forget a serious safety incident"
- Be prepared to work in shifts and in the nights
- Document the work thoroughly

Strategy for Process Development & ...

- Working on a right project/problem is more important than working rightly on the problem – the answer is not always known
- Millions could be spent with no potential for return on the investment if working on a wrong problem
- Rigorous internal debate on alternate technology evaluation process for development
- Technology evaluation covering all important aspects of the proposed process must be done
- Early stage capital cost estimate has low accuracy but variable cost estimate is usually more accurate

Strategy for Process Development & ...

- Aim for the simplest possible process
- Batch or continuous process? Or a combination of the two?
- Map out the data required to design the commercial plant
- Pilot or not?
- If yes, what scale?
- Impact of recycle on build up of impurities
- Samples Analysis capability
- High quality physical properties data (VLE, LLE etc.) are essential for successful process development
- Larger portion of a typical plant cost is in the non-reactor unit operations such as separation

Strategy for Process Development & ...

• Model the process as much as reasonably possible

"All models are wrong, some are useful!" - Unknown

• Search literature thoroughly

"Six months in a lab can save a day in a library!!!"

– George Quarderer

• Document the work in detailed scientific reports – helps the report writer too by "seeing the data in a different light!"

Finally

• Our processes must be safe and environmentally sustainable - minimize environmental impact, favor use of less toxic materials and synthesis pathways, minimize potential for exposure to toxic chemicals - It's the new way of life for us

• Great satisfaction comes from seeing a project/solution implemented commercially standing for years

• Persist, do not give up till you achieve the goal

• Question "it didn't work before"
  - Changing only one parameter can make a difference
  - The devil is in the details

Acknowledgements

- John Beckerdite
- Paul Beckett
- Phil Carlberg
- Rui Cruz
- Clark Davis
- Joe Henry
- Steve King
- Joe Longoria
- Charles Lipp
- Sammy Masters
- George Quarderer
- Siegfried Unger
- Dan Tirtowidjojo
- David Trent
- David West

all former or current Dow Chemical colleagues, mentors & seniors

INGENERO
Excellence Through Insight

Dehydration and Purification of Spent THF Solvent

Presented by

Sameer Thorat

at

7th AIChE Southwest Process Technology Conference 2015

Process Development and Chemical Manufacturing Session

Confidential & Proprietary 1

Tetrahydrofuran (THF)

Recovering spent tetrahydrofuran (THF) is attractive to the pharmaceutical industry

Tetrahydrofuran (THF)

- A solvent used in pharmaceutical industry for Grignard reaction
- In process gets diluted with water and contaminated with impurities
- Spent THF can be reused after dehydration and removal of impurities
- Realizable salvage value of spent THF makes recycling an economically attractive option

INGENERO

Confidential & Proprietary 2

THF Recovery Process - Concerns

Dehydration and removal of impurities are needed in order to recycle but both are difficult

Concerns in effective dehydration of THF process

- THF forms azeotropic mixture with water; which restricts purity of product THF
- THF reacts readily with oxygen (from air) to form peroxides
- Peroxides have caused violent explosions during past attempts

INGENERO

THF Dehydration – Process Evaluation

Existing processes for THF dehydration have limitations and serious drawbacks

Brine Extraction
- Merits: Lower Raw Material Cost, Easy Process
- De-Merits: Final Moisture High, Less Yield, Effluent Generation

Pressure Distillation
- Merits: No Raw Material Cost, Negligible Solvent Loss
- De-Merits: Purity, Effluent and Peroxide Formation, High Pressure

Solvent Extraction
- Merits: Use of Existing Solvent Recovery System
- De-Merits: Low Yields, Additional Cost of Solvent Recovery

Alkali Treatment
- Merits: Less Raw Material Cost, Easy Process
- De-Merits: Final Moisture Content High, Effluent Generation

Adsorption (Molecular Sieve)
- Merits: Low Final Moisture Content, No Peroxide, No Effluent
- De-Merits: High Capital Cost, Feed Water Content < 1-1.5%

Pervaporation
- Merits: Low Moisture, No Peroxide, No Effluent, High Yield
- De-Merits: High Capital Cost

INGENERO

THF Dehydration – Process Objectives

Provide a safe, efficient and economic THF recovery process to enable its reuse as a solvent

Criteria for evaluation and selection of suitable process:

- Reduction of moisture content in recovered/dried THF to desired levels of less than 0.05 w%
- Safe Process; Peroxide formation should be minimum if not zero
- Minimal effluent generation; effluent disposal is cumbersome and difficult
- High yields/recovery of THF
- Low CAPEX

INGENERO

Confidential & Proprietary 5

THF Dehydration – Process Comparison

Comparison of processes from Literature showed Alkali treatment and Molecular sieve adsorption to be best options

Parameters	Brine Extraction + Molecular Sieve	Alkali Treatment + Distillation	Alkali Treatment + Molecular Sieve	Pervaporation
Purity of THF (w%)	99.50%	99.95%	99.95%	99.95%
Water in Rec.THF (w%)	0.50%	≤ 0.05%	≤ 0.05%	≤ 0.05%
THF Yield (wt basis)	~ 80 - 85%	~ 90%	~ 90%	~ 95%
Safety	Peroxide formation possible	Peroxide formation possible (stabilizer addition needed)	No Peroxide formation	No Peroxide formation
Time requirement	10 - 12 hrs	10 - 12 hrs	10 - 12 hrs	~ 10 hrs
Capital Investment (INR)	Medium (0.5x)	Medium (0.45x)	Low (0.35x)	Very High (4.0x)
Raw material cost (INR)	High 1.5x /batch	Medium 1x /batch	Medium 1x /batch	Very minimum
Operating cost (INR)	High 1.5x /batch	Medium 1x /batch	Medium 1x /batch	Very minimum
Remarks	-	-	Dehydration of other solvents may be possible	Dehydration of other solvents can be done

Basis : Water content in dilute THF=10 w%

INGENERO

Confidential & Proprietary 6

Initial Lab trials: Alkali treatment + Adsorption

Lab trial was conducted to see result of doing Alkali treatment first and then molecular sieve treatment

1A) Alkali Treatment

- Addition of required quantity of Alkali pellets to dilute THF till mixture got saturated, followed by separation of aqueous and organic layers
- Dilute THF analysis : Water content = 19 wt%, Peroxide = Not detected (ND)

Test No.	Dilute THF	Alkali pellets	Aqueous layer	Organic layer	Analysis of Organic layer	
					Water %	Peroxide
UOM	ltr.	lb	ltr.	ltr.	wt%	
Test 01	1.0	216	0.08	0.92	0.20	ND
Test 02	1.0	278	0.11	0.89	0.27	ND
Test 03	1.0	282	0.11	0.89	0.20	ND
Test 04	2.0	507	0.18	1.82	0.27	ND
Test 05	2.0	503	0.16	1.84	0.28	ND
Test 06	2.0	494	0.19	1.81	0.30	ND

1B) Liquid phase adsorption using molecular sieves

- Semi-dried THF after alkali treatment passed through bed of Molecular Sieves
- Dilute THF analysis : Water content = 0.28 w%
- Recovered THF Analysis Results
  - Water content < 0.03%, Peroxide = Not detected (ND)
  - THF yield (wt basis) : 89%

INGENERO

Further Lab trials: Alkali treatment + Adsorption

After impressive results, additional larger scale Lab trials were conducted

- Impact on process was checked for variation in following parameters
  - Batch Size
  - Alkali types and addition rates
  - Temperature
  - Molecular Sieves type
  - Height and L/D ratio of Molecular Sieve column
  - Flow rates in Molecular Sieve column
- Results were found to be consistent with earlier trials
- Final moisture content in recovered THF was less then 0.05 w%
- Peroxide content was < 0.0018 w%
- Suitability of Recovered THF for Grignard reaction was checked

INGENERO

THF Purification – The Distillation Step

Organic impurities (if any) need to be removed for improving Dry THF purity to 99.9 w%

- Distillation was determined as the efficient process to follow
- Batch distillation intended to run occasionally when the organic impurities in dry THF exceed 2 w%
- In the Distillation operation
  - Low boiling impurities are removed in "First Cut"
  - Pure THF with min. purity of 99.9 w% is obtained as "Second Cut"
  - High Boiling Impurities are removed as Residue
- Process safety against Peroxide formation
  - High temperature distillation operation increases risk of peroxide formation
  - Adequate safety precautions considered for risk free operation
    - » Addition of suitable and optimum quantity of stabilizer
    - » Safe operating procedures and parameters

INGENERO

Optimal Distillation Design To Purify THF

Extensive simulation study was conducted to determine optimal process design of distillation system

- Several simulation runs were taken to determine
  - Optimum concentration of low boiling impurities in reactor residue after First Cut
    - » Higher low boiling impurities after first cut operation increases reflux requirement during second cut operation significantly
  - First Cut and Second Cut Reflux ratios
    - » First Cut operation is sensitive to reflux ratio; has minimal impact of change in no. of theoretical stages
    - » Second cut operation is sensitive to variation of theoretical stages
  - Optimum theoretical stages
- Simulation results were used to determine the column dimensions and column internal design
- Simulation results were successfully verified in Lab trials

INGENERO

Final Pilot Scale Trials To Demonstrate Recovery

After Lab trials, Pilot plant trials were undertaken and THF recovery process was demonstrated successfully

- Pilot plant scale trials
  - ✓ Development of process scheme
  - ✓ Specification of hardware requirements
  - ✓ Conduction of trials of alkali treatment and molecular sieve adsorption processes
  - ✓ Conduction of trials of distillation process for removal of organic impurities

INGENERO

Engineering And Commissioning The Recovery Unit

Process development, engineering of THF Recovery Unit was undertaken followed by commissioning

- Process development and Engineering
  - ✓ Basic Process Engineering Package
  - ✓ Mechanical Design and Data Sheet preparation
  - ✓ Piping Engineering
  - ✓ Instrumentation Engineering
  - ✓ HAZOP
- Identification and coordination with vendors
- Plant erection and process commissioning support
  - ✓ Checking of plant erection & suggesting corrective actions
  - ✓ Support for process commissioning
  - ✓ After erection of commercial scale THF distillation unit at site, a joint test run was conducted successfully in plant spanning ten batches

INGENERO

Commercial Validation Of the Recovery Process

THF recovery unit is running steadily resulting in significant reduction in fresh THF consumption

- Combination of Dehydration and Distillation process developed by Ingenero is running successfully in one of the pharmaceutical units in India
- Addressing of safety concerns related to peroxide formation makes this process a unique one
- This invention meets almost all the purity and THF specification requirements
- It provides a safe and economical option for THF recovery
- This THF recovery process has potential to be used in most of the pharmaceutical units all over the world

INGENERO

Confidential & Proprietary 13

Questions?

INGENERO

Confidential & Proprietary 14

INGENERO

Excellence Through Insight

4615 Southwest Freeway, Suite 320,
Houston, Texas, 77002, USA

www.ingenero.com info@ingenero.com

INGENERO

Confidential & Proprietary 15

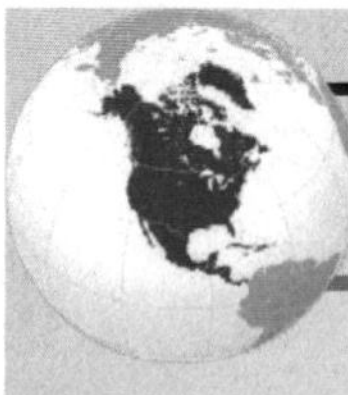

Rebalancing Global Crude Flows

John Mayes
Director of Special Studies

AICHE Technology Conference
Galveston, Texas
October 2, 2015

AIChE

Turner, Mason & Company
CONSULTING ENGINEERS

Annual U.S. Crude Production

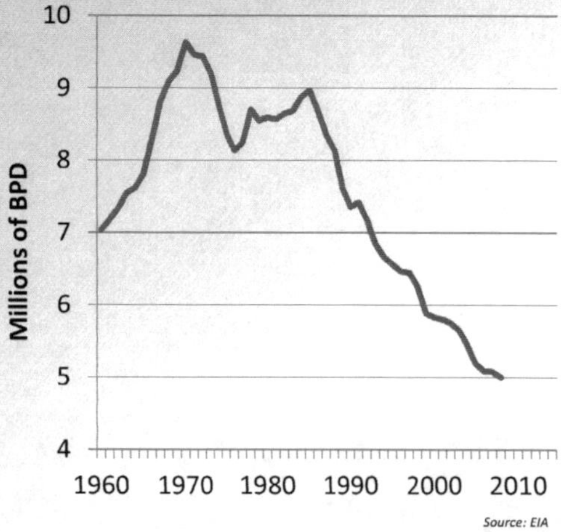

- U.S. crude oil production has been declining since 1985
- Output peaked in November of 1970 at 10.0 million BPD
- Output in 2008 fell to 5.0 million BPD

AIChE

Page 2

Turner, Mason & Company
CONSULTING ENGINEERS

U.S. Crude Production

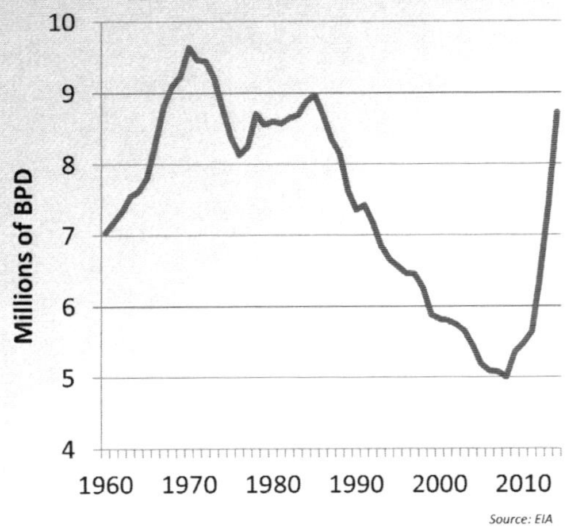

- U.S. crude oil production rose to 8.7 million BPD in 2014
- Estimated output in 2015 is 9.3 million BPD
- Recent high output was 9.6 million BPD in April

North American Production Gains

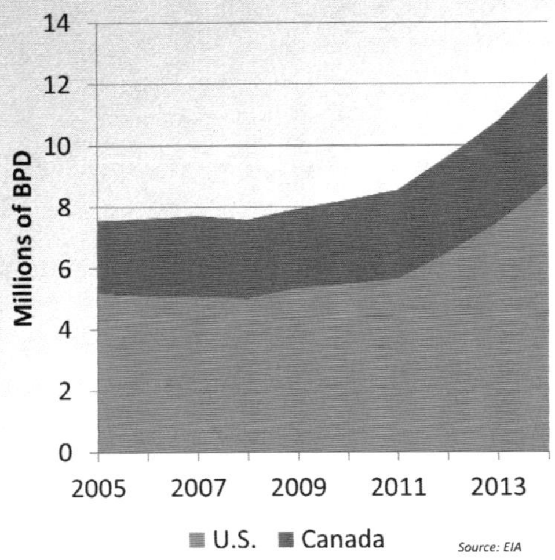

- U.S. crude oil production grew by 3.7 million BPD between 2008 and 2014
- The U.S. had its greatest oil growth in 2014 at 1.3 million BPD
- Canadian production grew by 1.0 million BPD between 2008 and 2014

Global Crude Production Growth from 2008 - 2014

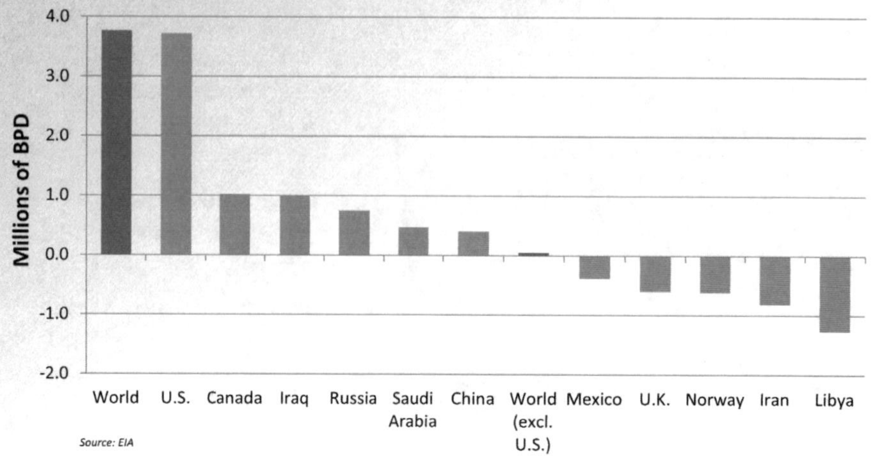

Brent Prices Are Sharply Lower

- Brent prices peaked in June of 2015 at $115.32
- Prices hit recent lows in Jan and August of 2015
- Brent had yearly lows in 4Q in 2008 and 2014

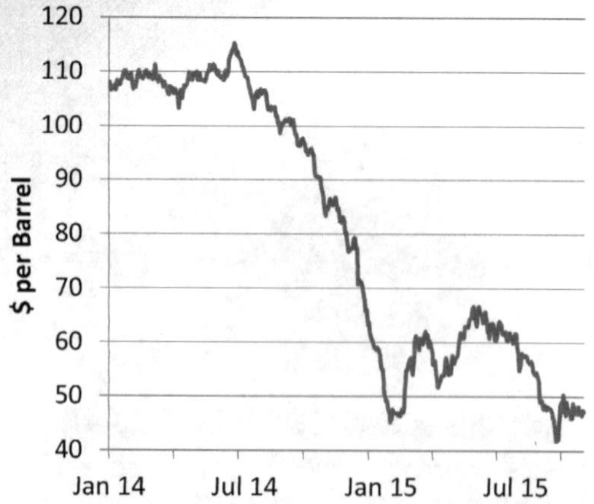

U.S. Dependence on Foreign Oil

- U.S. dependency peaked in 2005-2007 at more than 60%
- Dependency has recently fallen below 23%
- Decline a result of U.S. production, renewable fuels and moderating demand

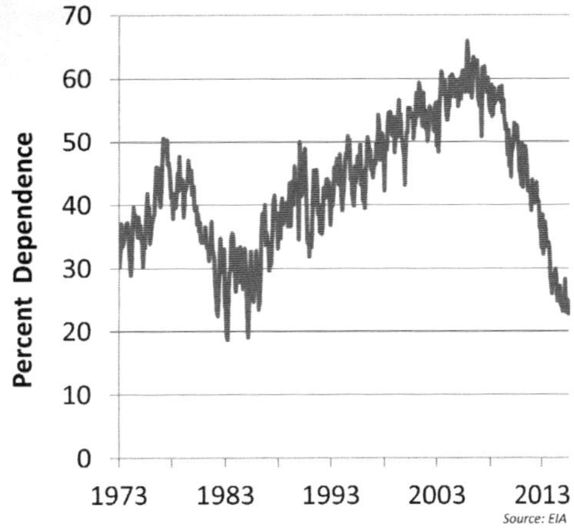

U.S. Oil Rig Count Also Down

- Oil rig count peaked in October of 2014 at 1,609
- Rig count bottomed out in June of 2015 at 628
- Recent rebound has faltered

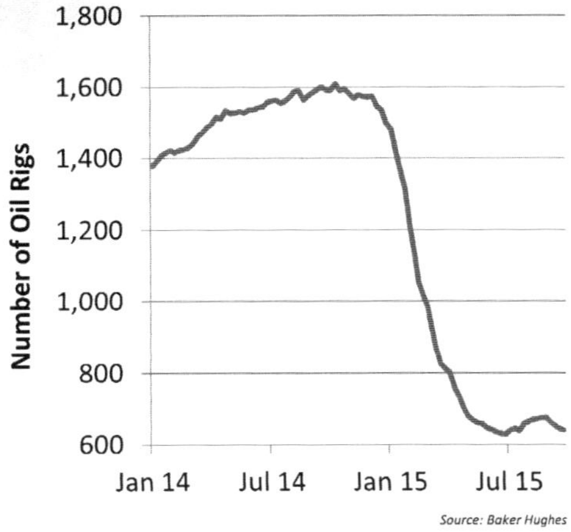

2015 U.S. Crude Production

- Production has declined in recent months
- Output is lower by 243 MBPD from the April peak
- Current production is 639 MBPD above the 2014 average

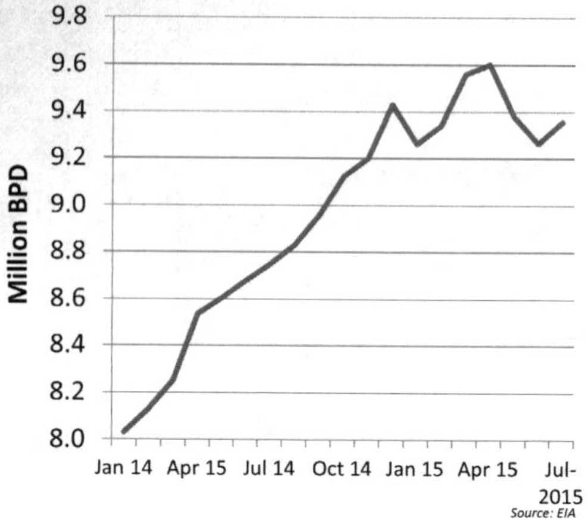

U.S. Crude Inventory (Excluding SPR)

- U.S. crude inventories rose by over 100 million barrels in the last year
- Inventories have moderated in recent months

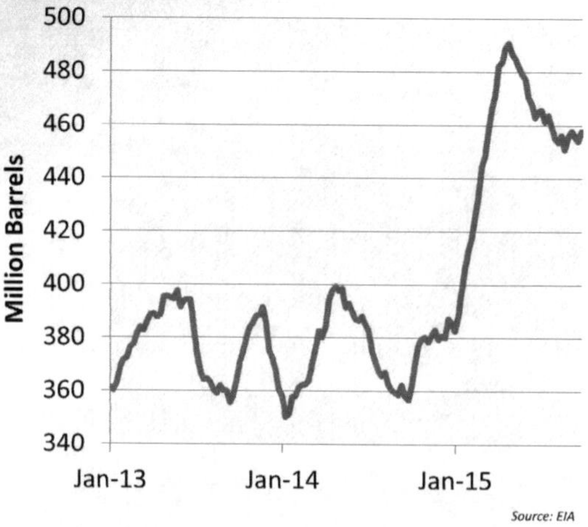

U.S. Product Inventories (Including Intermediates)

- U.S. product inventories have risen by over 110 million barrels since March
- Stocks have jumped by 187 million barrels since March of 2014

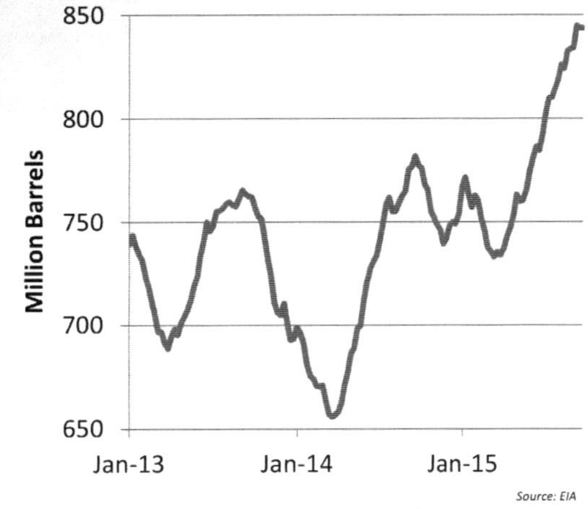

AIChE | Page 11 | Turner, Mason & Company CONSULTING ENGINEERS

U.S. Crude & Product Inventories (Excluding SPR)

- Total oil stocks have risen by 269 million barrels since early 2014
- The stock build is at a rate of 462 MBPD

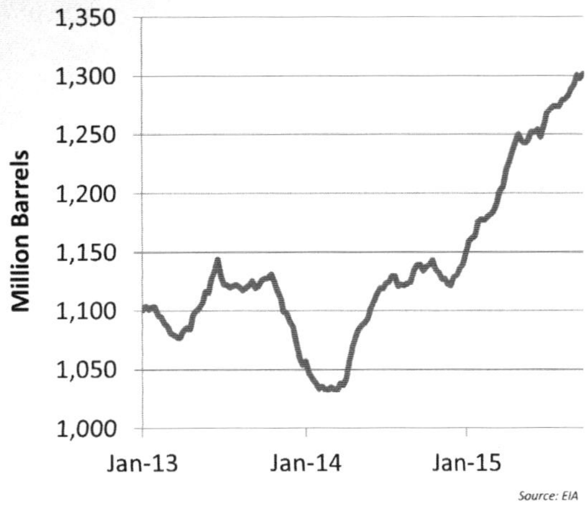

AIChE | Page 12 | Turner, Mason & Company CONSULTING ENGINEERS

Global Demand is Rising

- IEA projects 2015 demand growth at 1.7 million BPD
- 2015 to have the strongest growth since 2010

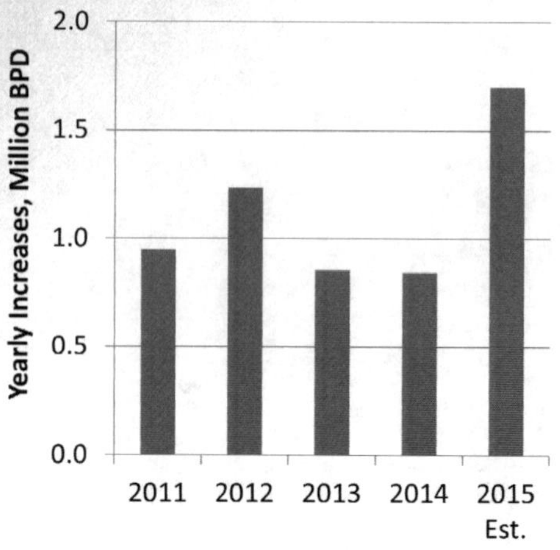

Source: IEA and EIA

The Growing Crude Oil Glut

- IEA calculates oil stocks have risen since 1Q2014
- 2Q2015 build is the greatest at 2.6 million BPD
- The total build in 2014 and YTD 2015 is 875 million barrels

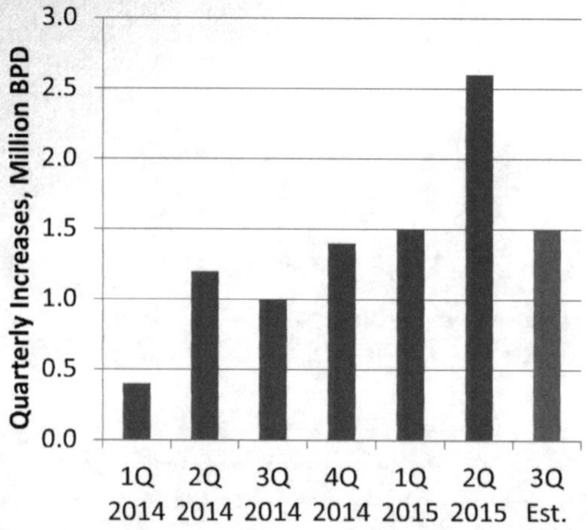

Source: IEA

Change In Crude Production 2Q2014 vs 2Q2015

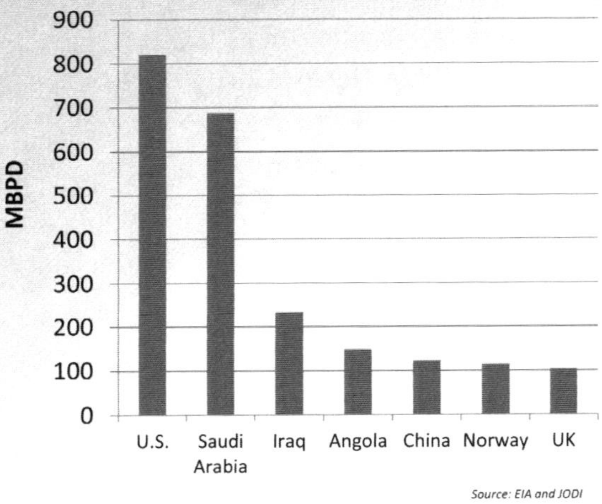

- Many countries are attempting to maximize revenues
- U.S. has greatest growth at 821 MBPD
- Saudi Arabia has increase production by 687 MBPD

AIChE | Page 15 | Turner, Mason & Company CONSULTING ENGINEERS

New Oil World Order

- The "New Oil World Order" is a multi-client study which calculates future Brent prices based on the convergence of crude oil supply and demand
- Production costs were evaluated for 45 regions
- Model is an iterative process which computes production rates in balance with crude demand
- Current Brent price would be in low $60s if supply were in balance with demand

AIChE | Page 16 | Turner, Mason & Company CONSULTING ENGINEERS

Inventory Solution Scenarios

- Political event which would reduce global output
- OPEC production reductions
- Wait until global demand increases to above crude supply

AIChE | Page 17 | Turner, Mason & Company CONSULTING ENGINEERS

Recent Saudi Production Levels

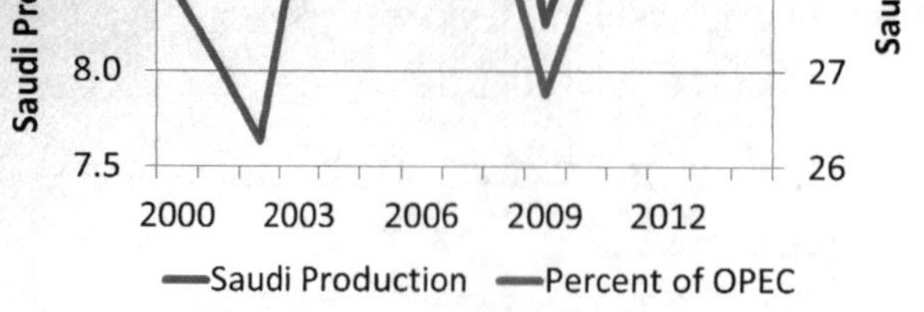

Source: EIA

- Low of 7.6 million BPD in 2002 and a high of 9.8 million BPD in 2012
- Saudi market share of OPEC at low in 2009 at 26.8% and a higher in 2014 at 30.5%

AIChE | Page 18 | Turner, Mason & Company CONSULTING ENGINEERS

OPEC Crude Production 1980 - 1985

	Production MBPD		Decline MBPD	
	1980	1985	MBPD	Percent
Saudi Arabia	9,900	3,388	6,512	66
Other OPEC	15,483	11,979	3,504	23
	25,383	15,367	10,016	

Source: EIA

The Long Road to a Supply/Demand Balance

- IEA projects oil stocks to increase through 3Q2016
- Total oil build will be over 1.2 billion barrels
- Forecast is based on OPEC output of 31.7 million BPD

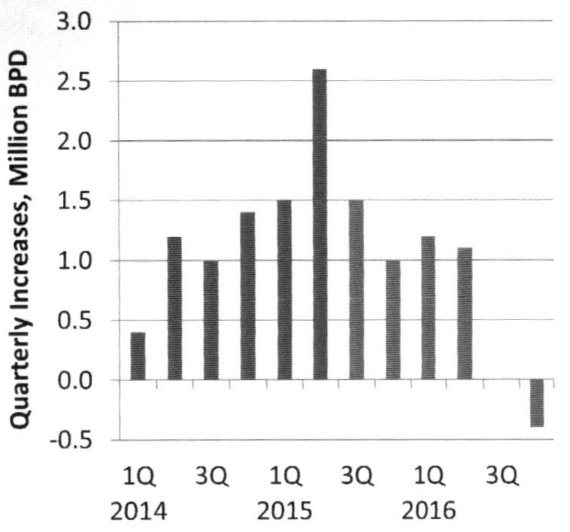

Source: IEA

Conclusions

- The U.S. has been the dominant source of incremental crude since 2008
- Surplus crude production has forced sharply lower crude prices
- OPEC (Saudi Arabia) has sought to defend its market share rather than crude prices
- OPEC (Saudi Arabia) is not likely to set output quotas in 2015
- Prospects for a broader production agreement in 2016 are declining

Conclusions (cont.)

- Many producing nations have increased output to minimize revenue losses
- Rising crude and product inventories will continue to depress prices
- The IEA is forecasting inventory builds until 3Q2016
- Surplus oil stocks will reach 1.2 billion barrels by mid-2016
- Barring a political event, a sustained price recovery is not likely before 2H2016

Conclusions (cont.)

- Brent prices in a balanced supply and demand market are in the low $60s

Presenter

John M. Mayes jmayes@turnermason.com
Director of Special Studies

Turner, Mason & Company
2100 Ross Ave., Suite 2920
Dallas, Texas 75201

turnermason.com

214-754-0898

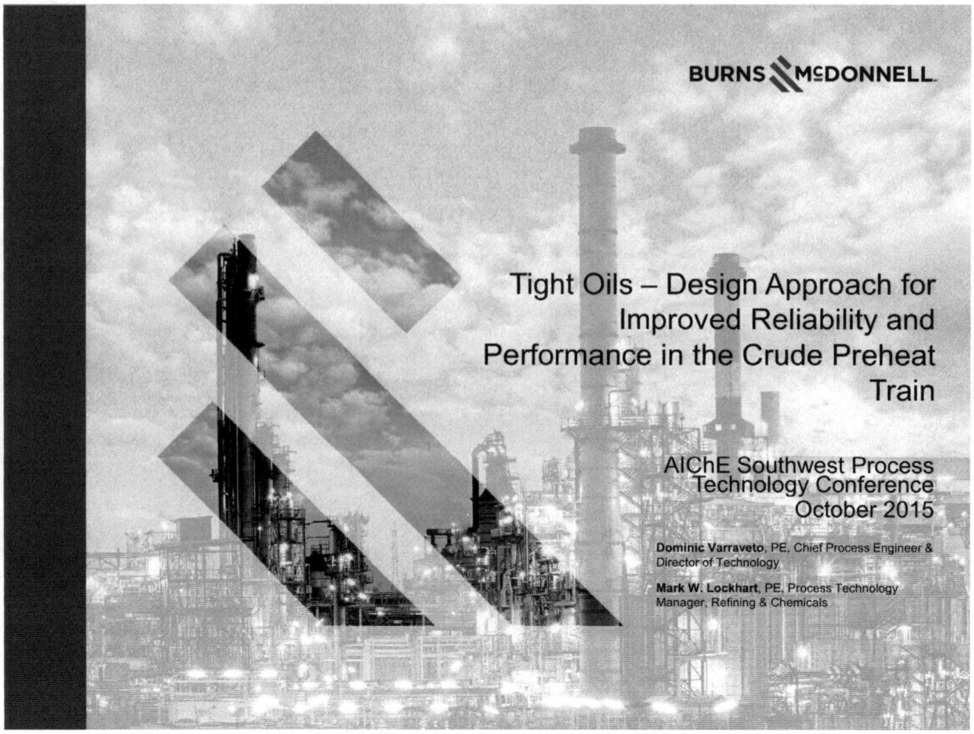

Introduction / Overview

- Shale plays - locations and current production
- Tight oil and condensate - composition
- Design approach for crude preheat train design
  - Exchanger fouling and design
  - Desalter design and operation
  - Flash drum placement and design
  - Crude charge heater fouling and design
- Conclusions

BURNS McDONNELL 2

Shale Plays Location & Current Production

- **Shale plays - lower U.S. 48 map:**
  - Niobrara
  - Bakken
  - Monterrey
  - Haynesville-Bossier
  - Utica
  - Wolf Camp & Bone Springs
  - Eagle Ford

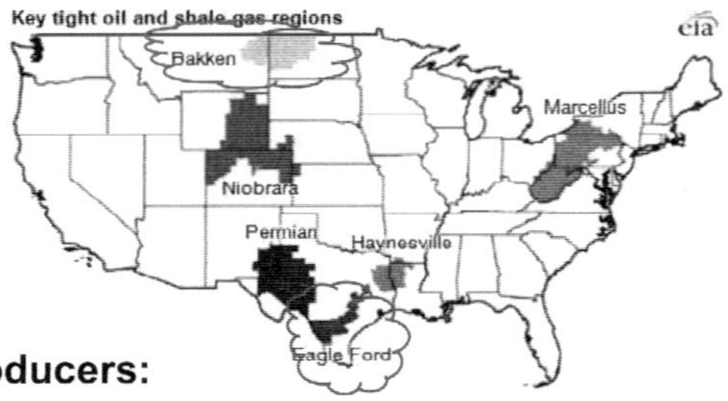

- **Largest liquid "Tight Oil" producers:**
  - Bakken in Montana / North Dakota
  - Eagle Ford in South Central Texas
- **Production:**
  - Bakken **2014E: 0.94 MMBPD**
  - Eagle Ford **2014E: 1.21 MMBPD**

Tight Oil & Condensate Composition

- **Tight oil and condensate are:**
  - Light and sweet
  - Sulfur: < 0.25 wt%
  - High naphtha content
  - Paraffinic
  - Heavy metals are low
  - High filterable solids

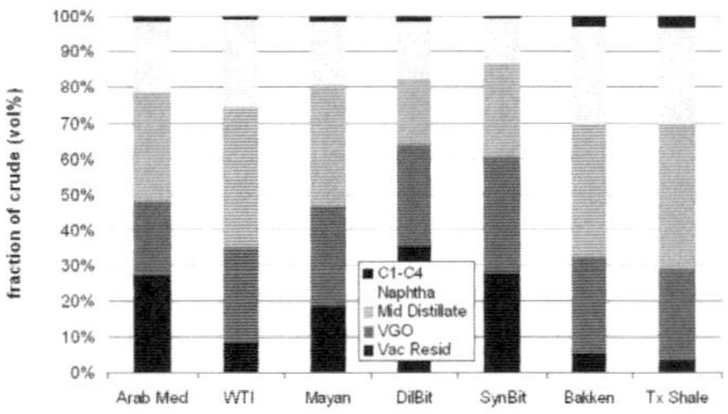

Source: AFPM AM-14-17, 2014

- **Tight Oil: ~ 40 – 55 API gravity**
- **Condensate: ~ 55 – 65 API gravity**
- **Processing Challenges from:**
  - Paraffinic nature – **wax deposition, asphaltene precipitation when blend crudes, exchanger and charge heater fouling**
  - Filterable solids – **exchanger and fired heater fouling**
  - **Refiners lack of ability to clean equipment on-line**

Crude Preheat Train - Design Approach for Reliability & Performance

- Exchanger Fouling and Design
- Desalter Design and Operation
- Flash Drum Placement and Design
- Charge Heater Fouling and Design

BURNS & McDONNELL 5

Crude Preheat Train Overview

Design Considerations:

- Exchanger Fouling and Design

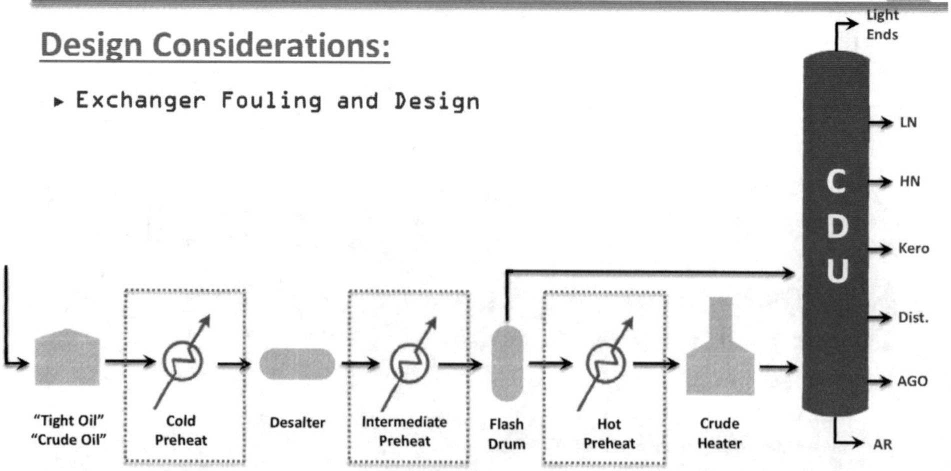

BURNS & McDONNELL 6

Exchanger Fouling & Design

- Tight oil and condensate considerations:
  - High paraffin content - wax deposition
  - High filterable solids – solids deposition
  - Asphaltene precipitation potential - crude blending
  - Desalter design and operation – upsets and carryover
- Exchanger Types:
  - Shell and tube
  - Helical baffle
  - Refer to figures on next two slides

BURNS & McDONNELL 7

Exchanger Types – Shell and Tube (S&T)

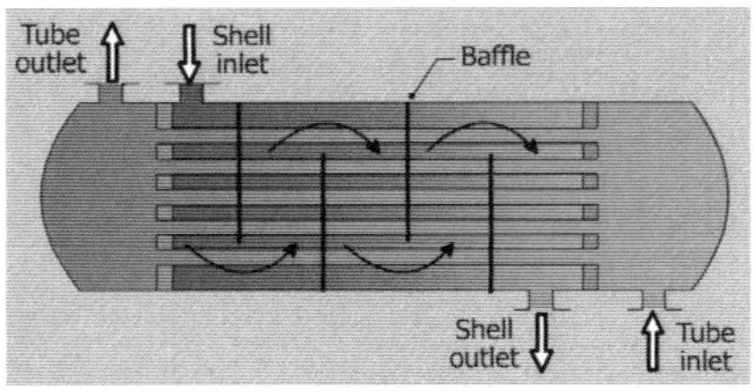

- Tube side – high velocity
- Shell side:
  - Lower velocity vs helical baffle exchanger (HBE)
  - Higher pressure drop (same velocity) vs HBE
  - Segmental baffles - potential for deposition

BURNS & McDONNELL 8

Exchanger Types – Helical Baffle (HBE)

- Tube side – high velocity
- Shell side:
  - Can achieve higher velocity vs S&T
  - Can achieve lower pressure drop with high velocity (without vibration)
  - Reduced dead zones – extended run times

Exchanger Selection Methodology

- Exchanger Selection – to start:
  - Optimization / process simulation is complete
  - Configuration / number of heat exchangers defined
  - Hot and cold streams, flowrates, and exchanger duties defined
- Methodology
  - Assess causes of fouling in the cold, intermediate and hot train
  - Based on causes, assess: can hot and cold fluids be placed on the shell and/or the tube side for S&T and Helical baffle exchangers (HBE)?
  - Use exchanger modeling tool (i.e. HTRI, EDR, etc.) with prescribed design requirements and conduct comparison

Exchanger Selection Methodology

- Methodology (continued)
  - Apply design margins, fouling factors, exchanger cleaning requirements
  - Tabulate results for S&T and HBE's while placing the hot and cold fluids on each of the applicable sides of the exchanger
  - Compare area, pressure drop, on-line cleaning considerations and installed capital and/or lifecycle costs
  - Based upon results - select exchanger type and fluid placement
- Other considerations – bigger picture
  - Exchanger pressure drop
  - Pumping horsepower
  - Equipment design pressure
  - Metallurgy
  - Seek existing exchanger fouling data

BURNS & McDONNELL 11

Exchanger Fouling & Design – Cold PHT

EXAMPLE

- Causes of Fouling:
  - Fouling is more likely to occur on cold crude side
  - Cause: Wax build-up, high filterable solids deposition
- Design Considerations:
  - Maintain high fluid velocity (>5 fps)
  - Minimize dead spots on crude side (i.e. NOT place on shell side for S&T option)
  - Recognize potential to reduce ΔP for crude if placed on the shell side.
  - Other (include design margins, fouling factors, exchanger cleaning requirements)
- Exchanger Selections:
  - Tabulate and compare results
  - Continue through all cold PHT exchanger services and make selections
  - Apply "Other considerations" for final selections

BURNS & McDONNELL 12

Exchanger Fouling & Design – PHT Exchangers

- Causes of Fouling:
  - Cold Train:
    - Fouling is more likely to occur on cold crude side
    - Cause: Wax build-up, high filterable solids deposition
  - Intermediate Train
    - Fouling potential for the cold crude and hot stream sides
    - Cold (crude) side - desalter operation/upsets with carryover of emulsions or solids, possible asphaltene precipitation
    - Hot side – potentially higher fouling streams (AGO, AR, VR, etc.)
  - Hot Train:
    - Highest temp in train - fouling potential for the cold crude and hot stream sides
    - Cold (crude) side - desalter operation/upsets with carryover of emulsions or solids, higher temps result in higher fouling with asphaltene precipitation potential
    - Hot side – higher fouling streams at high temps, in particular AGO, VGO, Atm and Vac resid

BURNS & McDONNELL 13

Exchanger Fouling & Design - Summary

- To achieve the most cost-effective, safe, and reliable PHT design - Evaluate and determine a balance between:
  - Exchanger fouling
  - Fluid velocity criteria
  - Exchanger pressure drop
  - Placement the cold and hot fluids
  - Exchanger metallurgy
  - On-line cleaning
  - Pumping head
  - Equipment design pressures

BURNS & McDONNELL 14

Crude Preheat Train Overview

Design Considerations:

- Exchanger Fouling and Design
- Desalter Design and Operation

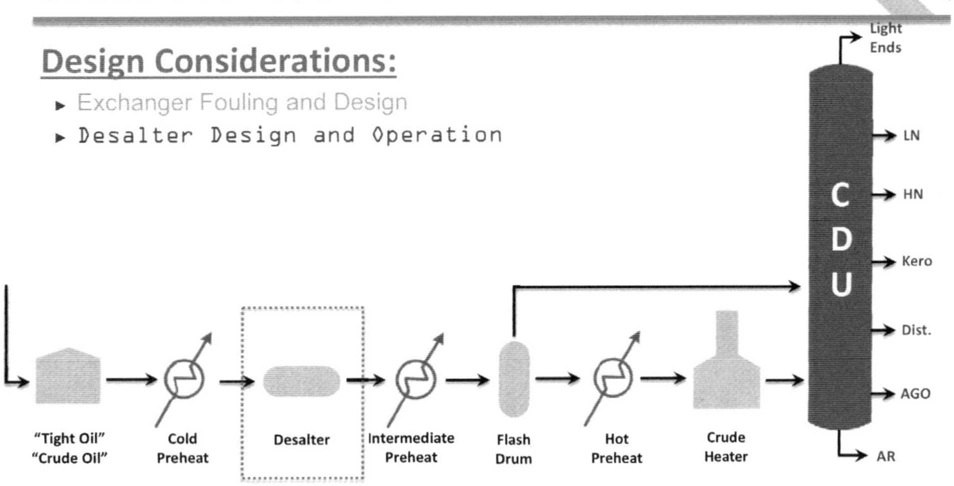

Desalter Design & Operation

Purpose & Function of Desalter:

- Reduce salt content and remove solids
- Wash Water mixed with Oil. Separates by differential gravity enhanced by electric field
- Resulting Oil is clean and dry. Brine contains salts, sediment

Design Considerations:

- Feed characteristic including paraffinic-asphaltene compatibility
- Temperature to reduce viscosity, limited by water solubility in oil & mechanical equipment
- Corresponding Pressure needed to maintain liquid phase and prevent vaporization
- Tendency to form emulsions and need for addition of chemical dispersants and demulsifiers
- Solids handling capacity internal and external oily brine processing

Tight Oil Desalter Features

- Two-Stage has higher removal efficiency than Single Stage. Reduces chance of carryover, Can be serviced w/o total shutdown
- Robust Mudwash system for high solids and stabilized emulsions.
- [illegible] control to assist in removal of amine added during [illegible] content in tight oil)

Desalter Design & Operation

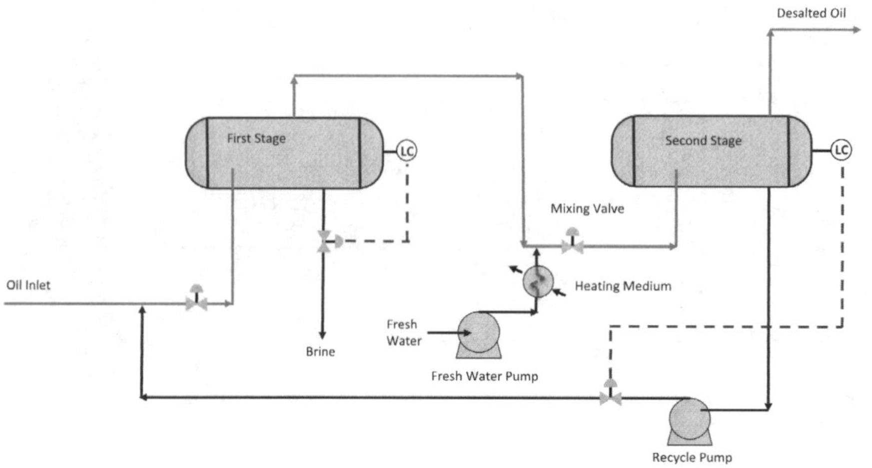

Source: Piping Engineering

Crude Preheat Train Overview

Design Considerations:

- Exchanger Fouling and Design
- Desalter Design and Operation
- Flash Drum Placement and Design

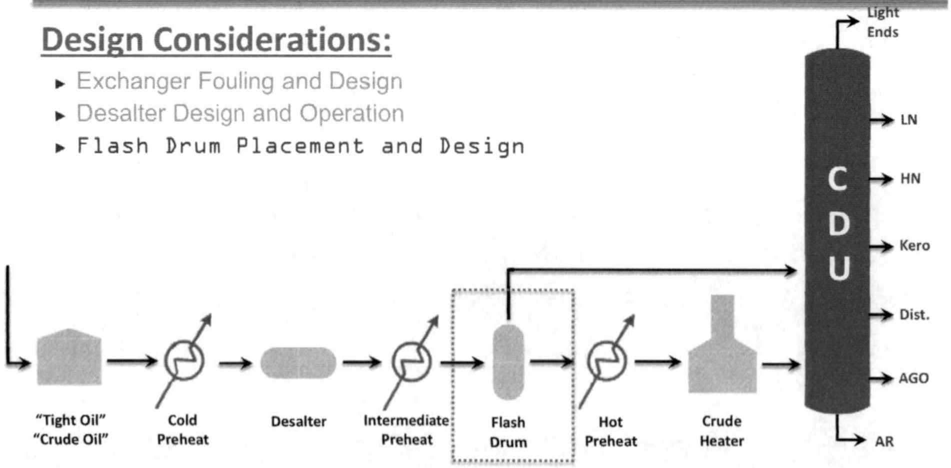

Flash Drum Placement & Design

- Location of vapor feed to column
  - Option 1: Flash zone where fired heater transfer line connects to the crude tower
    - Results in a larger crude tower bottom section diameter.
    - Vapor stream in flash zone acts as a quench requiring a higher heater outlet temp. to maintain the desired flash zone temp.
  - Option 2: Taken to an appropriately higher section in the tower where temp. and composition of the flash drum vapor is similar.
    - For light oil flashed at 400°F, vapor return location would be above the distillate section and below the naphtha section.
    - Required heater duty and crude tower [illegible] this scenario.

Flash Drum in CPHT

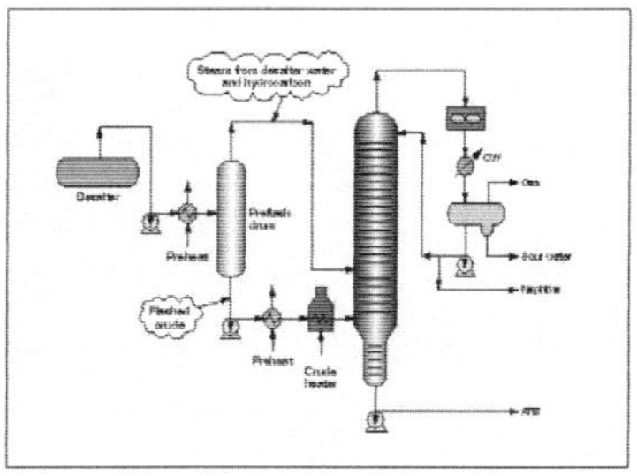

Source: Process Consulting Services

Flash Drum Placement & Design

- Option 3: Flash tower with trays and overhead system
  - Does NOT return any flashed vapor to the crude tower.
  - More complex and costly than a flash drum
  - Typical for a revamp due to:
    - Limitations on existing crude tower
    - Requirement of additional throughput and improved separation

Flash Tower in CPHT

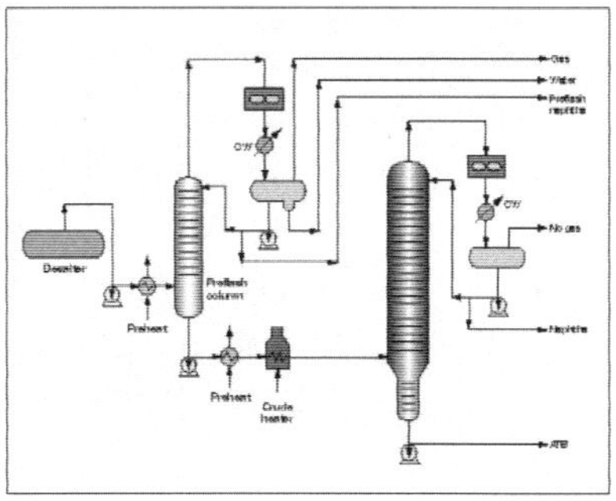

Source: Process Consulting Services

Crude Preheat Train Overview

Design Considerations:

- Exchanger Fouling and Design
- Desalter Design and Operation
- Flash Drum Placement and Design
- **Charge Heater Fouling and Design**

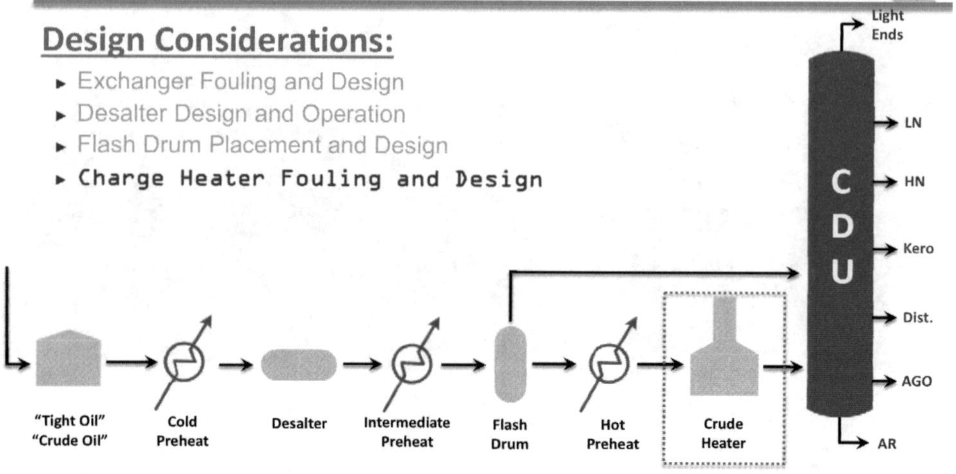

BURNS & McDONNELL

Charge Heater Fouling & Design

- Causes of fouling:
  - Asphaltene deposition from blending crudes
  - Desalter carryover of emulsions and solids can also contribute
  - High tube metal temperatures (TMT)
  - High residence times
  - High percent vaporization (tube running dry)
  - → Cracking reactions with coke laydown
- Design Considerations:
  - Even radiant heat flux – avoid localized hot spots
    - See figures to right
  - High mass velocity – lower residence time
    - Consider velocity steam injection at multiple locations

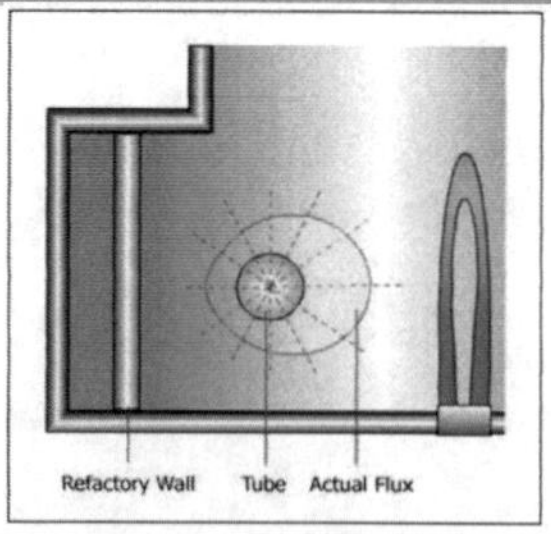

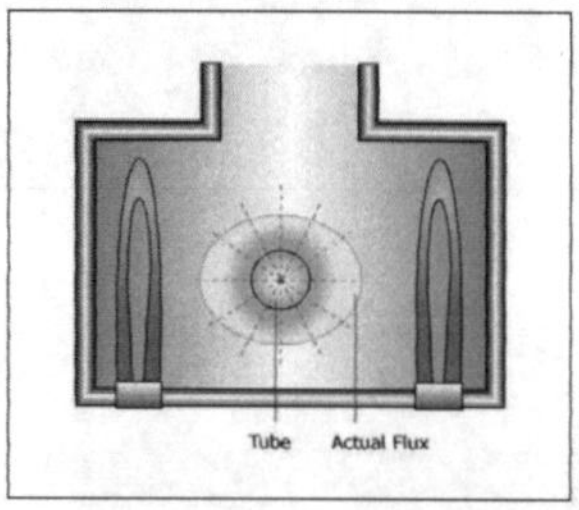

BURNS & McDONNELL

Charge Heater Fouling & Design

- Design Considerations (cont'd):
  - Max percent vaporization (50 – 60%)
    - Lights content of tight oil exacerbates
    - Can recirculate atm resid
  - Include on-line decoking or pigging of tubes
- Charge Heater type:
  - Double fired – even heat flux
    - All floor or wall and floor burners
    - Floor burners alone can accomplish even heat flux and is lower cost
  - Multiple cells – for on-line cleaning

BURNS & McDONNELL 23

Charge Heater Fouling & Design

- Charge Heater Type:
  - Figure to right features:
    - Two-cell, box cabin with floor burners
    - Ability to isolate each cell for on-line decoking and/or pigging (smart pigs)
    - Design may or may not include:
      - ID, FD fans, with SCR and common stack

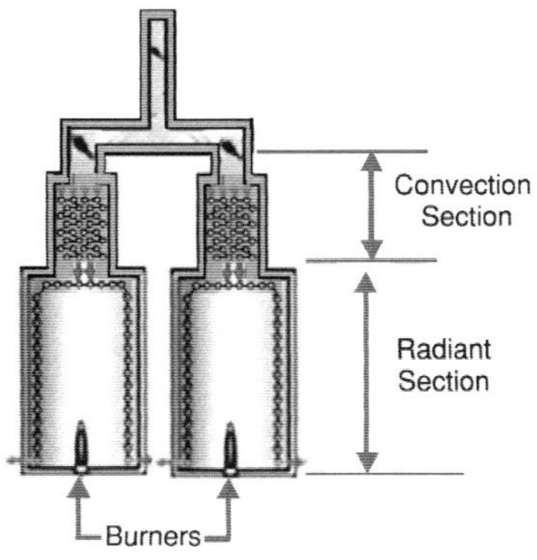

BURNS & McDONNELL 24

Charge Heater Fouling & Design

- Other Design Considerations:
  - Carefully consider charge heater design margin
    - Allows operation at higher capacity when cleaning one cell
  - Fouling in fired heater tubes is less desirable than fouling in heat exchanger
    - Strive for max heater inlet temperatures
  - Vaporization at heater outlet affected by
    - Flash zone temperature and pressure
    - Consider recirculation of atm resid

BURNS & McDONNELL 25

Conclusion

- Tight oils and condensates are presenting unique challenges for design of new CPHT's and in operating a existing CPHT's
- Existing crude units are experiencing high fouling in heat exchangers and charge heater tubes leading to unplanned outages and loss of production
- Crude unit reliability can be increased in new unit design, in crude unit retrofits and/or in adjusting operating parameters in existing units

BURNS & McDONNELL 26

Conclusion

- This paper has provided specific design approaches for the CPHT to address the unique challenges when processing tight oil and condensates, including:
  - Exchanger fouling and design
  - Desalter design and operation
  - Flash drum placement and design
  - Fired heater fouling and design

Questions and Answers

Dominic Varraveto
dvarraveto@burnsmcd.com
816.822.4282

Mark W. Lockhart
mlockhart@burnsmcd.com
832.214.1975

Abyar Aejaz
aaejaz@burnsmcd.com
832.214.1947

BURNS McDONNELL

The Petrochemical FCC Unit

de Graaf, B.,

Originally presented on: 10/2/2015 15:00:00 - 15:30:00

Propylene is a key building block in the petrochemical industry. The vast majority of propylene supply comes from steam crackers (approximately 60%). Fluid Catalytic Cracking (FCC) units provide approximately 35% and the balance is made up by propane dehydrogenation units (PDH). Feedstock diet of steam crackers is changing with the advent of shale oils to lighter feeds. These lighter feeds, and especially natural gas liquids change the yield pattern of steam cracker products towards more ethylene and less propylene. Typically FCC yields 5 wt% of propylene from a heavy feed, but this can be easily increased by the use of ZSM-5 additives. Petrochemical FCC units can achieve yields of typically 9 to 12 wt% propylene on feed using ZSM-5 additives. Where addition of small amounts of ZSM-5 additive to a base catalyst do not change FCC chemistry dramatically, adding substantial amounts of ZSM-5 additive (10 to 25 wt% to inventory) have a profound effect on the chemistry of gasoline and LPG as they can now react over two very different shape selective zeolites. Hydrogen transfer reactions over ulstrastiblized Y (main zeolite in FCC base catalyst), which are very dominant in FCC, are now 'diluted' by alternative cracking routes over ZSM-5. A new network of reactions appears in which FCC olefins are now no longer only a desired final product but an intermediate as well. The effect of ZSM-5 on light and heavy fractions of the gasoline is different. In the light gasoline the effects of de-alkylation and recombination of LPG olefins is more pronounced than in the heavy gasoline. In both fractions the change in hydrogen transfer over the catalytic system are very apparent. Changes in the gasoline and LPG composition show that these are a function of the process and catalyst both.

Coke Drum Feed Entry Design Considerations

Hughes, G.,
Adams, J., Adams Project Managers, Inc.
Originally presented on: 10/2/2015 15:30:00 - 16:00:00

Automated unheading valves are becoming the norm for both new and modernized delayed cokers, as these unheading devices have a proven track record of improving safety, operability, and reliability of the delayed coker.

But the installation of automated valves has created new challenges in the design of coke drum piping and the coke drum itself. Historically, with manual coke drum unheading, the coke drum feed was inject directly into the bottom head of the coke drum and the feed naturally flowed upwards with a relatively even distribution.

The use of bottoms, automated unheading devices requires the flow be injected into the side of the coke drum cone, an operation which has caused coke drum distortion in cases and maldistribution of the feed in some instances.

This paper presents engineering analysis and proven operating designs, showing that improved technology will assure even upward distribution of coker feed; the improved design must aims the feed upward and toward the center of the coke drum.

Successful coker feed injection can result in reliability and stream factors improvements which flow to the refiners bottom line. Technologies considered must include flow analysis, dual side entry nozzles, nozzle orientation, retractable center feed injection devices, and pipe design.

AUTHOR INDEX